AF615354

# Physiological Aspects of Anaesthetics and Inert Gases

# Physiological Aspects of Anaesthetics and Inert Gases

A. G. MACDONALD AND K. T. WANN
*Physiology Department*
*Marischal College*
*University of Aberdeen*
*Aberdeen AB9 1AS, UK*

1978

*ACADEMIC PRESS*
LONDON NEW YORK SAN FRANCISCO
*A Subsidiary of Harcourt Brace Jovanovich, Publishers*

ACADEMIC PRESS INC. (LONDON) LTD.
24/28 Oval Road
London NW1

*United States Edition published by*
ACADEMIC PRESS INC.
111 Fifth Avenue
New York, New York 10003



Library of Congress Catalog Card Number: 77-81385
ISBN: 0-12-464150-4

Printed in Great Britain by
The Garden City Press Limited
Letchworth, Hertfordshire SG6 1JS

# Acknowledgements

We are indebted to numerous colleagues for their comments on various aspects of this book; in particular to Dr V. Flook, Dr J. Kimura and Dr T. Stone. We are also extremely grateful to those authors who allowed us to reproduce their figures.

# Units of measurement

We have found it impractical to adopt S.I. units in this book; for those who wish to convert any of the units we have used we recommend Handy Matrices by Pennycuick, C.J., 1974, Arnold, London.

# Preface

This book attempts to set out in an orderly scientific way the multitude of effects which anaesthetics, inert gases and similar substances exert in living cells and organisms. It is intended for students of biological, and biophysical sciences and of medicine, but could prove useful to the specialist worker in anaesthetics research. The fields of study and established terminology with which we are concerned are often rather nebulous so we considered it necessary to devote much of the first chapter to an introduction of what the book is about.

Anaesthetics affect almost any and every biological function, and yet they are not normally included in systematic courses of physiology, biochemistry or other biological sciences. We describe here only a selection of anaesthetic effects. We have concentrated on cellular aspects, and although our treatment is not intended to be comprehensive, we hope this book will render the fascinating world of anaesthetics accessible to students.

For those who begin the study of anaesthetic effects there is the immediate prospect of gaining specific knowledge about medically important compounds. However the ubiquitous nature of anaesthetic effects presents the student with a greater scientific challenge, that of learning more about biological systems in general.

ALISTER G. MACDONALD
KENNETH T. WANN
*June, 1978*

# Contents

## CHAPTER III

## Biochemical effects of anaesthetics and inert gases



## CHAPTER IV

## Inexcitable membranes

## CHAPTER V

## Excitable membranes



## CHAPTER VI

## Synapses

CHAPTER I

# Introduction



## 1 Objectives

Anaesthetics and inert gases constitute a fascinating group of physiologically active substances. This book deals with their physiological actions, which are fundamentally important to both the scientist and the anaesthetist. It came to be written because the authors, who are physiologists and not specialists in anaesthetics research, found the subject lacking a basic text appropriate to the needs of biology or medical students. Accordingly, we have attempted to provide an introductory text which describes the basic, and diverse, effects of anaesthetics in a coherent physiological frame of reference.

In recent years a considerable amount of new information on anaesthetic effects at all levels of organisation has appeared and now presents a body of knowledge with which biology students may wish to become familiar. Although several books are already available which deal with certain aspects of the material discussed here, their approach is either clinical in outlook or their style too restricted and advanced to serve as an introductory text. A comprehensive account of anaesthetic actions is not the aim of this book but it is hoped that sufficient breadth of treatment is given to provide a biologically balanced introduction to the subject without undue sacrifice of depth. Anaesthetists have always been aware that the effects of anaesthetics are complex and medical students should know about the basic physiological problems which anaesthetics present. A discussion of anaesthetic effects set in a biological context should help the medical student and would-be anaesthetist to appreciate these problems.

This book is therefore intended to be introductory and informative and it should provide the reader with sufficient background information to stimulate further exploration of the subject at a more advanced level.

## 2 Contents

In discussing the physiological actions of anaesthetics we have considered two questions. What do anaesthetics do? What do these actions tell us about the properties of the physiological systems with which they interact? Anaesthetics are instructive probes of physiological function and this view will be seen to be a recurrent theme in the book. It enables us to form a bridge, albeit tenuous at times, between the well documented physical-chemical properties of these agents and the wealth of physiological actions which they possess. It might well be argued that this approach avoids questions concerning the basis of clinical anaesthesia. Indeed this is true and no attempt is made here to describe how anaesthetics act in the central nervous system to cause anaesthesia. It becomes increasingly difficult to link the physical-chemical properties of anaesthetics to their physiological effects at such a complex level of organisation. Their scientific usefulness as probes of physiological function is clearly limited, and this largely determines the selection of material for this book.

Chapter II deals with the effects of anaesthetics on the structure and organisation of cells in general. Chapter III deals with the biochemical actions of anaesthetics which were once mistakenly thought to be biochemically inert. As highly evolved homeostatic systems, cells and tissues have developed defence mechanisms which guard them against invasion by foreign substances. Cells are bounded by, and contain, membranes and other structures whose hydrophobic nature renders them highly susceptible to penetration by anaesthetic molecules. Consequently, anaesthetics affect the structure and organisation of cells and set in motion their defence mechanisms.

Chapters IV and V consider the special problem of anaesthetic effects on inexcitable and excitable membranes, respectively. In membranes, anaesthetics are particularly effective probes of physiological function, a fact which has been exploited in recent years.

The actions of anaesthetics on synapses are discussed in Chapter VI, although, as emphasised above, the discussion does not venture to explain the nature of the anaesthetised state, but simply draws attention to the stages in the synaptic process which are susceptible to anaesthetics.

The final chapter deals with some non-narcotic aspects of anaesthetics and inert gases in the whole animal (see p. 4 for a definition of non-narcotic and other terms). Here the physical properties of some of the agents are seen to make them effective probes in pulmonary function and gas exchange processes, and in thermal balance.

## 3 Definitions of the different physiological effects of anaesthetics

### 3.1 GENERAL ANAESTHESIA

Anaesthetics are chemically diverse substances which can reversibly depress the functioning of part or whole of an organism's nervous system. They exert this distinctive effect with little initial disturbance to metabolism although many useful anaesthetics are secondarily potent biochemical agents. General anaesthesia is that state in which the normal functioning of the CNS is temporarily suspended and sensory input, motor output and reflexes are variously inhibited. In humans, consciousness or the specific sensation of pain may be abolished, yet pulmonary ventilation and cardiac function continue. There is thus a hierarchy of systems the constituent parts of which are progressively affected as the anaesthetic dose is increased and anaesthesia "deepens". In a practical case an experimental animal or patient enters into equilibrium with the particular constant dose of anaesthetic which is selected for the purpose of the experiment or operation. In short, general anaesthesia is a non-progressive, readily reversible inhibition of CNS function which can be brought about by any one of a wide range of substances which share a number of physical-chemical properties.

### 3.2 NARCOSIS

Anaesthetics exert reversible, inhibitory effects in many different physiological and biochemical preparations, for example in cell division and oxidative phosphorylation. Following convention (see Butler, 1950), we use the term narcosis to describe such "anaesthetised" systems. The concept of a non-progressive and highly reversible depression of function caused by chemically unreactive substances has been recognised since the nineteenth century. Anaesthesia is the clinically useful form of CNS "narcosis".

We should not equate a cell whose division is inhibited by a clinical anaesthetic with an animal anaesthetised with the same substance. What is of scientific interest, however, is that clinical anaesthetics are so effective in disturbing or inhibiting many physiological targets, additional to the nervous system. What the narcotised cell and the anaesthetised animal have in common is primarily a vulnerability to anaesthetics, which implies they have certain molecular targets in common, and, secondarily, an organisation which can sustain temporary dislocation without incurring serious damage. Simpler narcotised systems, such as enzymes, obviously only possess the former property, the molecular target, in common with the anaesthetised animal.

### 3.3 NARCOTIC-LIKE EFFECTS

There are a number of effects which anaesthetics and related substances can exert and for which the term narcosis may not be strictly appropriate. The ability of inert gases to offset some of the physiological effects caused by high hydrostatic pressure or the way anaesthetics induce liver enzymes are examples. These effects are derived, in part, from the physical properties of the anaesthetics, but their physiological consequences are not necessarily inhibitory.

### 3.4 NON-NARCOTIC PROPERTIES

Anaesthetics and inert gases also exert a variety of physiological effects which are not related to their narcotic potencies, although they are intimately related to some of their physical properties. For example, the physical properties of anaesthetics determine how they distribute themselves in the lungs and tissues of the body. Inert gases can provide an artificial atmosphere for men and test animals in which thermal balance may be altered in a way determined by the choice of gas.

We therefore distinguish four types of effect: anaesthetic, narcotic, narcotic-like and non-narcotic.

## 4 Substances

The terminology in the field of anaesthetics will at first appear confusing. We have used the following terms in what we believe to be a widely accepted sense, but to avoid ambiguity we attempt to define them.

A *general anaesthetic* is a substance which can bring about general anaesthesia. This statement may seem trite, but a substance may be properly referred to as an anaesthetic even in the context of narcosis, thus: "the effect of general anaesthetics such as chloroform on cell division . . .".

Another definition is: an anaesthetic is a drug capable of blocking the conduction of an action potential without affecting the cell's resting potential (Seeman, 1972). This definition embraces many substances which are incapable of causing anaesthesia, such as detergents. As a broad definition, it is only surpassed by "an anaesthetic is a substance which, when injected into an animal, produces a scientific paper"! It is useful to distinguish between nerve-blocking agents such as detergents and general anaesthetics.

A *local anaesthetic* reversibly blocks axonal conduction and can thereby abolish the perception of pain without affecting the individual's consciousness. General anaesthetics can, in principle, act in this way if by experimental ingenuity they can be administered locally at a suitably high concentration.

In the present context, a *narcotic* substance brings about narcosis, previously defined. Our practice is to avoid the use of the word narcotic as a noun and to use it as an adjective.

*Inert gases* are a special group of substances some of which are good general anaesthetics and many of which have narcotic, narcotic-like and, indeed, non-narcotic effects. The inert gases occupy an important place in certain sections of this book and in narcosis theory generally. Some of the earliest observations on general anaesthesia were made using an inert gas, nitrous oxide, although practical anaesthesia began with diethyl ether. The substance which probably comes closest to the ideal, chemically unreactive general anaesthetic is the inert gas xenon, but it has major practical disadvantages.

"*Related substance*" is a term which has already been used, and its meaning is best clarified by example. Oxygen is not an anaesthetic, nor an inert gas, yet in special experimental conditions its toxic effects may be suppressed to reveal anaesthetic properties entirely consistent with some of its physical properties. It is therefore a "related substance". So too are benzene and other hydrocarbons, and many other substances which have some narcotic effects but about which we have insufficient physiological knowledge to say more.

## 5 Chemical diversity

Table I.1 may clarify the situation, but its main purpose is to emphasise the range of chemical diversity (a) within the group of good general anaesthetics and (b) within the range of substances with narcotic properties. The main unifying feature of the substances listed in Table I.1 is primarily their physiological action and secondly their high solubility in hydrophobic solvents and apparent chemical inertness. It is quite extraordinary that many of the substances in the Table should be useful anaesthetics, and

TABLE I.1

Anaesthetics and related substances

| Chemical | Example | State |
|---|---|---|
| 1. Inorganic inert gases | Xenon[a], nitrous oxide[a], nitrogen | Gaseous |
| 2. Hydrocarbons | | |
| straight chain | Ethylene[a], propylene, methane | Gaseous |
| cyclic | Cyclopropane[a] | Gaseous |
| 3. Halogenated hydrocarbons | Chloroform[a], halothane[a] | Volatile liquid |
| 4. Ethers | Diethyl ether[a], divinyl ether[a], methoxyfluorane[a] | Volatile liquid |
| 5. Alcohols | Methanol, ethanol, benzyl alcohol | Liquid |
| 6. Barbiturates | Pentobarbitone, phenobarbitone, thiamylal[a] | Solid |
| 7. Secondary/tertiary amines | Procaine, lidocaine, tetracaine (local anaesthetics) | Solid |
| 8. Steroids | Alphaxalone (3α-hydroxy-5α-pregnane-11,20-dione) | Solid |
| 9. Phenothiazines | Chlorpromazine (tranquilliser) | Liquid |
| 10. Miscellaneous | Urethane (hypnotic) | Solid |
| | Chloral hydrate (hypnotic) | Solid |
| | Paraldehyde (hypnotic) | Liquid |
| | Morphine (analgesic) | Solid |
| | Sodium laurylsulphate (detergent) | Solid |
| | Trimethadione (anticonvulsant) | Solid |

*N.B.* The chemical divisions used are not entirely exclusive, e.g. chloral hydrate is a halogenated hydrocarbon.

[a] Good general anaesthetics.

that all should exert reversible perturbations in various physiological and biochemical preparations. Unfortunately, a full range of substances has not been used in the majority of physiological systems described in the book. Inevitably, different fields have different emphasis; for example, experiments with membranes frequently involve local anaesthetics or alcohols, but rarely inert gases. It is hoped that such biases do not obscure the coherence of the subject.

## 6 "Theories of anaesthesia"

A close correlation exists between the potency of the anaesthetics listed in Table I.1 and their olive oil–water partition coefficient (Fig. I.1). There are also good correlations between other closely related physical properties and anaesthetic potency (see Wulf and Featherstone, 1957). The implication is clear; in anaesthesia or narcosis the anaesthetic "dissolves" in a

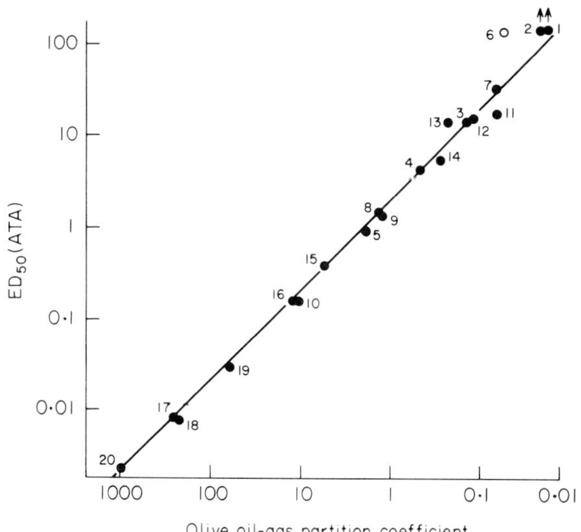


FIG. I.1 The correlation between the partial pressure of anaesthetics required to block the righting reflex in mice and the olive oil–water partition coefficient of the anaesthetics at 37°C. (From Miller *et al.*, 1972)
1, He; 2, Ne; 3, Ar; 4, Kr; 5, Xe; 6, $H_2$; 7, $N_2$; 8, $N_2O$; 9, $C_2H_4$; 10, $cC_3H_6$; 11, $CF_4$; 12, $C_2F_6$; 13, $C_3F_8$; 14, $SF_6$; 15, $CCl_2F_2$; 16, $CHClF_2$; 17, $CHCl_3$; 18, halothane; 19, ether; 20, methoxyfluorane.

molecular site with solvent properties similar to that of olive oil. Different anaesthetics thus impair the normal functioning of a target site, which we can imagine in a membrane preparation to be perhaps a specific ion-channel protein, its bilayer environment or indeed *any* or *all* of a number of targets. Chemically inert narcotic substances, are likewise hydrophobic, their potency being related to their solubility in a hydrophobic solvent.

The description of the molecular interactions which occur between anaesthetics and their hypothetical target molecules or sites is an elegant extension of solution chemistry. It has attracted successive generations of chemists who have gradually refined the dogma. The correlation between hydrophobicity and potency suggests that equal degrees of anaesthesia should be caused by equal local concentrations of anaesthetic molecules, or better, by equal activities of anaesthetics (Ferguson, 1939). One school of thought (Mullins, 1954) is that the presence of anaesthetic molecules is sensed by the system because of their molecular volume, and hence equal degrees of anaesthesia or narcosis should occur with the same molar volume expansion of the target site. The expansion can be envisaged as loosening of the molecular packing in a bilayer structure, following the intrusion of anaesthetic molecules, or as a displacement of solvent water by a conformational change in a protein brought about by anaesthetic-binding.

Whatever picture of anaesthetic–target interaction one prefers, it is evident that if the molecular volume expansion idea is correct, then anaesthesia should be abolished if the expansion can be prevented. This is thought to be what happens when high hydrostatic pressure is applied to an anaesthetised animal or a narcotised cell. Cases are known in which anaesthesia and narcosis are abolished by pressure and restored when the pressure is removed. There are other cases known in which pressure does not have this effect, and these are not pursued particularly vigorously by those arguing the volume expansion case.

It is not our intention to involve ourselves with the physical-chemical arguments which have been used to describe hypothetical anaesthetic–target interactions, sometimes called molecular theories of anaesthesia(!). Impressive though these arguments are, they have contributed little to our understanding of how anaesthetics actually inhibit nerve or CNS function. From the physiological viewpoint, real progress in understanding anaesthesia or narcosis can only be made in conjunction with increased knowledge of the target system or systems—be they excitable membrane, synapse, enzyme or structural protein. There are two rather different reasons for this. If we seek a physical description of an anaesthetic–target interaction, then clearly we must specify the target, even though many of the physical-chemical arguments appear to cope perfectly well with an imaginary one. The most detailed knowledge of an anaesthetic–target interaction is the case of xenon complexing with myoglobin (Schoenborn *et al.*, 1965; Schoenborn and Nobbs, 1966). The structure of this protein was known before its interaction with xenon was examined, and it is curious that the nature of the interaction seems to stand apart from that predicted from solution chemistry. It is also unfortunate that the xenon–myoglobin interaction is physiologically rather unimportant. Other "relevant" anaesthetic–target interactions are under investigation with every prospect of forging a link between solution-chemistry theories on the one hand and physiology on the other.

The second reason for identifying the target sites of anaesthetic action is physiological. If there is to be a full explanation of anaesthesia, then it will grow up with a much greater understanding of integrated polysynaptic pathways, and of mechanisms of consciousness itself. The extraordinary selectivity with which anaesthetics progressively inhibit more pathways as their dose is increased reveals an underlying organisation which is not based on molecule–molecule interactions, but on cell–cell or higher unit interactions. This is a difficult and neglected area with which, as we have indicated earlier, this book is not concerned. The physical-chemical treatments of anaesthesia and narcosis action are good rigorous examples of reductionist thinking and are thus severely limited.

Table I.2 lists a number of references which have contributed to our thinking about mechanisms of anaesthetic action. This book undoubtedly leans heavily on them from time to time, but the emphasis is on physiological rather than the physical aspects.

TABLE I.2

Physical-chemical aspects of anaesthesia and narcosis

| | |
|---|---|
| Meyer (1937) | Draws attention to the parallel between potency and lipid solubility. The concentration (mole fraction) in the lipid phase is the important feature |
| Ferguson (1939) | Thermodynamic activity of the anaesthetic emphasised |
| Mullins (1954) | Narcosis depends on the anaesthetic attaining a certain volume in the membrane (i.e. mole fraction × molecular volume) |
| Johnson *et al.* (1945)<br>Johnson and Flager (1951)<br>Brauer and Way (1970)<br>Miller *et al.* (1973)<br>Stern and Frisch (1973)<br>Eyring *et al.* (1973) | Pressure modifies anaesthesia and narcosis, emphasising the significance of molecular volume changes |

## 7 Concept of environmental physiology

Anaesthesia and narcosis stem from certain molecular interactions within an organised, physiological system. The non-narcotic effects of anaesthetics and inert gases (previously defined) likewise comprise a reversible physical interaction between an anaesthetic and an organised system, the outcome of which is potentially highly instructive. All four types of physiological effects discussed in this book, anaesthetic, narcotic, narcotic-like and non-narcotic, have in common a subtle, specific disturbance of the molecular machinery which drives a physiological function. Seen in the widest context these effects are really part of environmental physiology. Cells are vulnerable to anaesthetics because membranes and other hydrophobic molecular targets are "tuned" to function in an aqueous environment (Chapter II. 8 and Chapter IV). The gas-transporting systems of animals and the thermal balance of warm-blooded terrestrial animals are also shown to be "tuned" to the physical properties of the environment by the substitution of air by an appropriate volatile anaesthetic or inert gas (Chapter VII). Thus, despite the wide range of physiological systems and the variety of chemicals dealt with, this book is concerned with a body of knowledge which becomes coherent when seen as part of environmental physiology.

## References


Brauer, R. W. and Way, R. O. (1970). Relative narcotic potencies of hydrogen, helium, nitrogen, and their mixtures. *J. Appl. Physiol.* **29**, 23–31.

Butler, T. C. (1950). Theories of general anaesthesia. *Pharmacol. Rev.* **2**, 121–160.

Eyring, H., Woodbury, J. W. and d'Arrigo, J. S. (1973). A molecular mechanism of general anaesthesia. *Anesthesiol.* **38**, 415–424.

Ferguson, J. (1939). The use of chemical potentials as indices of toxicity. *Proc. Roy. Soc. B.* **127**, 387–404.

Johnson, F. H., Eyring, H., Steblay, R., Chaplin, H., Huber, C. and Gherardi, G. (1945). The nature and control of reactions in bioluminescence. *J. Gen. Physiol.* **28**, 462–537.

Johnson, F. H. and Flagler, E. A. (1951). Activity of narcotized amphibian larvae under hydrostatic pressure. *J. Cell. Comp. Physiol.* **37**, 15–25.

Meyer, K. H. (1937). Contributions to the theory of narcosis. *Trans. Faraday Soc.* **33**, 1062–1068.

Miller, K. W., Paton, W. D. M., Smith, E. B. and Smith, R. A. (1972). Physicochemical approaches to the mode of action of general anesthetics *Anesthesiol.* **36**, 339–351.

Miller, K. W., Paton, W. D. M., Smith, R. A. and Smith, E. B. (1973). The pressure reversal of anaesthesia and the critical volume hypothesis. *Molec. Pharmacol.* **9**, 131–143.

Mullins, L. S. (1954). Some physical mechanisms in narcosis. *Chem. Rev.* **54**, 289–323.

Schoenborn, B. P. and Nobbs, C. L. (1966). The binding of xenon to sperm whale deoxymyoglobin. *Molec. Pharmacol.* **2**, 495–498.

Schoenborn, B. P., Watson, H. C. and Kendrew, J. C. (1965). Binding of xenon to sperm whale myoglobin. *Nature*, **207**, 28–30.

Stern, S. A. and Frisch, H. L. (1973). Dependence of inert gas narcosis on lipid "free volume". *J. Appl. Physiol.* **34**, 366–373.

Wulf, J. and Featherstone, R. M. (1957). A correlation of van der Waals constants with anaesthetic potency. *Anesthesiol.* **18**, 97–105.

CHAPTER II

# Cell structure, movement and division



## 1 Introduction

No biologist can fail to be moved by the spectacle of a living cell seen with high-powered light microscopy. The cytoplasm is seen to stream, cilia flicker, membranous surfaces bulge, and particles jump, revealing a host of fundamental biological problems.

Anaesthetics and inert gases affect the structure, movement and many other visible properties of cells. This chapter examines the proposition that these effects illuminate some of the general principles of cellular structure and organisation. The fact that anaesthetics and inert gases have such practical importance adds a certain urgency to the subject.

## 2 Cytoplasmic structure

### 2.1 BULK CYTOPLASM

The bulk structure of living cytoplasm, its viscosity or consistency, was a property much studied by the early cell physiologists, and it remains an interesting and instructive parameter.

A dramatic and reversible solidification of the cortical cytoplasm of the giant Amoeba *chaos* occurs on immersion in an aqueous solution of diethyl ether (about 200 mM) and halothane (about 2 mM). The cortex becomes sufficiently solid to be sliced with a microdissecting tool and cell movement ceases. The interior of the cell remains fluid and contains many granules, whereas the solid cortex is free of visible inclusions. An electrical potential of 65 mV, with the cell interior positive, is present across the cortex, and in the absence of a membrane as judged by electron microscopy, the potential is thought to arise at a phase boundary. The tough cortex is probably a tightly cross-linked gel, providing a remarkable example of the ability of clinical anaesthetics to affect the structure of cytoplasm (Bruce and Christiansen, 1965).

Comparable doses of anaesthetics seem to fluidise the normally stiff cortical cytoplasm in *Amoeba proteus*, in the eggs of *Chaetopterus* (a worm), and the sea urchin *Arbacia* (Heilbrunn, 1920; Dougherty, 1937; Wilson and Heilbrunn, 1952). Here the measure of cytoplasmic consistency is the time which a standard centrifugal force had to be applied to the cells to shift intracellular granules a certain distance. It is difficult to reconcile these effects with those seen in *Chaos*. The slime mold *Physarum* is like *Chaos*, forming solid "islands" of cytoplasm when exposed to diethyl ether. Liquid cytoplasm persists in the cell as well, although under the influence of cyclopropane or chloroform the cytoplasm as a whole solidifies (Seifriz, 1941).

What are we to make of these rather conflicting results? First, it is important to know the dose of anaesthetic acting on the cytological target of interest. Should the reader consult the original papers, he will find that the experiments were not always carried out in a way which makes this possible, and the objectives of the earlier experimenters were more often exploratory than analytical. Anaesthetics at low dose can exert effects opposite to those seen at high dose. This is true of the effects on whole organisms as well as on cells. According to Heilbrunn it also applies to the effect of anaesthetics on plant cell cytoplasm, the outer cortex of which is liquefied by a low "anaesthetic" dose and solidified by a high dose. These points emphasise the importance of working with the physiological target of interest in equilibrium with the known anaesthetic dose. Often the anaesthetic may be satisfactorily measured as a partial pressure in a gas phase, but the chemical concentration of the aqueous phase in which the cell is immersed is also satisfactory.

Anaesthetics partition between aqueous and hydrophobic phases to an extent which depends on the physical properties of the individual anaesthetic. Solubility is generally increased by a reduction in temperature and hence partition coefficients change with temperature (Allott *et al.*, 1973). Thus, to define the dose of anaesthetic which a cell or any other physiological target is receiving, it is necessary to know (1) the ambient partial pressure or concentration of the anaesthetic, (2) the temperature, and (3) that the system is fully equilibrated. Doubts about some of these points confuse the comparison of the effects which anaesthetics have on the state of cortical cytoplasm, but there is no doubt that the effects are both profound and reversible.

Although cytoplasmic viscosity may seem a rather simple property to study, in fact it is peculiarly sensitive and by no means simple. Within most eukaryote cells there exists a cortical gel with more fluid internal cytoplasm which often streams. As we have only rudimentary knowledge of the basis for this intracellular differentiation or of the factors which determine the physical state of the cytoplasm, it is perhaps hardly surprising that anaesthetics produce apparently conflicting effects in different cells. For many years Heilbrunn expounded the view that $Ca^{2+}$ ions influence the state of cytoplasm and that anaesthetics affected the interior of cells by altering the binding of $Ca^{2+}$ ions. For reasons which the reader may wish to review elsewhere for himself (Berwick, 1951; Heilbrunn, 1956), a rigid cell cortex was thought to be achieved by $Ca^{2+}$ ions binding to the constituent molecules, and liquefaction of the cortex caused by a high dose of anaesthetic is to be explained as an anaesthetic-induced dissociation of $Ca^{2+}$ ions. It was thought that liquid, interior cytoplasm might also behave in this way and would therefore set firm when dissociated $Ca^{2+}$ ions diffuse in from the peripheral gel. This early view of the importance of calcium in cytoplasmic organisation and anaesthetic action persists and at least one contemporary hypothesis of the action of anaesthetics on synaptic transmission envisages a role for $Ca^{2+}$ ions (Chapter VI).

## 2.2 ULTRASTRUCTURE

The bulk properties of cytoplasm are the product of its macromolecular constituents, in particular the filaments and microtubules which confer mechanical strength on bulk cytoplasm. These structures have become major topics of research in recent years. Anaesthetics, in common with drugs and other probe molecules, can affect certain kinds of microtubules although we do not know in all cases if they do so directly or indirectly.

A spectacular effect is seen in the microtubular array in the protozoan *Actinosphaerium nucleofilum* (Fig. II.1). When this cell is exposed to anaesthetics the axopods retract or collapse, due to the disruption of the supporting microtubules (Allison *et al.*, 1970). Electron microscopy of the treated cells reveals little structure in the axopods although occasionally distinctive fibrous material is discernible. Following the removal of the anaesthetic the axopods grow to their original length as their microtubule support reconstitutes itself. Anaesthetics such as halothane, chloroform, methoxyfluorane, divinyl ether and cyclopropane were found to be effective in disorganising the microtubules at a dose twice that used clinically, whereas nitrous oxide was relatively more potent and diethyl ether had virtually no effect. The interesting question of molecular specificity therefore arises in this case, but so far remains unresolved. In the similar cell *Actinophrys*, macrotubular structures appeared after exposure to anaesthetics disrupting the normal microtubules.

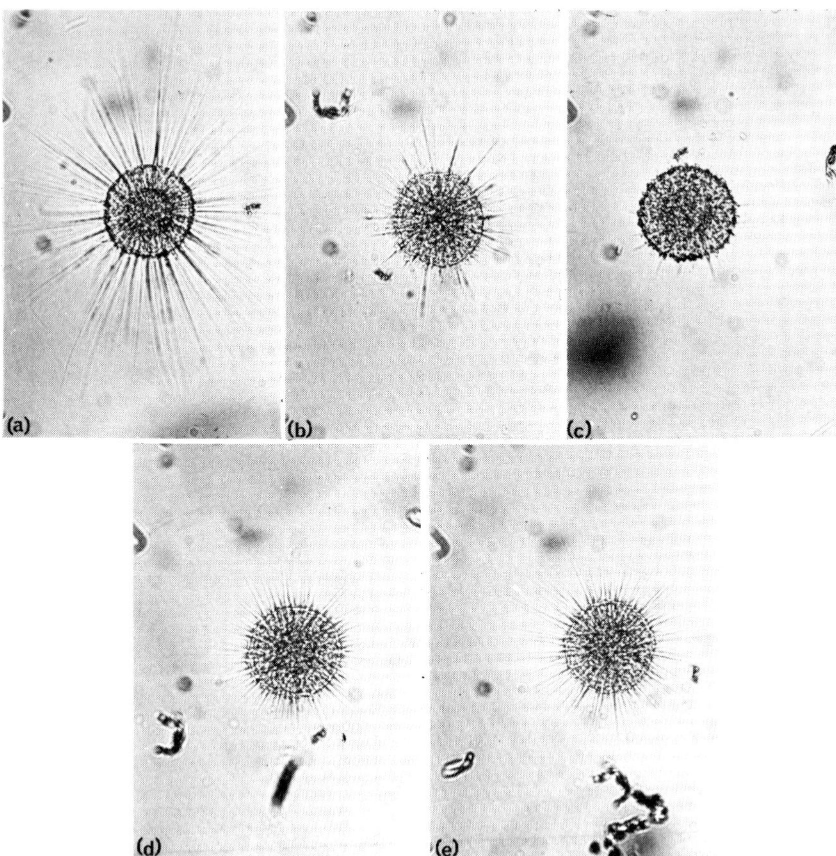


FIG. II.1. Effect of 0·45 per cent methoxyfluorane at room temperature on the axopods of *Actinosphaerium nucleofilum*. (a) before treatment; (b) after 3 min exposure; (c) after 8 min exposure; (d) 6 min after withdrawal of methoxyfluorane; (e) 15 min after withdrawal. (Allison *et al.*, 1970)

Although it is tempting to think that the anaesthetics act directly on the microtubules, it is possible that they affect the surrounding membrane or the cytoplasm, thereby altering the internal environment which favours microtubular stability. Other agents cause the microtubules in *Actinosphaerium* and *Actinophrys* to collapse, such as colchicine, low temperature and high hydrostatic pressure. Indeed, the microtubules are highly labile in the normal life of the cell.

The susceptibility of the microtubules to anaesthetics is, however, no oddity without general significance. Other microtubules, for example in the mitotic spindle of cleaving cells (see p. 29), are disrupted by the same agents. Clearly Nature has good reason for this, and it may be a consequence of the design of readily assembled and disassembled protein polymers. Microtubules and microfilaments contribute to cytoplasmic viscosity and in those cases in which anaesthetics liquefy a rigid cortex, it seems likely that a reversible disorganisation of the ultrastructural constituents takes place.

## 3 Motility

Cellular motility is bulk cytoplasm and its ultrastructural components in action. Table II.1 distinguishes the main forms of cellular movements. In Part A, the two main intracellular types of mechanical activity are listed and in Part B the ways in which whole cells move along are summarised.

TABLE II.1

Forms of cell motility (After Sleigh, 1973)

A. *Intracellular movement*
- 1. Cytoplasmic streaming
  - (a) Microfilaments, e.g. *Neurospora*, *Physarum*, *Nitella*
  - (b) Microtubules, e.g. axoplasmic flow
- 2. Movement of fibrous structures
  - (a) With change in length of structure: microfilament contraction, e.g. *Spirostomum* (protozoa); microtubular movement, e.g. chromosomes in mitosis
  - (b) With no change in length of structure: cilia, flagella

B. *Locomotion of whole cells*
- (a) Extrusion and flow of cytoplasm, e.g. amoeba
- (b) Traction by lamellae, e.g. leucocytes
- (c) Gliding, shear, e.g. blue-green algae
- (d) Ciliary and flagella propulsion, e.g. *Paramecium*, sperm

Cytoplasmic streaming may be inhibited by agents which gel or which further liquefy the cytoplasm. Unfortunately, we lack simultaneous measurements of cytoplasmic motion and viscosity during the exposure of a particular cell to anaesthetics. The slime mold *Physarum* normally exhibits vigorous streaming which, we have noted, stops under the influence of chloroform or cyclopropane which solidifies the cytoplasm. In the mold *Neurospora* the streaming is reversibly depressed by nitrous oxide, krypton and xenon, all at a partial pressure of 0·8 atm, but we do not know if the mechanism of inhibition involves solidification or liquefaction (Fig. II.2; Table II.1). Certain types of microtubular-based movement are sensitive to anaesthetics and the case of chromosome movement is separately discussed on p. 30. Axoplasmic transport and the susceptibility of neuronal microtubules to anaesthetics have attracted a number of experimenters, stimulated by the *Actinosphaerium* experiment of Allison *et al.* (1970).

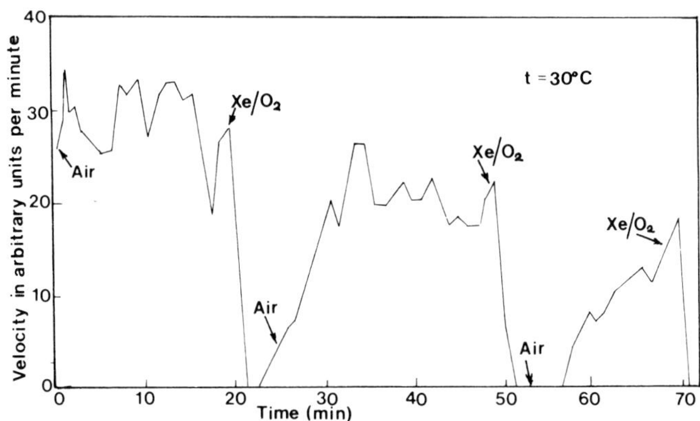


FIG. II.2. Effect of xenon on cytoplasmic streaming in *Neurospora crassa* at 30°C. The cell was exposed alternately to air and to a gas mixture of Xe 80 per cent and $O_2$ 20 per cent, at atmospheric pressure. (Doebbler *et al.*, 1967)

Axoplasmic transport is the term given to the rapid movement of particles (vesicles, mitochondria) or soluble substances such as enzymes, RNA and transmitters, along the length of an axon. The movement can occur either towards or from the cell body and at widely different rates. An underlying structure of filaments and microtubules seems to be necessary for axonal transport and some of these structures seem to be sensitive to anaesthetics. The situation is confusing and specific cases have to be considered in some detail.

The microtubules in the vagus nerve of the rabbit *increase* their number, judged from electron micrographs, when isolated nerves are incubated in the presence of 3 mM or 10 mM halothane at room temperature (Hinkley and Green, 1971). An anaesthetic dose of halothane or pentobarbitone has no effect on the microtubules in the optic tract in mice, and in the same nerve in rabbits, halothane fails to affect the transport of labelled leucine, although when combined with mild hypothermia some reduction in the rate of transport was apparent (Fink and Kennedy, 1972; Sauberman and Gallagher, 1973).

However, the microtubules in the isolated ventral nerve cord of the crayfish are disrupted by halothane. Initially, halothane treatment causes a new types of filament to appear in the form of a C. Over a 3 hr period the C filaments decline in number and are replaced by macrotubular structures. 420 Å in diameter. These are similar to those seen when microtubules are destroyed by vinblastine. Macrotubules were discovered by Tilney and Porter (1967) when disaggregating the microtubules of *Actinosphaerism* at low temperature and, as we have noted, similar large tubules were seen in halothane-treated *Actinophrys*. The conversion of a microtubule of 220 Å diameter to a macrotubule with a diameter much bigger is probably brought about by the linear array of protofilaments relaxing to a 45° spiral. The suggestion is therefore that halothane disrupts the normal cross-linking between linear protofilaments and favours an abnormal spiral-assembly. It may do this by binding directly to the tubulin or by less direct means. The macrotubules are merely a stable end-product and give no clue to the mechanism of halothane's action. Hinkley and Samson (1972) argue that the C-shaped filaments are an intermediate stage and that halothane acts initially to disrupt microtubules, and from the debris (C-filaments) macrotubules assemble by stacking.

While we await unequivocal evidence that halothane and other general anaesthetics can react with microtubules directly, considerable progress is being made in the study of how non-valotile hydrophobic substances affect microtubules and bind with tubulin.

The local anaesthetic lidocaine has been shown to inhibit reversibly the transport of labelled compounds in rat nerves both *in vivo* and *in vitro*, an effect which is only loosely coupled with the disruption of the microtubules (Byers *et al.*, 1973; Fink and Kish, 1976) and which might therefore be caused by some other metabolic disturbance. Whatever the details of lidocaine's action in axons are, it possesses the property of binding with purified tubulin in a manner which blocks polymerisation (Haschke *et al.*, 1974). Barbiturates also bind with tubulin and block its polymerisation, and they also disrupt the microtubules in isolated frog sciatic nerves in a way which inhibitors of aerobic metabolism fail to mimic. Barbiturates such as methohexitone, thiopentone, pentobarbitone, etc. block polymerisation in proportion to their hydrophobicity, with the exception of hexobarbitone whose inhibition is anomalously weak (Edström *et al.*, 1975). The binding of chlorpromazine (a tranquilliser) to brain tubulin is probably the best understood example of an interaction between tubulin and a hydrophobic molecule. On the tubulin molecule there appears to be two binding sites, one of high affinity which binds one chlorpromazine molecule and a lower affinity site which binds eight to nine molecules (Hinmann and Cann, 1976). These authors speculate that as chlorpromazine inhibits axonal transport in fairly high concentrations it might do so by way of the weaker interaction with tubulin. Its ability to block mitosis at much lower concentrations implicates the high-affinity tubulin binding site.

A cautionary comment on relating *in vitro* experiments to anaesthetic effects *in vivo* is called for. Isolated components, such as purified tubulin, may be more labile than functional polymerised tubulin in a "living" microtubule. One reason is chemical, that isolation removes stabilisers, and the other reason is physiological, that repair and regulatory processes may counteract the potential disturbance caused by a probe molecule. It is also clear from the different susceptibilities of neuronal microtubules to anaesthetics that important differences between microtubules exist.

The locomotion of whole cells is affected by anaesthetics in ways which are far from simple.

The effects seen by Goldacre (1952) in *Amoeba discoides* demonstrate some of the complexity of cell movement and the way anaesthetics may upset it. The first obvious response of this species of amoeba to anaesthetics is an increased sensitivity to touch. Normally an advancing pseudopod responds to a prod from a micro-manipulator by a localised contraction or gelation. In the presence of a low dose of chloroform (about 5 mM aqueous concentration) a prod elicits an exaggerated response from a pseudopod. A higher dose of chloroform (20 mM) causes the cell surface to "blister", leading to a clear zone around the periphery, separating the now expanded cell surface from the cortex. (Compare with *Chaos*.) A prod from the micro-manipulator elicits no response in these conditions, unless it is sufficient to cause the surface membrane to touch the underlying gel, in which case a response is seen.

Although an amoeba exposed to a high dose of chloroform is undoubtedly experiencing a uniformly distributed partial pressure of anaesthetic, the cell exhibits some polarity in the appearance of the clear layer which comes between membrane and cortical gel. The "contractile" tail region is reluctant to produce the clear layer and, according to Goldacre, this insensitivity to the anaesthetic derives from the membrane being compressed laterally, which renders it resistant to the insertion of anaesthetic molecules. Amoeboid progression in *Amoeba proteus*, in the slime molds *Physarum* and *Dictyostelium*, and in mammalian lymphocytes are all inhibited by anaesthetics. Wiklund and Allison (1972) found the movement of *Dictyostelium*, which is powered by microfilaments rather than microtubules, is inhibited by clinical doses of halothane, chloroform, divinyl ether and others, but not by diethyl ether. Lymphocytes cultured from the mammalian lung and thymus gland were stopped by a clinical dose of halothane, yet macrophages from lung appeared unaffected (Nunn *et al.*, 1970). Whatever intracellular mechanism of cell movement is involved, traction between substrate and cell is required, and this may be impaired by the anaesthetics.

Clear blisters appear on the surface of many cells exposed to moderate doses of anaesthetics, on, for example, the lymphocytes but not the macrophages studied by Nunn and his colleagues, and blisters were also seen on the surface of *Dictyostelium*. In short, when considering cell movement we cannot ignore the action of anaesthetics on the cell surface even when an apparently internal cellular function is uppermost in our minds. A remarkable demonstration of this was given by Marsland over 40 years ago when he demonstrated the phenomenon of "cut-off".

The narcotic potency of the homologous series of alkanes increases with the length of the carbon chain up to about $C_{10}$. Longer chain paraffins are weaker narcotics, although intensely hydrophobic. "Cut-off" refers to the abrupt change in potency with increased carbon chain length.

The cause of "cut-off" will become apparent from Marsland's experiments. *Amoeba dubia* was exposed to the longer chain alkanes in three different ways: (1) by immersion in an aqueous solution of alkanes, (2) by micro injection, and (3) by depositing a droplet on to the external surface of the cell (Marsland, 1934).

Amoebae immersed in an aqueous solution of heptane or octane became immobilised, but in solutions of duodecane or tetradecane they were unaffected for up to 5 days. They became immobilised within 5 hr, however, on receiving an injection of duodecane. Tetradecane injections took longer to act, and the shorter chain alkanes exerted their effect very quickly. The application of a droplet of duodecane to the external surface of the amoeba caused the cell to become immobile much more quickly than an injection. This proved to be generally true of all the alkanes used. These observations suggest two things. First, the higher paraffins are insufficiently soluble in water to gain access to their hydrophobic site of action in the cell. This is the basis of "cut-off" in this cell, although in other preparations "cut-off" may have other explanations. Secondly, the cell surface is a crucial target in the narcosis of amoeboid movement.

Marsland was also able to demonstrate lateral diffusion in the cell surface, by placing a drop of olive oil on to the surface of an amoeba already narcotised by a surface application of duodecane. The cell showed partial recovery of movement, almost certainly due to the diffusion of some of the duodecane from the cell surface into the olive oil droplet. The effectiveness of injected droplets of duodecane and tetradecane suggests that hydrophobic diffusion paths may connect the cell interior with the cell surface but it is difficult to explain why the final narcosis is not so severe as in the case of surface applications. The alkane may be primarily acting, albeit feebly, on a different intracellular component of the cell's motile system.

In the interaction between the cell surface and the movement of the whole amoeba which both Marsland and Goldacre demonstrated, we see a parallel with the membrane of a muscle and its internal contractile machinery. It is likely that anaesthetics (including here the alkanes) affect the control which the membrane of an amoeba exerts over the contractile machinery within the cell. Actomyosin filaments have been described in amoeboid cells and the sensitivity of movement in one such cell, *Physarum*, to $Ca^{2+}$ ions closely resembles that of muscle (Komnick *et al.*, 1973).

Returning to the point that anaesthetics affect the locomotion of cells in different ways (we have already noted that lymphocytes are immobilised by clinical agents whereas macrophages are not), it is important to be clear on the criterion used to measure movement. A cell which loses directional movement may cease to make headway because it tends to move simultaneously in all directions. An example would be a fibroblast treated with the microtubule poison colcemid (Vasiliev *et al.*, 1970). We need to distinguish between this effect and the specific inhibition of the machinery which generates the motile force. It is by no means clear which of these two possibilities apply to lymphocytes.

Movement of neutrophil leucocytes into the peritoneum of mice is depressed by halothane anaesthesia, but here the situation offers even more plausible sites of anaesthetic action. In addition to the possibilities already mentioned, the cell's movement into the peritoneum might, for example, be blocked by changes in the walls of blood vessels. This is of clinical interest because it has been shown that the diminished population of neutrophils in the peritoneum of anaesthetised mice leads to an increased susceptibility to experimental infections (Bruce, 1966, 1967). The ability of anaesthetised neutrophils to carry out phagocytosis is also suspect, and indeed Graham (1911) demonstrated that anaesthetics depressed leucocyte phagocytosis both *in vitro* and in human subjects. Hamburger (1916), using leucocytes from the horse, found that a low dose of chloroform actually stimulated phagocytosis and a higher dose depressed it. We do not know the sites at which phagocytosis is affected by anaesthetics, but it is clear that effects on a sensory or receptor stage and a motile stage should be distinguishable.

## 4 Cell division

### 4.1 EARLY OBSERVATIONS

In animal cells division is a contractile process and in plant cells it more often involves the deposition of a new cell wall, but in the division of both, anaesthetics exert a number of interesting effects. In 1902, Wilson described how diethyl ether caused the astral rays in fertilised sea urchin eggs to fade in a few minutes, leaving in their place clear vacuolated cytoplasm. In such cells, although cytoplasmic cleavage was blocked, nuclear division proceeded to produce binucleate cells. Prolonged exposure to ether thus resulted in multinucleate eggs. Wilson (1902) also emphasised the different effects of ether according to the stage of the mitotic cycle when it was applied. First there is the case in which ether was applied to cells in late anaphase, when the chromosomes have separated into daughter nuclei and the asters have passed their maximal development. Ether causes the asters to fade, mitosis is arrested and when the cell is removed from the solution of ether, each nucleus resumes where it left off and promptly enters into the next division to yield four cells. In contrast, when ether is administered to cells earlier in anaphase the asters are suppressed, and on removal of the ether they partially recover but cleavage fails to take place. A binucleate cell is thus formed, but restores itself to normal when it divides into four uninucleate daughter nuclei at the next division. The distinction between the two cases probably arises from the state of the centrosome, which is divided prior to inhibition in the first case but undivided in the latter. These observations are good examples of anaesthetics' reversible and stage-dependent effects within the cell cycle. Wilson's observations failed to stimulate much work at the time, although in 1914 Lillie extended his observations and more recent investigations have taken over where Lillie left off.

We will consider first the ways in which anaesthetics affect the cell cycle before returning to the astral rays, now known to be labile microtubules, which Wilson saw.

## 4.2 THE CELL CYCLE

### 4.2.1 *Sea urchin egg*

Cells which are progressing through their cell cycle are either preparing for division, committed to it, or in the mechanical process of dividing. Figure II.3 summarises the conventional terminology in this field. In the study of division-inhibitors such as anaesthetics, the concept of the transition point is important, marking distinct stages in the cel cycle when abrupt changes occur. Transition points may mark the onset of mitosis or of some metabolic change. Transition points have been defined in prokaryotes such as *E. coli*, the mold *Physarum*, and in eukaryotes such as the sea urchin egg and the protozoan *Tetrahymena*.

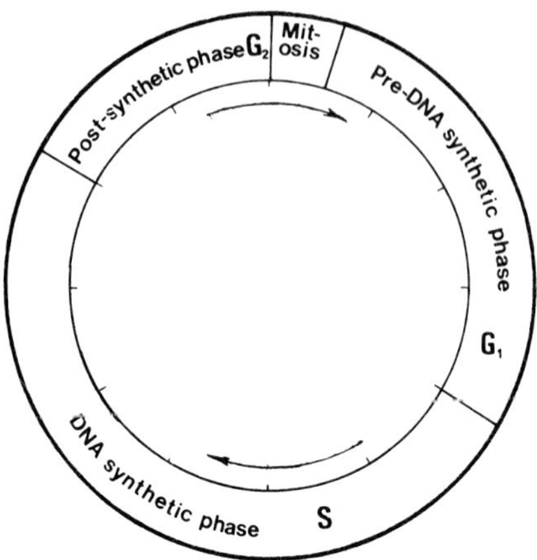


FIG. II.3. The cell cycle. The life span of a cell is shown as a circle with DNA synthesis (S) occupying a mid-portion, the periods of no DNA synthesis as $G_1$, and $G_2$, and mitosis as a small portion of the whole.

In the case of the sea urchin egg (*Psamechinus miliaris*) the cytochrome poison carbon monoxide reveals a transition 35 min after fertilisation, that is at the start of mitosis. When carbon monoxide is applied to the fertilised cell before the 35 min stage, the ensuing delay in cleavage is equal to the duration of exposure. If applied after 35 min, then the subsequent cleavage is not delayed but the next-but-one is, and by an amount equal to the exposure to carbon monoxide. Thus the first part of the cell cycle of the sea urchin egg involves aerobic metabolism which prepares the cell for division, and the second part of the cycle is independent of aerobic metabolism. Diethyl ether (7 mM in seawater) causes no division-delaying effect in the first part of the cycle, but when applied after the carbon monoxide transition point it causes a cleavage delay equal to the period of treatment (Swann, 1954). Ether slows the cells' progression through mitosis by blocking the assembly of the mitotic figure. This may be seen with polarised light microscopy, which reveals the integrity or otherwise of the mitotic apparatus in the living cell. Similar doses of ether have no effect on the rigidity of the cell cortex, which can be measured with a cell elastimeter, a device consisting of a suction micropipette which deforms and thereby measures the stiffness of a cell cortex. A higher dose of ether (15 mM) applied before the transition point slows the cells' progress through the first part of the cell cycle. When applied after the transition point it causes a cleavage delay rather longer than a similar exposure to the low dose of ether.

Slowing the first part of the cycle might be an effect on aerobic respiration; blocking the assembly of the mitotic apparatus and disrupting the preformed structure might be a direct anaesthetic-aster interaction. Generally, anaesthetics are not very effective respiratory inhibitors and progress through the cell cycle is unlikely to be affected much by a slight depression of respiration. In *Tetrahymena* or sea urchin eggs whose division cycle is totally blocked by urethane or chloral hydrate, for example, aerobic respiration is reduced by only 25–30 per cent. Also, the functioning of the mitotic apparatus is unaffected by a cessation of aerobic metabolism.

### 4.2.2 *Experiments with other animal cells*

The eggs of sea urchins are a favourite cell for division studies, but they have special properties, being primed for rapid division and negligible growth, and clearly differ from tissue cells which have to grow before dividing. Cultured tissue cells provide the experimenter with cells which are also convenient to use but which both grow and divide. Investigations with such cells have shown additional stages in the cell cycle to be sensitive to anaesthetics. Nunn and colleagues (Sturrock and Nunn, 1975) have used hamster lung fibroblasts in both suspension and monolayer culture. Clinical anaesthetic doses prolong these cells' mean generation time. For example, a partial pressure of 0·02 atm (2 per cent) halothane at 37°C doubled the generation time in both monolayer cultures, in which time-lapse photography was used to record growth, and in suspension cultures which were monitored by cell counts. As the cells approach mitosis so they assume a distinctive rounded form, the duration of which provides a measure of the mitotic stage within the cell cycle. Halothane prolongs mitosis thus measured by 40 per cent. The mitotic index is depressed to 60 per cent of the control value over a 3 hr exposure. Analysis of the occurrence of the different stages of mitosis established that cells progressed through $G_2$ into prophase more slowly in the presence of halothane than in the controls. Halothane's effect on mitosis is unlike that of the drug colchicine, which specifically arrests cells only at metaphase. In the presence of halothane, metaphase is little affected, but multinucleate cells arise, following nuclear division without cytoplasmic cleavage and, additionally, by cell fusion.

A very unusual synergism between low doses of nitrous oxide and halothane also produces multinucleate fibroblasts (Sturrock and Nunn, 1976). Nitrous oxide also upsets mitosis in cultured heart cells (Kieler, 1957), and HeLa cells cultured in both monolayers and in suspension show rather specific effects when exposed to nitrous oxide (Rao, 1968). Metaphase was specifically inhibited, the most effective dose being 5 atm. $G_1$, S and $G_2$ were not prolonged and DNA, RNA and protein synthesis were all unaffected. This is a colchicine-like effect (Table II.2). It would be interesting to expose fibroblasts to hyperbaric nitrous oxide and HeLa cells to halothane, to see if the apparent differences between these experiments are caused by the cells or by the anaesthetics.

TABLE II.2

Effect of nitrous oxide on the progression of HeLa cells through the mitotic cycle (Rao, 1968)

| Duration of treatment (hr) | $N_2O$ at 5·1 atm | | Control with colcemid | | Control without colcemid | |
|---|---|---|---|---|---|---|
| | M.I. | Labelled (%) | M.I. | Labelled (%) | M.I. | Labelled (%) |
| 1·0 | 0·03 | 35·0 | 0·03 | 39·0 | 0·02 | 36·0 |
| 2·0 | 0·08 | 51·0 | 0·06 | 47·5 | 0·04 | 49·5 |
| 4·0 | 0·11 | 55·0 | 0·12 | 55·0 | 0·03 | 58·0 |
| 8·0 | 0·27 | 78·0 | 0·26 | 78·0 | 0·03 | 79·5 |
| 16·0 | 0·62 | 80·0 | 0·65 | 81·5 | 0·03 | 83·0 |

M.I.: mitotic index; labelled: percentage of cells labelled with tritiated thymidine.

According to the reports of Bruemmer *et al.* (1967) and Ueda (1974), anaesthetics and inert gases impair the adhesion of cells to a glass substrate. The division of cells cultured in monolayers may require the continued presence of a substrate, so it is conceivable that certain types of cell division might be affected by anaesthetics which impair adhesion. Table II.3 summarises the effect of xenon on the way HeLa cells come out of suspension to attach to the walls of the culture tube, and on pre-existing attached

TABLE II.3

Effect of xenon on HeLa cell adhesion (after Bruemmer *et al.*, 1967)

| 1. Xenon applied prior to adhesion to glass<br>Gas mixture (atm) | % Cells not attached after 24 hr treatment |
|---|---|
| 0·75 $N_2$–0·2 $O_2$ | 9 |
| 4·2 Xe–0·2 $O_2$ | 36 |
| 2. Xenon applied to attached cells | |
| 0·75 $N_2$–0·2 $O_2$ | 0·5 |
| 4·2 Xe–0·2 $O_2$ | 6·7 |
| 7·2 Xe–0·2 $O_2$ | 69 |

cells. The results show that attachment (cell adhesion) is inhibited and that the multiplication of previously attached cells is also reduced, either by detachment or by more direct inhibitory means. It does seem important to know the extent to which anaesthetics and inert gases can affect division by such an indirect means as cell adhesion and by the more direct interactions with the cell cycle.

In tissues growing *in vivo* the cell cycle is not usually accessible to the experimenter, but cells in the crypts of Lieberkühn in the intestinal epithelium are an exceptional case. They divide approximately once a day and may be pulse labelled with tritiated thymidine injected into the peritoneum of a suitable experimental animal such as a rat (Bruce and Traurig, 1969). Those cells which are undergoing DNA synthesis in the presence of the labelled precursor incorporate it and subsequently reveal the labelled DNA in chromosomes which appear in mitosis. (Halothane has no effect on the

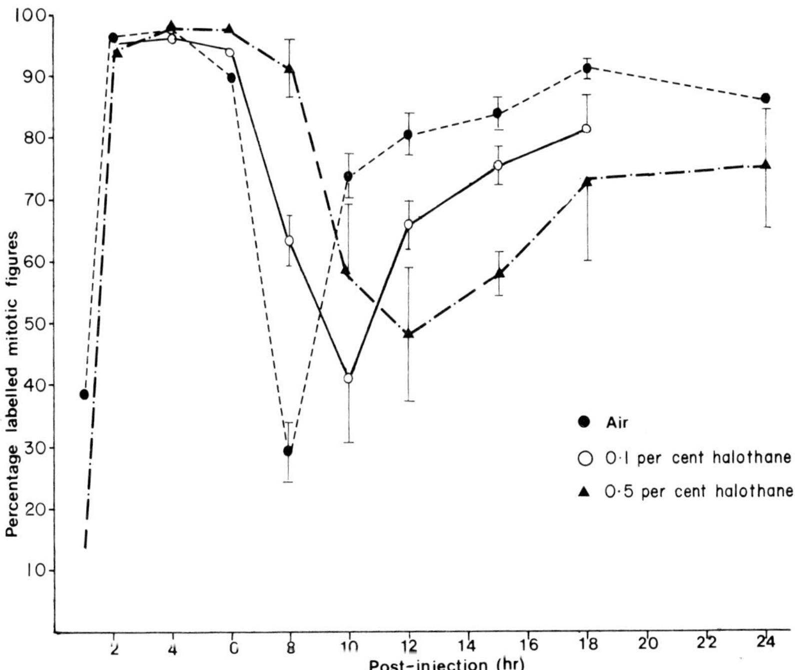


FIG. II.4. The effect of halothane on the cell cycle in the epithelial crypt cells of the small intestine. Percentages of epithelial crypt cell mitotic figures which contained a radioactive label at intervals following injection of 250 $\mu$C tritiated thymidine and subsequent inhalation of air or 0·1 per cent or 0·5 per cent halothane in air. Bars indicate $\pm$SE. DNA synthesis (S) is the time interval between ascending and descending curves crossing the line of 50 per cent labelled mitotic figures. (Bruce and Traurig, 1969)

uptake of thymidine (Bruce, 1972a).) A graph of the incidence of labelled mitotic figures against time shows that control and halothane-treated cells obtain peak labelling at the same time (Fig. II.4). Not shown in the figure is the fact that the mitotic index is unaffected by halothane. These two observations suggest that the prolonged high incidence of labelled mitoses in the treated group is due to an extension of the period of synthesis. A closer inspection of Bruce and Traurig's data also suggests that $G_2$ may be prolonged, a point made by Grant *et al.* (1974), whose anaesthetic experiments on plant cell mitosis are described below.

It is worth emphasising that in addition to finding no effect on the mitotic index of the epithelial cells, no abnormal mitotic figures were seen. In this *in vivo* experiment halothane is not acting as a mitotic inhibitor but merely slows down the progression through S phase.

Another aspect of the cell cycle which anaesthetics affect is the transformation of a cell from a resting state to an actively "cycling" condition. Lymphocytes transformed ("activated") by phytohemaglutinin (PHA) are susceptible to halothane, which blocks the surge of synthesis which such cells normally exhibit (see Section 7.1).

### 4.2.3 *Plant cells*

In 1878 Bernard showed ether inhibited the growth of seedlings, an effect which implied that cell division was susceptible to anaesthetics. In fact the root cells of the broad bean *Vicia fabia* are particularly convenient tissues for studies of mitosis and have featured in a number of modern anaesthetic experiments. Conventional techniques of visualising the chromosomes readily provide data on how the mitotic cycle is affected. Halothane at a partial pressure of 0·01 atm at room temperature (a) depresses the mitotic index (Fig. II.6) and (b) it slows the progression of cells through both S, $G_2$ and all the phases of mitosis (Fig. II.5). Note, for example, in Fig. II.5 how halothane differs from colchicine. In halothane the cells do not *preferentially* accumulate in metaphase, because progress through the preceding stages of the cycle is impeded. Some cells do exhibit metaphase arrest, however. When the anaesthetic gas ethylene is applied to the same preparation the mitotic index is also depressed by a mild dose but metaphase arrest does not occur. A higher partial pressure (1·3 atm at room temperature) induces metaphase arrest (Fig. II.6(b)), which suggests that ethylene is capable of disorganising spindle microtubules.

*Allium cepa* (onion) root cells exhibit mitotic arrest in metaphase when exposed to the inert gases propane (1·5 atm), nitrous oxide (6 atm), methane (40 atm), argon (75 atm), nitrogen (80 atm) and hydrogen (200 atm) (Ferguson *et al.*, 1950). Polyploid cells usually arise from these disturbances to mitosis.

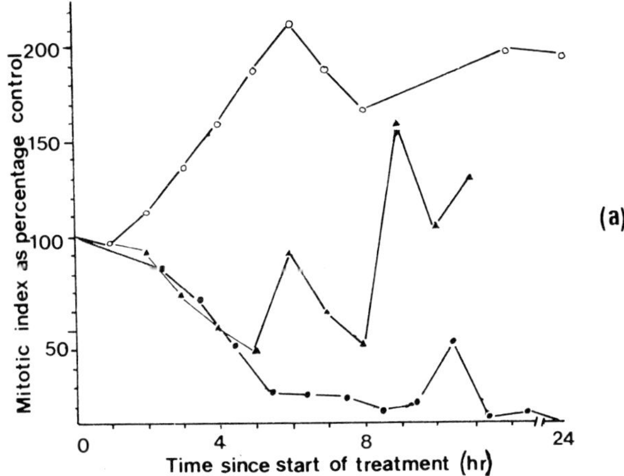


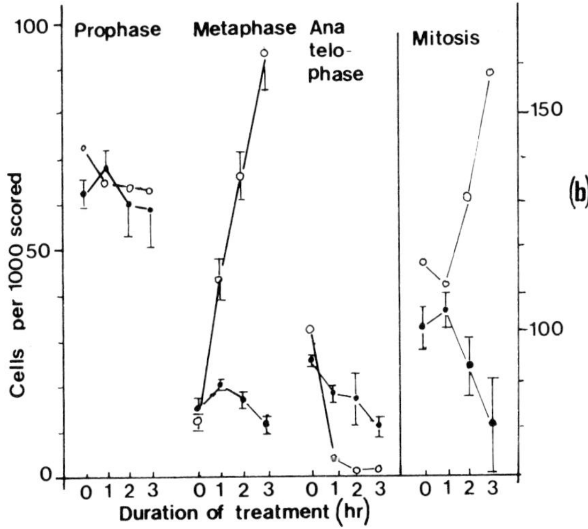


FIG. II.5. The effect of halothane on the cell cycle in the root cells of the bean *Vicia fabia*. (a) Mitotic index (●, ▲) including a comparison with colchicine (○). (b) Mitotic stages (●), also including a comparison with the effects of colchicine (○). (Grant *et al.*, 1974)

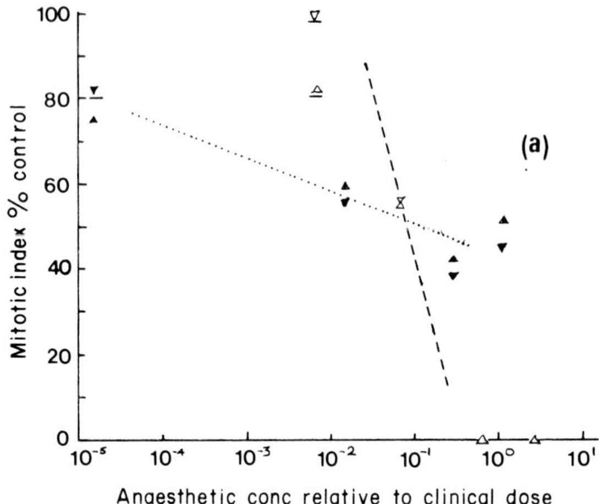


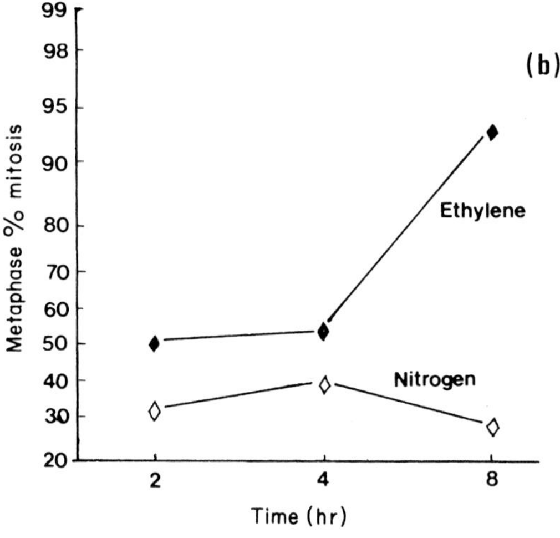


FIG. II.6. The effect of ethylene on the cell cycle in the root cells of the bean *Vicia fabia*. (a) Compared with the effects of halothane. Closed symbols are ethylene, open are halothane; △ = 24 hr; ▽ = 48 hr exposures. (b) Higher dose of ethylene (0.86 atm). (Powell *et al.*, 1973)

The hydrogen experiment reported by Ferguson and his colleagues is incomplete because it fails to distinguish between hydrostatic pressure and gas partial pressure, and at 200 atm the former is likely to disturb mitosis. It will be shown in a later section that inert gas pressure and pure hydrostatic pressure exert interesting interactions in dividing cells.

*Summary:* The effects of anaesthetics on the cell cycle in plant and animal cells show general similarity with the effects of sedatives, tranquillisers and many other oil-soluble agents. Mitosis is often, but not always, susceptible and in those cases in which mitosis is inhibited, it is not necessarily the only part of the cell cycle to be affected. Anaesthetics are rather unspecific cell cycle inhibitors. Table II.4 provides a representative list of anaesthetic-like agents which affect mitosis; the reader seeking a more comprehensive list should consult the old reviews by Östergren (1944) and Andersen (1966).

TABLE II.4

Anaesthetics, inert gases and related substances which inhibit mitosis in various cells at concentrations broadly comparable to anaesthetic doses (Compiled from Andersen (1966) and other sources)

| | |
|---|---|
| Halothane | |
| Diethyl ether | |
| Chloroform | Ethyl alcohol |
| Nitrous oxide | Chloral hydrate |
| Cyclopropane | Pentobarbitone |
| Ethylene | Nitrobenzene |
| Propane | Chloretone |
| Methane | Ethyl urethane |
| Xenon | Phenyl urethane |
| Krypton | |
| Argon | |

### 4.2.4 *Mitosis*

The susceptibility of mitosis to anaesthetics is a consequence of the labile nature of the microtubules involved. The mitotic apparatus (MA) of the egg of the sea urchin, *Arbacia punctulata* (and of cells in general), consists of radiating microtubules. Transverse sections of the MA in metaphase, passing through the mitotic spindle, reveal in the electron microscope some 2000 microtubules (Cohen and Rebuhn, 1970). Estimates of the number of aster microtubules are somewhat lower at 1200. The microtubules are composed of the protein tubulin, organised in protofilaments lying side by side, each protofilament being a polymer with a 33 Å repeat unit. Tubulin comprises only a small part of the total MA protein with estimates ranging from 5–15 per cent.

It is characteristic of the tubulin dimer that it binds the antimitotic alkaloids colchicine, podophyllotoxin and vinblastine at different *loci*, and as we have noted, neuronal tubulin also binds lidocaine and chlorpromazine. Polymerised tubulin exists in dynamic equilibrium with the depolymerised form, and at a morphological level microtubules are known to exist in equilibrium with precursors which are not readily visualised by light or electron microscopy. Protein synthesis inhibitors exert no effects of interest during mitosis because the dynamic polymer–monomer equilibrium exists between preformed constituents. Colchicine, in contrast, blocks mitosis at metaphase. Tubulin isolated from brain tissue, which is a rich source of the protein but not of the mitotic apparatus, undergoes polymerisation *in vitro*, which is thermally driven in the presence of ATP, the appropriate $Mg^{2+}$ ion concentration and the absence of $Ca^{2+}$ ions, to yield microtubules.

The MA is intimately associated with the mechanical act of chromosome separation which is the ultimate purpose of mitosis. We have seen that anaesthetics variously disturb the formation and function of the MA, but we do not know if they do so by acting on tubulin, or by affecting the monomer–polymer equilibrium in some less direct way by affecting the intracellular $Ca^{2+}$ concentration, for example. And we do not know if the agents listed in Table II.4 all act in the same way. They probably do not. In the case of mitosis in *Haemanthus* endosperm cells, for example, methanol seems to disorientate the spindle without inhibiting the motive force generated on or around it (Östergren, 1960). This seems to be a rather special effect and much less drastic than that seen with volatile anaesthetics, which totally disorganise the spindle.

General anesthetics can only exert molecular disturbances by weak, usually hydrophobic, forces. It is therefore reasonable to consider whether such forces are likely to lead to an interaction with tubulin *in vivo*. In view of what is known about the polymerisation of the MA of certain cells, the answer is fairly clear. Inoué *et al.* (1975) have developed quantitative polarised light microscopy to the stage where it is possible to estimate the mass of MA microtubules assembled in a dividing cell. From their measurements of birefringence changes throughout mitosis at different temperatures and hydrostatic pressures, thermodynamic data have been derived. These show that polymerisation proceeds with an increase in entropy and enthalpy and with a small free energy change, broadly consistent with hydrophobic interactions being primarily involved. In particular, the increase in entropy can be envisaged as a release of ordered water surrounding hydrophobic groups where mutual adhesion binds sub-units together. Colchicine binding also involves a hydrophobic interaction, so there is little doubt that the MA has prospective anaesthetic binding sites, but a direct general anaesthetic–MA tubulin interaction has yet to be demonstrated.

#### 4.2.5 *Cell synchrony*

Many techniques for converting a population of growing cells, whose division cycle is asynchronous, into a population of cells which divides in synchrony use inhibitors which act to arrest cells at specific stages of the cell cycle. Since anaesthetics disturb metabolism in a relatively mild way they might be rather useful synchronising agents, but their efficiency will depend on the specificity with which they inhibit the cell cycle. Although anaesthetics are not highly specific cell cycle inhibitors, Rao (1968) found some division synchrony in HeLa cells treated with 5 atm of $N_2O$, although no synchrony in macromolecular synthesis was obtained (Table II.2).

#### 4.2.6 *Cleavage*

The constriction of the cytoplasm of a dividing cell is a special form of cellular contractility. Agents which can selectively block it give rise to abnormally large and multinucleate cells. In the case of the sea urchin egg the contractile force is thought to be generated by a band of microfilaments whose specific susceptibility to anaesthetics has not been studied but which is affected by the cell inhibitor cytochalasin *B* (Schroeder, 1972). One possibility is that the microfilaments are unaffected by both anaesthetics and colchicine, which generally act more effectively on microtubules. Thus by disturbing the mitotic apparatus of dividing cells, anaesthetics might upset the reactions which initiate cleavage and not actually affect the contractile machinery, but from what we have seen of the effects of anaesthetics on cytoplasmic viscosity and cell movement it would be surprising if they did not more directly affect the microfilaments.

## 5 Growth and division in tissues

### 5.1 BLOOD CELLS

In the mid 1950s a few selected patients suffering from severe tetanus were lightly anaesthetised with $N_2O$, and also received other drugs such as *d*-tubocurarine, sodium pentobarbitone and antibiotics (Lassen *et al.*, 1956). Over several weeks a decline in the number of granulocytes, leucocytes and thrombocytes (platelets) was noted (Fig. II.7). Fewer erythrocytes were also proliferated from the bone marrow. Other cases suggested that it was primarily the nitrous oxide which was responsible for inhibiting cell growth. Confirmation of nitrous oxide's potency against human white

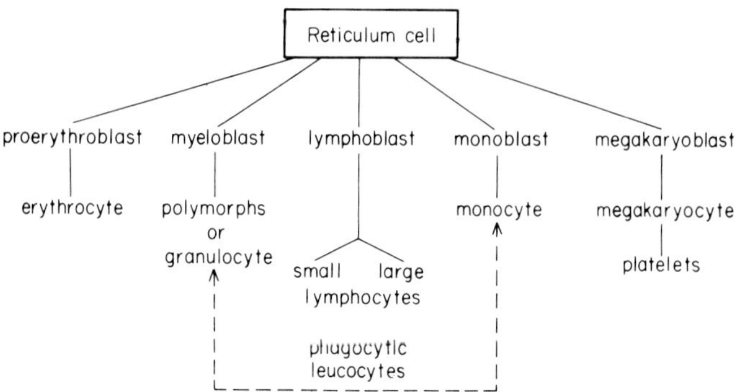


Fig. II.7. Simplified scheme showing the terminology of red and white cells.

cells emerged in two case histories reported by Eastwood *et al.* (1963). The study began with experiments on rats which, when exposed to 80 per cent $N_2O$–20 per cent $O_2$ for 6 days, showed a white cell count of less than 10 per cent normal (Green and Eastwood, 1963). Lymphocytes seem to be less susceptible than granulocytes, and megakaryocytes were relatively unaffected by the treatment. Two patients suffering from myelogenous leukemeia were then given nitrous oxide in a breathing mixture and both showed a markedly reduced white cell count after several days (Eastwood *et al.*, 1963).

Further animal experiments by other workers using weaker gaseous anaesthetics and inert gases produced consistent data, summarised in Table II.5 from Aldrete and Virtue (1967a, 1967b). Erythrocytes were shown to be relatively unaffected by the inert gases, which is also consistent with Green and Eastwood's findings. However, a more recent experiment in which rats were exposed to nitrous oxide, also for a period of days, revealed a depression in the numbers of both leucocytes and erythrocytes. The normal increase in reticulocytes in response to exogenous erythropoietin was also depressed by nitrous oxide (Table II.6, Johnson *et al.*, 1971).

A reduced number of leucocytes (leucopenia) also occurs in animals and men exposed to other anaesthetics and related substances, including sodium pentobarbitone, urethane and benzene. The latter can be an industrial hazard to health (Ert and Rhoades, 1939; Usenik and Cronkite, 1965).

The term leucocyte is applied to several types of cell: polymorphonuclear granulocytes, lymphocytes and monocytes (Fig. II.7). The previously mentioned studies showed that inert gases affected granulocytes in particular, and halothane appears to have a similar selective effect (Bruce and Koepke, 1966). Figure II.8 shows the decline in the leucocyte count of rats anaesthetised with halothane.

TABLE II.5

Changes in the blood cell count of rats exposed to anaesthetics and inert gases for 6 days at atmospheric pressure. Figures are percentage of control counts (animals in air) or of counts obtained from the animals before treatment (Aldrete and Virtue, 1967a, 1967b)

| | Cyclo-propane 6% (+30% $O_2$ 64% $N_2$) | Acetylene 50% (+50% $O_2$) | Ethylene 60% (+40% $O_2$) | Xenon 65% (+35% $O_2$) | Argon 80% (+20% $O_2$) | Helium 80% (+20% $O_2$) |
|---|---|---|---|---|---|---|
| Erythrocytes | 86 | 82 | 89 | 82 | 98 | 99 |
| Thrombocytes = platelets | 87 | 78 | 81 | 96 | 97 | 96 |
| Leucocytes | 44 | 57 | 52 | 70 | 94 | 99 |

TABLE II.6

Effect of nitrous oxide on the response of rats to exogenous erythropoietin (Johnson *et al.*, 1971)

| | Conditions | Reticulocytes, % blood count |
|---|---|---|
| Control | 96 hr, air | 4·4 |
| | 96 hr, air + erythropoietin | 10·1 |
| Experimental | 96 hr, air + 80% nitrous oxide | 1·4 |
| | 96 hr, air + 80% nitrous oxide + erythropoietin | 2·0 |

Rats injected with tritiated thymidine showed labelled dividing cells whose rate of maturation could then be followed. In normal conditions, the labelled cells progressed to maturity in about 20 hr, but in the presence of halothane (0·45 per cent) very few cells matured in that time. Halothane thus acts on dividing cells to cause granulocytopenia. The proliferation of antibody producing lymphocytes in the rat spleen is also depressed by halothane treatment (Wingard and Humphrey, 1969).

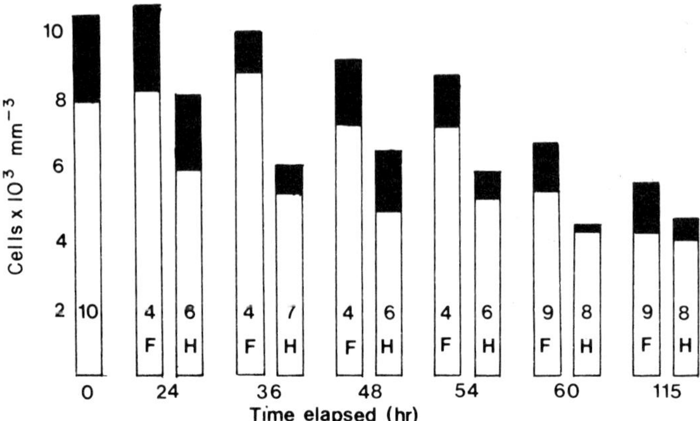


FIG. II.8. The effect of halothane on the leucocytes of rats. The histogram shows mean peripheral blood counts of fasting (F) and halothane anaesthetised rats (H) over a period of time. The black portion of each bar represents granulocytes (98 per cent of which are neutrophils) and the remainder of the bar shows lymphocytes and monocytes. The number of animals is indicated in each bar. (Bruce and Koepke, 1966)

## 5.2 EMBRYO TISSUES

The expectation that anaesthetics would upset the normal development of embryos by interfering with cell division has been confirmed by a variety of studies. The chicken embryo has been a popular choice of test organism because it lacks the complications of a placental mammal, whose embryological development can be affected by agents acting either on the placental circulation or directly on the foetal tissues.

Nitrous oxide, cyclopropane, halothane and diethyl ether have all been shown to cause abnormalities in developing chicken eggs (Andersen, 1968; Snegireff and Andersen, 1971). In the case of 20 per cent cyclopropane applied for 12 hr to $1\frac{1}{2}$-day-old embryos, cells in the neural tube showed an increase in the metaphase index, whereas 50 per cent cyclopropane caused a decrease in both prophase and mitotic indices. The former suggests a more stage-specific inhibition and the latter a generalised slowing down of the cell cycle.

One of the disadvantages of using chick eggs is that they take a long time to equilibrate with the ambient atmosphere. The effective partial pressure of cyclopropane in the neural tube was probably about half that applied externally, so the lower dose mentioned falls into the clinical range. Other experiments with halothane and nitrous oxide produced results similar to those obtained with cyclopropane. Thus, 36 hr exposures to 80 per cent $N_2O$ increased the metaphase count in the neural tube without altering the mitotic index. An intermediate dose of halothane (1·2–1·6 per cent) reduced the neural tube mitotic index, but a higher dose did not, although development was impaired at both doses. The developmental consequences of the cytological disturbances caused by the anaesthetic are similar to those elicited by many other teratogenic compounds.

Andersen (1968) demonstrated that chicken embryos were most susceptible to anaesthetics on the fourth day of development (Table II.7), and the common abnormalities included incomplete development of the caudal vertebrae (rumplessness) and malformed ribs and skull.

TABLE II.7

Frequency as percentage control of abnormalities in chicks exposed to cyclopropane (from Andersen, 1968)

| Day | Cyclopropane (% applied for 6 hr) 20–40 | 40–60 | 60–75 |
|---|---|---|---|
| 3 | 3·9 | 12·8 | 6·8 |
| 4 | 8·0 | 6·2 | 38·0 |
| 6 | −0·1 | 7·8 | — |
| 8 | 6·8 | 0·4 | 3·7 |

Care has to be taken to ensure proper incubation conditions in these experiments because the chick embryo is highly sensitive to changes in its environment. The introduction of alien gases can exert unexpected effects. Weiss *et al.* (1965) studied the effect of helium on chick development and were able to detect interesting changes. Despite all the precautions taken, survival of the embryos growing in helium was about half normal, and the emerging chicks were lighter than the controls. Young chicks surviving the helium environment showed no abnormalities other than eating more and having higher than normal heart rates. These effects are in line with helium's negligible narcotic potency at atmospheric pressure and its high heat transfer capacity, which imposed a thermal stress on the developing animals (see Chapter VII).

Mice and rats exposed to anaesthetics during gestation also give birth to young with abnormalities. For example, mice injected with sodium pentobarbitone during the first 2 weeks of gestation failed to produce young in about half the cases, and many of the treated young had incompletely developed limbs or lacked limbs altogether (Setälä and Nyyssönen,

1964). Rats exposed to nitrous oxide or halothane during gestation produce offspring with complicated skeletal abnormalities (Basford and Fink, 1968: Shepard and Fink, 1968).

The development of insects involves a complicated embryogenesis and later development, which in the case of the wasp *Mormoniella* is sensitive to the inert gases nitrogen (10 atm), argon and sulphur hexafluoride in the usual potency series. Helium at 10 atm was found to be without effect (Frankel and Schneiderman, 1958).

Placental hypoxia is a potent teratogen, but in view of the relatively mild effects of anaesthesia on the mammalian cardiovascular system and of the teratogenic effects in chick embryos, it seems likely that mammalian embryonic abnormalities result from anaesthetic action in cell division. The implications for anaesthesia of pregnant women are not at all specific, but clearly caution must prevail. For a fuller treatment of the subject, including the difficult question of satisfactory control experiments, the reader should refer to the multi-author books edited by Fink (1968, 1972).

### 5.3 TUMOURS

The inhibitory effects of anaesthetics on dividing cells would be of special clinical (and scientific) interest if it could be used to arrest preferentially the division of tumour cells. Examination of this possibility by Parbrook (1967) revealed little that was encouraging and Setälä and Nyyssönen (1964) found that sodium pentobarbitone had less effect on the mitosis in malignant epidermal hyperplasia than on mitosis in benign epidermal hyperplasia in mice. The administration of nitrous oxide to patients suffering from myelogenous leukemia has already been mentioned. It achieved an inhibition of white cell division and some slight temporary beneficial effects (Eastwood *et al.*, 1963). Perhaps the most likely way of using a general anaesthetic to destroy tumour cells would be to use it in conjunction with specific cytotoxic drugs. The anaesthetic might help to prolong certain susceptible parts of the cell cycle in a way which imposes negligible stress on the patient, thereby exposing more tumour cells to a stage-specific drug (Bruce *et al.*, 1972b).

The carcinogenic properties of anaesthetics are rather obscure but the potential risk to medical personnel has been discussed by Cohen (1976).

## 6 Narcotic-like effects of inert gases on cell division

### 6.1 ANAESTHETICS, INERT GASES AND HYDROSTATIC PRESSURE

Some inert gases are sufficiently potent to be useful clinical anaesthetics, such as nitrous oxide, cyclopropane and xenon (the usefulness of the latter is primarily limited by its high cost). Other inert gases are very weak and only exert an anaesthetic effect at high partial pressure (e.g. argon, nitrogen and hydrogen), of which the last two are important in human diving. On theoretical grounds the weakest of all inert gases is helium, which is also used in diving (see Chapter VII). It is of considerable interest to establish whether helium is capable of exerting a narcotic or even a narcotic-like effect. Attempts to do so using whole animals are complicated and lead to equivocal results. Experiments with dividing cells, however, have technical advantages which have enabled decisive results to be obtained.

The problem of measuring the putative effects of helium or any other weak inert gas is to separate the effect of the gas from the hydrostatic pressure which is necessarily present. In the case of the relatively potent gas, nitrous oxide, which affects cell division at partial pressures in the range 1–10 atm, the hydrostatic pressure is of little consequence. When weaker agents are applied at tenfold partial pressures then hydrostatic pressure becomes important. The simplest way of distinguishing between the effects of hydrostatic pressure and gas partial pressure is to compare the effects obtained with hydraulic compression (free of gas phase) with those obtained in the presence of a suitably high partial pressure of gas. This has been done in the case of the cell division of the ciliated protozoan *Tetrahymena pyriformis*. This cell divides amitotically into two cells once every 3–4 hr and may be grown in tiny hanging drop cultures placed on the window of a microscope observation pressure vessel (Macdonald, 1975). Hydrostatic pressure is applied to the cells hydraulically by compressing liquid paraffin which fills the pressure vessel, and helium pressure may be applied to the cells by mounting the cultures in an air-filled, humidified pressure vessel and bleeding in helium from a high-pressure source. Microscope counts yield cell multiplication curves in either case, and these are compared.

A second method is to pressurise a cell suspension within a glass syringe and compare the resultant cell multiplication with that found in a similar volume of culture equilibrated with helium in a rocking pressure vessel. This method only permits cell counts to be made on a limited number of samples. As no serious discrepancy between the two methods has come to light only data obtained with the microscope pressure vessel will be dealt with here.

A comparison of the rates at which cell numbers increase at high hydrostatic pressure and in high pressures of helium shows a difference at 175 atm after 7 hr exposure (Fig. II.9); 250 atm hydrostatic pressure is particularly significant because it reversibly inhibits division with the cells otherwise unimpaired. In the same pressure of helium, cell division persists.

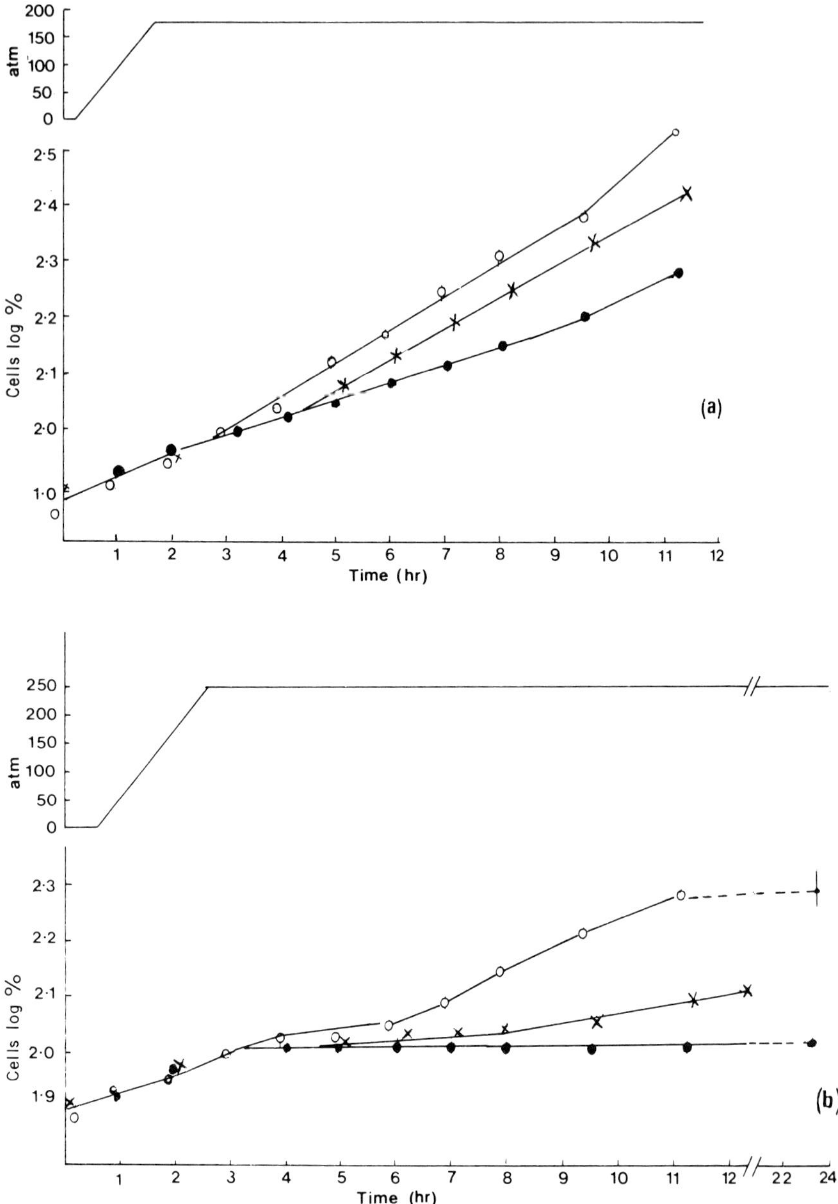


FIG. II.9. Effect of helium (×) and hydrogen (○) on the division of *Tetrahymana pyriformis* at 25°C, compared with the rate of division at comparable hydrostatic pressures (●). Vertical scale: number of cells normalised to 100 per cent. (a) 175 atm pressure of helium, 165 atm hydrogen (+10 atm helium) and 175 atm hydrostatic pressure. (*b*) Similarly, 250 atm helium and hydrostatic pressure, and 240 atm hydrogen (+10 atm helium). (Macdonald, 1975)

Helium thus counteracts the inhibitory effects of hydrostatic pressure, and similar experiments with hydrogen show that it too alters the effects of hydrostatic pressure, the critical partial pressure being 130 atm applied for $6\frac{1}{2}$ hr. The experiments provide little information about the sensitivity of different stages of the cell cycle to the gases. Cleavage is blocked by 250 atm hydrostatic pressure, and so He and $H_2$ must offset that effect, but the cells are also held in interphase by the same pressure, and so He and $H_2$ presumably act there too. He, and $H_2$ to a greater extent, probably counteract hydrostatic pressure by stabilising pressure-labile components which would otherwise be impaired by hydrostatic pressure. They probably do so by way of weak hydrophobic interactions (Macdonald, 1975). The act of cleavage in high-pressure helium proceeds at a relatively normal speed, suggesting that pressure-labile components involved in the process are fully stabilised by the helium molecules. Cells exposed to high pressures of hydrogen for many hours become somewhat amoeboid, with an apparently excessive area of membrane enclosing the cell body. Generally, cells in hydrogen are more active and more normal in appearance than cells in helium.

Should we regard the "anti-pressure" effects of these gases as comparable to the actions of the more potent agents which inhibit *Tetrahymena*'s division (Table II.8)? There are several similarities. Both show reversible

TABLE II.8

Hydrostatic pressure and partial pressures of inert gases and anaesthetics (atm) which influence the division of *Tetrahymena pyriformis* at 25°C (Macdonald, 1978)

| | Sub-effective dose | | Effective dose | |
|---|---|---|---|---|
| Halothane | 0·001 | No effect alone but combine with 100 atm hydrostatic pressure to inhibit division, i.e. additive interaction | 0·01 | Inhibitory alone, counteracted by 100 atm hydrostatic pressure |
| Chloroform | 0·001 | | 0·005 | |
| *n*-Propane | 0·25 | | 0·75 | |
| Nitrogen | 30 | | 60 | |
| Hydrostatic pressure | 100 | Indistinguishable from 1 atm | 140 325 | Inhibitory |
| Nitrogen | 100 | See text | 175–250 | Counteracts inhibition and lysis caused by hydrostatic pressure |
| Hydrogen | <130 | Indistinguishable from hydrostatic pressure | 130–400 | |
| Helium | <175 | | 175–250 | |

effects at partial pressures which, in whole animals, cause anaesthesia. The effective partial pressure of 130 atm $H_2$ in *Tetrahymena*, for example, is close to its narcotic dose in mice. Helium, and to a lesser extent $H_2$, also exert a concomitant hydrostatic pressure on animals, which causes motor disturbances and complicates measurements of potency. In the case of helium, we can only predict a theoretical narcotic dose from its physical properties. Using the righting reflex of mice this hypothetical narcotic dose is approximately 160 atm, which is very close to the critical pressure of 175 atm He in *Tetrahymena*.

The ability of helium and hydrogen to counteract pressure in dividing cells shows some similarity with the way inert gases and anaesthetics counteract the pressure-induced motor disturbances in whole animals. *Tetrahymena*'s division is inhibited by orthodox doses of clinical agents (halothane, chloroform) and also by *n*-propane (Table II.8). These inhibitory effects are themselves reversed by the application of 100 atm hydrostatic pressure, which on its own has no effect on division.

It is natural now to consider what effect nitrogen and other gases of intermediate potency might have on division. Does nitrogen, for instance, offset hydrostatic pressure more effectively than helium or hydrogen? Two different series of experiments have produced conflicting data. Figure II.10(a) shows 100 atm $N_2$ to be inhibitory, which one would expect from its clinical potency (Kirkness and Macdonald, 1973). Figure II.10(b) shows that in a subsequent series of experiments much higher pressures of $N_2$ allow a slow rate of division, i.e. counteract the inhibitory effects of hydrostatic pressure. Isobaric transfer to 250 atm hydrostatic pressure then blocks division. In this second series the cells were not inhibited by 100 atm $N_2$. These phenomena have been fully discussed elsewhere (Macdonald, 1978), with the conclusion that the sensitivity of the cells to nitrogen is dependent on unspecified culture conditions. Table II.8 summarises our knowledge of anaesthetic and inert gas effects on *Tetrahymena*'s division. It would be interesting to know the factors which determine whether nitrogen acts as an inhibitory agent or as a weak inert gas able to counteract hydrostatic pressure.

## 6.2 OTHER NARCOTIC-LIKE EFFECTS OF INERT GASES

A number of earlier studies on the effects of inert gases at high pressure failed to include a comparison with hydrostatic pressure, and are therefore inconclusive. An exception was the demonstration that helium and several other inert gases reduce the oxygen-dependent sensitivity to X-rays shown by root cells of the broad bean *Vicia fabia* and mouse Ehrlich ascites tumour cells. Table II.9 lists the gases which confer protection and it is apparent

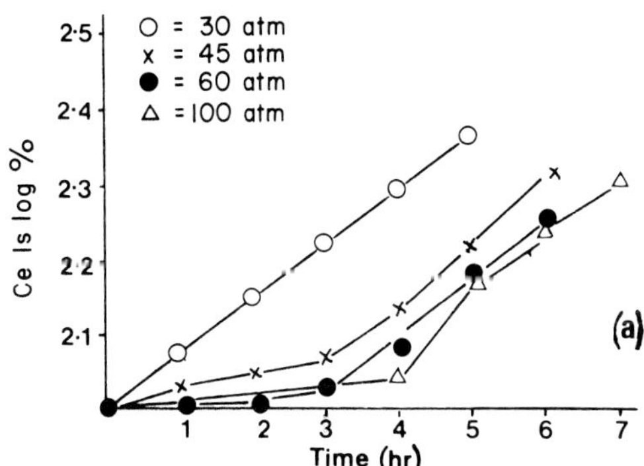


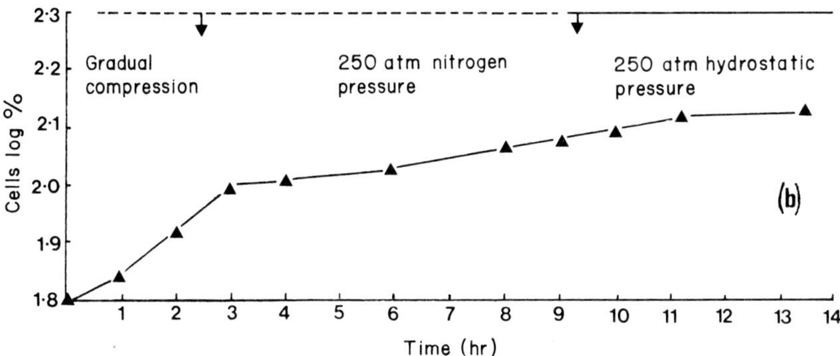


FIG. II.10. Effects of hyperbaric nitrogen on the division of *Tetrahymena pyriformis*. (a) From Kirkness and Macdonald (1973). (b) From Macdonald (1978). In (b), nitrogen gas is isobarically replaced with hydraulic pressure at 250 atm pressure.

that they conform to the narcotic potency series. Remarkably, this protective phenomenon is little studied. It may be related to the protection which nitrogen (33 atm) and, to a lesser extent helium, confer on cells cultured in hyperbaric oxygen (Stephens, 1969). Their effect may be to displace oxygen from a complex which is involved in the process of radiation or hyperbaric oxygen damage (Longmuir, 1963).

Some cellular effects of inert gases seem inconsistent with the narcotic potency series established with animals. The action of $N_2$ in *Tetrahymena* is possibly a case in point and so is the failure of nitrogen at 120 atm to inhibit the growth of *Neurospora* (Fig. II.11(a)). Perhaps if the hydrostatic

TABLE II.9

Effectiveness of various gases in reducing the oxygen-dependent X-ray sensitivity of *Vicia fabia* roots (from Ebert and Hornsey, 1958a, 1958b)

| Gas | Pressure (atm) for 50% reduction in sensitivity due to oxygen |
|---|---|
| Helium | 55 |
| Hydrogen | 55 |
| Nitrogen | 12·5 |
| Argon | 2 |
| Krypton | 2 |
| Cyclopropane | 2 |
| Xenon | 1·1 |
| Nitrous oxide | (more effective than xenon in ascites tumour cells) |

pressure effect were to be taken into account, $N_2$ would fall into its expected place between argon and neon, but that would raise new questions about the different pressure-sensitivities of the inert gas effects. HeLa cells are inhibited by the same gases in the orthodox sequence (Bruemmer *et al.*, 1967). It has been claimed that at pressures close to atmospheric helium potentiates anoxia in the seeds of rye and that it affects the growth of *Neurospora hyphae* (Buchheit *et al.*, 1966; Latterell, 1966; Schreiner *et al.*, 1962). Figure II.11(b) shows that helium (partial pressure approximately 1 atm) can hardly be distinguished from nitrogen, but xenon and krypton do seem to be inhibitory. The linear relationship between growth rate and diffusivity (molecular weight) shown in Fig. II.11(b) was held to imply that the inert gases influence growth by affecting some diffusion limiting factor. The small number of experiments hardly justifies this, but as the mold was grown in sealed, gas-tight containers in which the gases were recirculated, it is conceivable that some volatile growth-limiting factor diffuses from the hyphae at a rate influenced by the gaseous environment. On the whole, the provisional conclusion is that the data show no distinction between the weaker agents He, Ne, $N_2$, Ar, but xenon is probably slightly inhibitory. The same conclusion may be drawn from similar experiments with HeLa cells (Bruemmer *et al.*, 1967).

## 6.3 ETHYLENE

Ethylene's anaesthetic properties were well known over 50 years ago, and

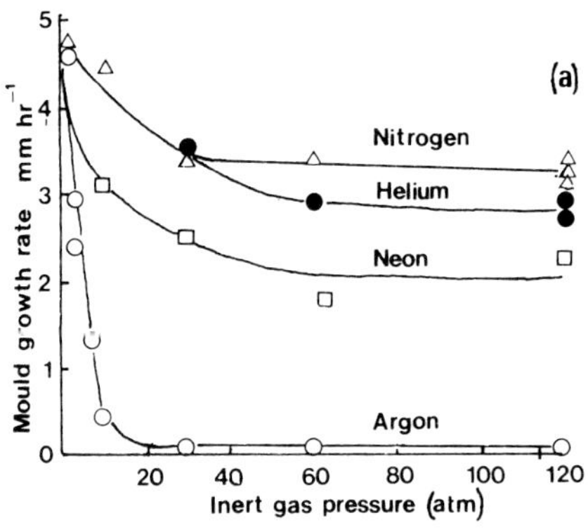


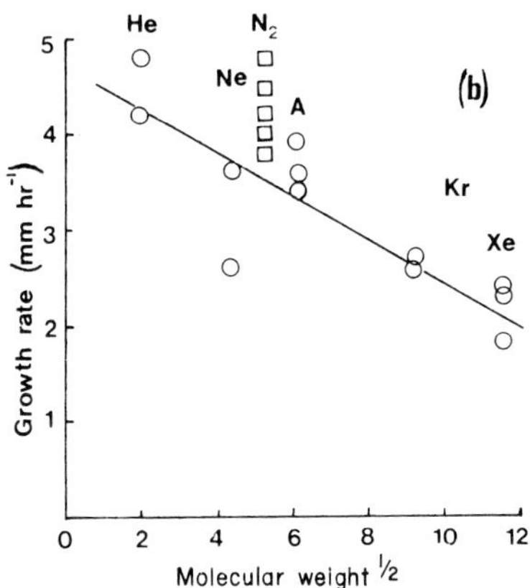


FIG. II.11. Effects of inert gases on *Neurospora*. (a) Hyperbaric inert gases on rate of growth of the hyphae. (b) The relation between growth rate of the hyphae and the molecular weight of the gas applied at 0·8 atm + 0·2 atm $O_2$ partial pressure. (Buchheit *et al.*, 1966)

in all respects ethylene is an orthodox anaesthetic gas. We have seen that ethylene inhibits the cell cycle, and mitosis in particular, in much the same way as other clinical agents. Additional to its narcotic properties, however, it has an important physiological role in plants, where it is synthesised and functions as a hormone.

The hormonal effects of ethylene include the shedding of leaves, differential growth in stems and leaves, and the inhibition of growth in stems and root tips (Abeles, 1973). It is thought that ethylene exerts its various hormonal actions in plant tissue by a common mechanism. Burg *et al.* (1971) summarise the ethylene–receptor interaction as a non-covalent binding reaction which involves oxygen. The receptor is probably a metal-containing protein with some affinity for carbon dioxide which can competitively inhibit ethylene. The effective concentration of ethylene the hormone is of the order of 1 part in $10^7$, in contrast to the effective dose of ethylene the anaesthetic, which is 1 part in 2. At a dose which is high for its hormonal action, 8 hr exposure of ethylene reduces the mitotic index in *Vicia* root cells and in the meristem of *Pisum* (10 ppm and 50 ppm, respectively). An equivalent dose of halothane has no effect (Fig. II.6). Other comparisons between ethylene and halothane have been made by Powell *et al.* (1973), who conclude that the narcotic and hormonal effects are quite separate phenomena.

The significance of ethylene in narcosis theory is that it demonstrates how a simple molecule can exert highly specific effects when the receptor is complex and presumably programmed to initiate certain processes. It raises the question, how direct are the effects of anaesthetics and inert gases? Do they fortuitously trigger set responses? In the present context, one might ask whether anaesthetics exert most of their effects by disturbing the cell membrane, whose natural role it is to mediate between "external" signals such as hormones, and "internal" processes such as synthesis and cell division. The following section deals with how anaesthetics influence a number of interactions between cells and provides more evidence for the susceptibility of membranes to anaesthetics.

## 7 Cellular interactions

### 7.1 LYMPHOCYTE ACTIVATION

Peripheral lymphocytes do not divide actively unless stimulated by highly specific antigens or by less specific stimulating agents called mitogens. In their normal condition lymphocytes are in the resting stage of the cell cycle, designated $G_0$. Mitogens are interesting and useful experimental tools, the most commonly used being PHA (phytohaemaglutinin) and Con-A (Concanavalin A), both of which are extracted from plants. Different cells exhibit different sensitivities to mitogens, for example, PHA seems to activate T-lymphocytes more than B-lymphocytes. The activation process involves some specificity and is complicated, but some features of the cells' entry into division are established. Within an hour or so of adding PHA to a culture of human lymphocytes there is an increase in the uptake of small molecules by the cell, an increase in RNA and protein synthesis, followed by a peak in DNA synthesis 72 hr later.

A clinical dose of halothane inhibits the activation of lymphocytes *in vitro*. DNA synthesis, measured as the rate of incorporation of tritiated thymidine, is inhibited and so is RNA and protein synthesis, also judged by the incorporation of labelled precursors (Bruce, 1972a, 1972b, 1975; Cullen *et al.*, 1972). It has been shown that halothane is without effect on the entry of thymidine into the cells, but a similar control experiment for RNA and protein precursors have not been reported. The earliest effect which halothane is known to have on activated lymphocytes comes 16 hr after the addition of PHA, but there is some evidence that RNA synthesis is inhibited only 2 hr after activation. To pinpoint the halothane effect, it may be necessary to expose the cells to the anaesthetic for very short periods.

Two negative, but interesting, effects shed some light on the action of the anaesthetic. First, it probably does not involve cyclic AMP (Bruce, 1976a). Secondly, since antimicrotubule drugs (e.g. colchicine) do not delay the DNA synthesis following activation by Con-A, it is plausible to argue that halothane does not exert its inhibitory effect by disturbing the skeletal microtubular structure situated on the inner face of the cell membrane (Betel and Matijnse, 1976).

Activation is best achieved in populations of cells rather than in individuals, which suggests that cell–cell interactions may be involved in the process, and by affecting these an anaesthetic might inhibit activation indirectly. It might therefore be rather misleading to say that halothane blocks the progression of the individual lymphocyte from $G_0$ to $G_1$, although it certainly inhibits activation in a cell population, which is of some clinical interest because it probably weakens the patient's immunological defence.

## 7.2 CELL–CELL INTERACTIONS

The local anaesthetic dibucaine has the interesting property of enhancing the lymphocyte agglutination which Con-A induces in untransformed cells (Post *et al.*, 1975). The suggestion is that it does so by fluidising the cell membrane and facilitating lateral diffusion of Con-A receptors in the membrane, which in turn cluster and bring about agglutination. If this simple hypothesis is correct, then general anaesthetics should exert the same effect (p. 108).

Aggregation of a different sort, namely the platelet aggregation which ADP causes without involving an immunological mechanism, is inhibited by general (and local) anaesthetics (Ueda, 1974) (Fig. II.12).

Another example of a cell–cell interaction being affected by anaesthetics is seen in the homotypic reaggregation of embryonic brain cells which have been previously disaggregated by trypsin. Using a method which does not require motility or substrate adhesion on the part of the reaggregating cells, Ungar and Keats (1973) showed that halothane, chloroform, diethyl ether, trichloroethylene, nitrous oxide, and methoxyfluorane, all at anaesthetic dose, inhibited reaggregation in a reversible way. They also caused recently reaggregated cells to separate. Local anaesthetics had no effect and it appears that the volatile general anaesthetics impair the recognition or adhesion mechanisms in the reaggregation process in some relatively specific way. It would be interesting to know if other tissue cells are similarly affected, or whether brain cells have particular surface features on which anaesthetics act preferentially.

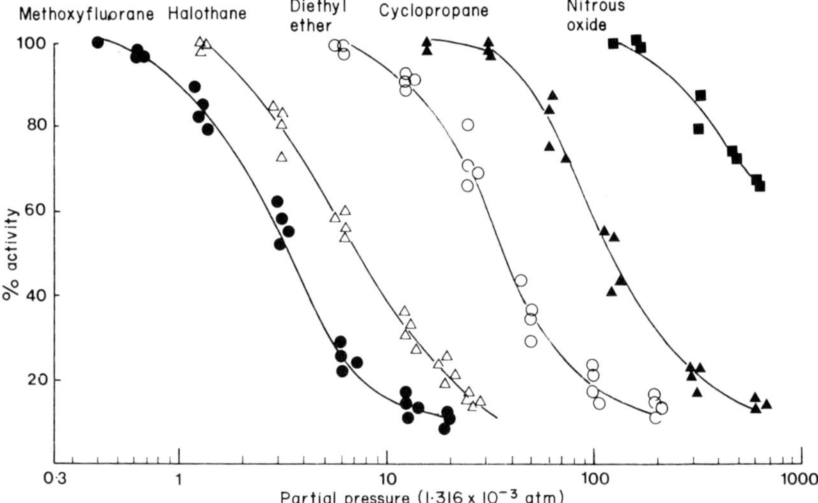


Fig. II.12. Effect of clinical anaesthetics on platelet aggregation. Aggregation was induced by adding ADP. Vertical scale: percentage activity compared with control cell suspension which was bubbled with air. (Ueda, 1974)

Certain lymphocytes possess the acquired property of destroying other cells whose surface carries a particular antigen. The lymphocytes are obtained by injecting a suitable mammal with target cells and subsequently collecting the immune lymphocytes from the peritoneal fluid. This procedure was adopted by Kemp and Berke (1973), who injected tumour cells into mice and thereby obtained lymphocytes which were cytotoxic to the tumour cells. The cytolysis of the tumour cells by the lymphocytes is a two-stage process involving, first, an intercellular interaction which is then followed by lysis. Kemp and Berke found that nerve-blocking concentrations of benzyl alcohol, methanol and butanol inhibited the cytolysis by acting during the first stage of cell interaction. Ethanol, procaine and chlorpromazine had very much less inhibitory effect. Halothane also inhibits lymphocyte cytotoxicity *in vitro*, but here the question of the inhibitory mechanism is wide open. A reduction in the locomotor activity of the cells, leading to fewer cell–cell contacts, could be responsible, or more specific effects in cell membranes might occur (Cullen *et al.*, 1976). This is another example of how anaesthetics may reduce the immunological defence mechanism in patients.

## 8 Conclusions

The multiplicity of the cellular effects of anaesthetics and inert gases arises from the ease with which they combine with target molecules, whose identity is unknown and whose relationship to the visible effects is obscure. In certain cases, e.g. microtubules, it is conceivable that an anaesthetic–target interaction leads straight to the visible symptoms, the disruption of structure. Most overt anaesthetic effects are more indirectly caused, probably many are mediated by the cell membrane. It is just conceivable that *all* effects stem from an alteration in the normal way the membrane regulates the cell's internal activities. Trigger effects may occur, as in the very special case of ethylene in plant cells.

Narcotic effects are sometimes biphasic, as in the responses of amoeba to anaesthetics and related substances. The narcotic-like effects of the weak gases $N_2$, $H_2$, He actually enhance cell division against an inhibitory background of high hydrostatic pressure, and protect against radiation damage.

Seen within the organisation of a whole animal, anaesthetics weaken immunological defences by inhibiting lymphocyte division, activation and cytolysis, and by affecting antibody production (see Bruce and Wingard (1971) for a good discussion of this aspect). Anaesthetics also disrupt the orderly development of some embryonic tissues and depress the turnover of some actively growing cells in the adult.

For substances which we have good reason for thinking act without forming covalent bonds and whose primary molecular effects are usually weak and reversible, this is an impressive performance.

Can we establish any biological order in these apparently miscellaneous effects? We need to see organisms as dynamic systems finely tuned to their physical environment, which is fundamentally aqueous. Molecular interactions, which are the moving parts of the living machine, proceed in, or close to, an aqueous phase. Life processes are primarily concerned with maintaining the system intact, balancing the needs of a cell to feed off its environment with the need to be protected from it. Anaesthetics and inert gases are physiologically potent probes because they seek out the nonaqueous. They disturb the balance between the aqueous phase and "biological" molecules and, at the cellular level, they disturb the interface between the cell and its aqueous environment. The ways in which they do this are considered in detail in subsequent chapters.

To return to the proposition mentioned in the introduction to this chapter, the cellular actions of anaesthetics can be seen to relate to general principles of cellular organisation. However, we are a long way from establishing a coherent cellular physiology of anaesthetics.

## References

ABELES, F. B. (1973). "Ethylene in Plant Biology", p. 302. Academic Press, New York and London.

ALDRETE, J. A. and VIRTUE, R. W. (1967a). Effects of prolonged inhalation of hydrocarbon gases on rats. *Anesthesiol.* **28**, 238–239.

ALDRETE, J. A. and VIRTUE, R. W. (1967b). Prolonged inhalation of inert gases by rats. *Anesth. Analag. Curr. Res.* **46**, 562–565.

ALLISON, A. C., HULANDS, G. H., NUNN, J. F., KITCHING, J. A. and MACDONALD, A. C. (1970). The effect of inhalational anaesthetics on the microtubular system in *Actinosphaerium nucleofilum. J. Cell. Sci.* **7**, 483–499.

ALLOTT, P. R., STEWARD, A., FLOOK, V. and MAPLESON, W. W. (1973). Variation with temperature of the solubilities of inhaled anaesthetics in water, oil and biological media. *Brit. J. Anaesth.* **45**, 294–300.

ANDERSEN, N. B. (1966). The effect of CNS depressants on mitosis. *Acta Anaesth. Scand.* **10**, Suppl. 22, 3–36.

ANDERSEN, N. B. (1968). The teratogenicity of cyclopropane in the chicken. *Anesthesiol.* **29**, 113–122.

BASFORD, A. B. and FINK, B. R. (1968). The teratogenicity of halothane in the rat. *Anesthesiol.* **29**, 1167–1173.

BERWICK, M. C. (1951). The effect of anaesthetics on calcium release. *J. Cell. Comp. Physiol.* **38**, 95–107.

BETEL, I. and Matijnse, J. (1976). Drugs that disrupt microtubuli do not inhibit lymphocyte activation. *Nature*, **261**, 318–319.

BRUCE, D. L. (1966). Effect of halothane anaesthesia on extravascular mobilization of neutrophils. *J. Cell. Physiol.* **68**, 81–84.

BRUCE, D. L. (1967). Effect of halothane anaesthesia on experimental *Salmonella* peritonitis in mice. *J. Surg. Res.* **7**, 180–185.

BRUCE, D. L. (1972a). Normal thymidine entry into halothane-treated lymphocytes. *Anesthesiol.* **37**, 588–591.

BRUCE, D. L. (1972b). Halothane inhibition of phytohemagglutinin-induced transformation of lymphocytes. *Anesthesiol.* **36**, 201–205.

BRUCE, D. L. (1975). Halothane inhibition of RNA and protein synthesis of PHA-treated human lymphocytes. *Anesthesiol*, **42**, 11–14

BRUCE, D. L. (1976a). Halothane action on lymphocytes does not involve cyclic AMP. *Anesthesiol.* **44**, 151–154.

BRUCE, D. and CHRISTIANSEN, R. (1965). Morphologic changes in the giant amoeba *Chaos chaos* induced by halothane and ether. *Exptl. Cell. Res.* **40**, 544–553.

Bruce, D. L. and Koepke, J. A. (1966). Changes in granulopoiesis in the rat associated with prolonged halothane anesthesia. *Anesthesiol.* **27**, 811–816.

Bruce, D. L. and Traurig, H. H. (1969). The effect of halothane on the cell cycle in rat small intestine. *Anesthesiol.* **30**, 401–405.

Bruce, D. L. and Wingard, D. W. (1971). Anesthesia and the immune response. *Anesthesiol.* **34**, 271–282.

Bruce, D. L., Lin, H.-S. and Bruce, W. R. (1972). Effects of anaesthetics on the response of normal and abnormal bone marrow cells to cytotoxic drugs. *In* "Cellular Biology and Toxicity of Anaesthetics" (B. R. Fink, ed.), pp. 251–261, 328. Williams and Wilkins Company, Baltimore.

Bruemmer, J. H., Brunetti, B. B. and Schreiner, H. R. (1967). Effects of helium group gases and nitrous oxide on HeLa cells. *J. Cell. Physiol.* **69**, 385–392.

Buchheit, R. G., Schreiner, H. R. and Doebbler, G. F. (1966). Growth responses of *Neurospora crassa* to increased partial pressure of the noble gases and nitrogen. *J. Bact.* **91**, 622–627.

Burg, S. P., Apelbaum, A., Eisinger, W. and Kang, B. G. (1971). Physiology and mode of action of ethylene. *Hort. Sci.* **6**, 359–364.

Byers, M. R., Hendrikson, A. E. and Fink, B. R. (1973). Effects of lidocaine on axonal morphology, microtubules, and rapid transport in rabbit vagus nerves *in vitro*. *J. Neurobiol.* **4**, 125–143.

Cohen, E. N. (1976). Anesthetics and cancer. *Anesthesiol.* **44**, 459–460.

Cohen, W. D. and Rebuhn, L. I. (1970). An estimate of the amount of microtubule protein in the isolated mitotic apparatus. *J. Cell. Sci.* **6**, 159–176.

Cullen, B. F., Duncan, P. G. and Ray-Keil, L. (1976). Inhibition of cell-mediated cytotoxicity by halothane and nitrous oxide. *Anesthesiol.* **44**, 386–390.

Cullen, B. F., Sample, F. and Chretien, P. B. (1972). The effect of halothane on phytohemagglutinin-induced transformation of human lymphocytes *in vitro*. *Anesthesiol.* **36**, 206–212.

Doebbler, G. F., Bruemmer, J. H. and Schreiner, H. R. (1967). Influences of high pressures of inert gases upon cell activity. *Proc. 3rd Symp. Underwater Physiology* (C. J. Lambertsen, ed.), 404–411. Williams and Wilkins Company, Baltimore.

Dougherty, K. (1937). The action of anaesthetics on amoeba protoplasm. *Physiol. Zool.* **10**, 473–483.

Eastwood, D. W., Green, C. D., Lamdin, M. A. and Gardner, R. (1963). Effect of nitrous oxide on the white cell count in leukemia. *New England J. Med.* **268**, 297–299.

Ebert, M. and Hornsey, S. (1958a). Effect of nitrous oxide on the radiosensitivity of mouse Erlich ascites tumour. *Nature*, **182**, 1240.

Ebert, M. and Hornsey, S. (1958b). Effect on radiosensitivity of inert gases. *Nature*, **181**, 613–616.

Edström, A., Hansson, H.-A., Larsson, H. and Wallin, M. (1975) Effects of barbiturate on ultrastructure and polymerization of microtubules *in vitro*. *Cell Tiss. Res.* **162**, 35–47.

Erf, L. A. and Rhoades, C. P. (1939). The hematological effects of benzene (benzol) poisoning. *J. Ind. Hyg. Toxicol.* **21**, 421–435.

Ferguson, J., Hawkins, S. W. and Doxey, D. (1950). *c*-Mitotic action of some simple gases. *Nature*, **165**, 1021–1022.

Fink, B. R. (ed.) (1968). "Toxicity of Anesthetics". Williams and Wilkins Company, Baltimore.

FINK, B. R. (ed.) (1972). "Cellular Biology and Toxicity of Anesthetics". Williams and Wilkins Company, Baltimore.

FINK, B. R. and KENNEDY, R. D. (1972) Rapid axonal transport: effect of halothane anesthesia. *Anesthesiol.* **36**, 13–20.

FINK, B. R. and KISH, S. J. (1976). Reversible inhibition of rapid axonal transport *in vivo* by lidocaine hydrochloride. *Anesthesiol.* **44**, 139–146.

FRANKEL, J. and SCHNEIDERMAN, H. A. (1958). The effects of nitrogen, helium, argon, and sulfer hexafluride on the development of insects. *J. Cell. Comp. Physiol.* **52**, 431–451.

GOLDACRE, R. J. (1952). The action of general anaesthetics on amoebae and the mechanism of the response to touch. *Symp. Soc. Expt. Biol.* **6**, 128–144.

GRAHAM, E. A. (1911). The influence of ether and ether anaesthesia on bacteriolysis, agglutination and phagocytosis. *J. Inf. Dis.* **8**, 147–175.

GRANT, C. J., POWELL, J. N. and RADFORD, S. G. (1974). Effects of halothane on DNA synthesis and mitosis in root tip meristems of *Vicia faba*. *Brit. J. Anaesth.* **46**, 653–657.

GREEN, C. D. and EASTWOOD, D. W. (1963). Effects of nitrous oxide inhalation on hemopoiesis in rats. *Anesthesiol.* **24**, 341–345.

HAMBURGER, H. J. (1916). Researches on phagocytosis. *Brit. Med. J.* 37–41.

HASCHKE, R. M., BYERS, M. R. and FINK, B. R. (1974). Effect of lidocaine on rabbit brain microtubular protein. *J. Neurochem.* **22**, 837–843.

HEILBRUNN, L. V. (1920). The physical effect of anesthetics upon living protoplasm. *Biol. Bull.* **39**, 307–315.

HEILBRUNN, L. V. (1956). "The Dynamics of Living Protoplasm", p. 327. Academic Press, New York and London.

HINKLEY, R. E. and GREEN, L. S. (1971). Effects of halothane and colchicine on microtubules and electrical activity of rabbit vagus nerves. *J. Neurobiol.* **2**, 97–105.

HINKLEY, R. E. and SAMSON, F. E. (1972). Anesthetic-induced transformation of axonal microtubules. *J. Cell. Biol.* **53**, 258–263.

HINMANN, N. D. and CANN, J. R. (1976). Reversible binding of chlorpromazine to brain tubulin. *Molec. Pharmacol.* **12**, 769–777.

INOUÉ, S., FUSELER, J., SALMON, E. D. and ELLIS, G. (1975). Functional organization of mitotic microtubules. Physical chemistry of the *in vivo* equilibrium system. *Biophys. J.* **15**, 725–743.

JOHNSON, M. C., SWARTZ, H. M. and DONATI, R. M. (1971). Hematologic alterations produced by nitrous oxide. *Anesthesiol.* **34**, 42–49.

KEMP, A. S. and BERKE, G. (1973). Inhibition of lymphocyte mediated cytolysis by the local anaesthetics benzyl and salicyl alcohol. *Eur. J. Immun.* **3**, 674–677.

KIELER, J. (1957). The cytoxic effect of nitrous oxide at different oxygen tensions. *Acta Pharmacol. Toxicol.* **13**, 301–308.

KIRKNESS, C. M. and MACDONALD, A. G. (1973). Effect of helium pressure on the narcotic potency of nitrogen in dividing cells. *Swedish J. Def. Med.* **9**, 310–313.

KOMNICK, H., STOCKEM, W. and WOHLFORTH-BOTTERMANN, K. E. (1973). Cell motility: mechanism in protoplasmic streaming and amoeboid movement. *Int. Rev. Cytology*, **31**, 169–249.

LASSEN, H. C. A., HENRIKSEN, E., NEUKIRCH, F. and KRISTENSEN, H. S. (1956). The treatment of tetanus. Severe bone-marrow depression after prolonged nitrous oxide anaesthesia. *Lancet*, April, 527–530.

LATTERELL, R. L. (1966). Nitrogen and helium-induced anoxia: different lethal effects on rye seeds. *Science*, **153**, 69–70.

LONGMUIR, I. S. (1963). The application of polarography to the study of tissue respiration. *Rec. Chem. Progr.* **24**, 3–25.

MACDONALD, A. G. (1975). The effect of helium and of hydrogen at high pressure on the cell division of *Tetrahymena pyriformis*. *J. Cell. Physiol.* **85**, 511–528.

MACDONALD, A. G. (1978). The reversal of the effect of hydrostatic pressure on dividing cells by inert gases. Sixth Symposium of Underwater Physiology. Fed, Am. Soc. Exp. Biol.

MARSLAND, D. (1934). The site of narcosis in a cell: the action of a series of paraffin oils on *Amoeba dubia*. *J. Cell Comp. Physiol.* **4**, 9–33.

NUNN, J. F., SHARP, J. A. and KIMBALL, K. L. (1970). Reversible effect of an inhalational anaesthetic on lymphocyte motility. *Nature*, **226**, 85–86.

OSTERGREN, G. (1944). Colchicine mitosis, chromosome contraction, narcosis and protein chain folding. *Hereditas*, **30**, 429–467.

PARBROOK, G. D. (1967). Experimental studies into the effect of nitrous oxide on tumour cell growth. *Brit. J. Anaesth.* **39**, 549–553.

POST, G., PAPAHADJOPOULOS, D., JACOBSON, K. and VAIL, W. J. (1975). Local anaesthetics increase susceptibility of untransformed cells to agglutination by conconavalin A. *Nature*, **253**, 552–554.

POWELL, J. N., GRANT, C. J., ROBINSON, S. M. and RADFORD, S. G. (1973). A comparison with halothane of the hormonal and anaesthetic properties of ethylene in plants. *Brit. J. Anaesth.* **45**, 682–690.

RAO, P. N. (1968). Mitotic synchrony in mammalian cells treated with nitrous oxide. *Science*, **160**, 774–776.

SAUBERMAN, A. J. and GALLAGHER, M. L. (1973). Failure of pentobarbital and halothane to depolymerize microtubules in mouse optic nerve. *Anesthesiol.* **38**, 25–29.

SCHREINER, H. R., GREGOIRE, R. C. and LAWRIE, J. A. (1962). New biological effect of the gases of the helium group. *Science*, **136**, 653–654.

SCHROEDER, T. E. (1972). The contractile ring II. Determining its brief existence, volumetric changes, and vital role in cleaving *Arbacia* eggs. *J. Cell. Biol.* **53**, 419–434.

SEIFRIZ, W. (1941). A theory of anesthesia based on protoplasmic behaviour. *Anesthesiol.* **2**, 300–309.

SETÄLÄ, K. and NYYSSÖNEN, O. (1964). Hypnotic sodium pentobarbital as a teratogen in mice. *Naturwissen*, **51**, 413.

SHEPARD, T. H. and FINK, B. R. (1968). Teratogenic activity of nitrous oxide in rats. *In* "Toxicity of Anesthetics", p. 332. Williams and Wilkins Company, Baltimore.

SLEIGH, M. A. (1973). Cell motility. *In* "Cell Biology in Medicine" (E. E. Bittar, ed.), p. 723. Wiley Interscience, New York.

SNEGIREFF, S. and ANDERSEN, N. B. (1971). The effect of cyclopropane on mitosis in chicken embryos. *Anesthesiol.* **34**, 157–162.

STEPHENS, N. G. (1969). Effects of increased partial pressures of oxygen, nitrogen and helium on cells in culture. *Cell. Tiss. Kinet.* **2**, 225–234.

STURROCK, J. E. and NUNN, J. F. (1975). Mitosis in mammalian cells during exposure to anaesthetics. *Anesthesiol.* **43**, 21–33.

STURROCK, J. M. and NUNN, J. F. (1976). Effects of mixtures of nitrous oxide and halothane on the nuclei of dividing fibroblasts. *Brit. J. Anaesth.* (Abstract), **48**, 267.

SWANN, M. M. (1954). The mechanism of cell division: experiments with ether on the sea urchin egg. *Exptl. Cell. Res.* **7**, 505–517.

TILNEY, L. G. and PORTER, K. R. (1967). Studies on the microtubules in Heliozoa II. The effect of low temperature on these structures in the formation and maintenance of the axopodia. *J. Cell. Biol.* **34**, 327–325.

UEDA, I. (1974). The effects of volatile general anesthetics on adenosine diphosphate-induced platelet aggregation. *Anesthesiol.* **34**, 405–408.

UNGAR, G. and KEATS, A. S. (1973). Inhibition of embryonic brain cell aggregation by general anaesthetics. *Anesthesiol.* **39**, 362–369.

USENIK, E. A. and CRONKITE, E. P. (1965). Effects of barbiturate anesthetics on leucocytes in normal and spleenectomized dogs. *Anesth. Analag. Cur. Res.* **44**, 167–170.

VASILIEV, J. M., GELFAND, I. M., DOMNINA, L. V., IVANOVA, O. Y., KOMM, S. G. and OLSHEVSKAJA, L. V. (1970). Effect of colcemid on the locomotory behaviour of fibroblasts. *J. Embryol. Exp. Morph.* **24**, 625–640.

WEISS, H. S., WRIGHT, R. A. and HIATT, E. P. (1965). Embryo development and chick growth in a helium-oxygen atmosphere. *Aerospace Med.* **36**, 201–206.

WIKLUND, R. A. and ALLISON, A. C. (1972). Effects on the motility of *Dictyostelium discoideum*. *Nature New Biol.* **239**, 221–222.

WILSON, E. B. (1902). Experimental studies in cytology. *Archiv für Entwickelungsmechanik der Organismen*, **13**, 353–395.

WILSON, W. L. and HEILBRUNN, L. V. (1952). The protoplasmic cortex in relation to stimulation. *Biol. Bull.* **103**, 139–144.

WINGARD, D. W. and HUMPHREY, L. J. (1969). Depression of antibody production by halothane: a dose response. *Anesthesiol.* **30**, 353.

CHAPTER III

# Biochemical effects of anaesthetics and inert gases



## 1 Introduction

Anaesthetised animals show gross changes in the level of metabolites within cells or circulating in the blood. These changes are certainly biochemical effects, but they may be caused by several mechanisms including, for example, the action of anaesthetics on the sympathetic nervous system, as in the case of the hyperglycemia induced by ether (Alexander *et al.*, 1972). This chapter is not so much concerned with a biochemical description of the anaesthetised animal as with an outline of anaesthetics as biochemically active substances, which, despite their inability to form covalent bonds, they most certainly are (Chapter 1.5).

## 2 Soluble enzymes

### 2.1 EFFECTS OF SUBSTANCES RELATED TO ANAESTHETICS ON ENZYMES AND OTHER PROTEINS

If anaesthetics do not form covalent bonds readily, how do they affect biochemical reactions? They interact with other molecules by a variety of weak reversible mechanisms, perhaps including hydrogen bonds, dipole–dipole interactions, hydrophobic interactions (entropic effects) or by clathrate formation (water-ordering). To many biologists, these terms convey rather mysterious binding reactions, quite unlike the familiar case of the ionic bond, by which some local anaesthetics interact, in conjunction with hydrophobic interactions (Chapter V). Ionic effects are, of course, extremely important in much physiological thinking, but so too are some of the other molecular interactions, as for example, between oxygen and haemoglobin or the interaction between the non-polar constituents of a lipid bilayer. Examples of the binding of hydrophobic molecules with the protein tubulin have been mentioned in the previous chapter.

Let us consider an example of a hydrophobic interaction between a soluble enzyme and a neutral molecule. Chymotrypsin is a hydrolytic enzyme whose activity in catalysing the hydrolysis of a hydrophobic substrate, an ester, can be conveniently followed in solution (Miles *et al.*, 1963). First, the enzyme-substrate association constant may be measured, and defined as a conventional equilibrium constant

$$E + S \rightleftharpoons ES, \quad K = \frac{(ES)}{(E)(S)}$$

It has been shown that $K$ is affected by the ionic strength of the reaction mixture. That is, the solubility of the hydrophobic substrate is decreased by an increase in ionic strength. This is a case where the substrate is apparently "salted out" into the enzyme's active site, with consequences for the catalytic reaction.

A hydrophobic substance such as benzene (true anaesthetics were not used in the study unfortunately) can interfere with the enzyme's function by competitive inhibition, and so can toluene, ethylbenzene, naphthalene and anthracene, none of which can form hydrogen bonds. The potency of these purely hydrophobic inhibitors is, in this simple case, determined by their ability to leave the aqueous phase and dissolve in (bind to) the enzyme's hydrophobic active site. Other compounds, chlorobenzene, nitrobenzene and azulene, all of which can interact by hydrogen bonding or dipole interaction, appear to reinforce their hydrophobic interaction by these additional means.

A good example of a hydrophobic interaction between alcohols and an enzyme, ribonuclease, is provided by Schrier *et al.* (1962). Ribonuclease undergoes thermal denaturation at 62°C, the conformational change in the molecule being detected by spectrophotometry. Because the change in the molecule is abrupt it is referred to as a thermal transition, and the change in the enzyme involves the exposure of hydrophobic groups which are normally buried within the molecule, away from solvent water. The presence of alcohols favours the thermal transition, which occurs at progressively lower temperatures as the alcohol concentration is increased. Straight chain aliphatic alcohols were more effective in lowering the transition temperature than equivalent branched chain alcohols, and among the former the longer chain alcohols were found to be more effective than the shorter alcohols.

The action of the alcohols may be visualised as follows. The thermally driven transition leads to the exposure of hydrophobic groups to the solvent water. In the presence of alcohols this state is energetically favoured by a hydrophobic interaction between the alcohol and the protein hydrophobic groups, a mutual dissolution. Branched chain alcohols are presumably less effective than straight chain alcohols in favouring the transition for steric reasons. A thermodynamic treatment of the transition yielded a simple equation from which it is possible to calculate the effect of a given alcohol in lowering the transition temperature.

A number of studies of how inert gases, including hydrocarbons, bind to proteins have been carried out. Bovine serum albumin and $\beta$-lactoglobulin in particular have been studied, revealing several effects which may be relevant to anaesthetic action. Butane, for example, interacts with an aqueous solution of albumin to cause a change in the pH and viscosity of the solution, indicating a shift in the equilibrium between distinct states of the molecule. Lactoglobulin is similarly affected by butane. In general, hydrocarbons are able to exert reversible effects on proteins by an interaction which we can simply describe as a hydrophobic effect, but whose complexity is determined by the detailed structure of the protein in question (Wetlauffer and Lovrien, 1964). The case of xenon binding to deoxymyoglobin is rather different. Crystallographic analysis of deoxymyoglobin crystals equilibrated with 2 atm xenon revealed that xenon binds in a specific cavity within the molecule (Schoenborn and Nobbs, 1966). The binding, envisaged as a dipole to induced-dipole interaction and not a simple hydrophobic effect, is the most precisely defined anaesthetic–target interaction known.

An apparently simple hydrophobic interaction takes place between hydrocarbon gases and sickle cell haemoglobin. Because of the substitution of the amino acid valine in the $\beta$-chains of haemoglobin, an interaction occurs with the $\alpha$-chains which leads to the formation of characteristic sickle-shaped cells when the haemoglobin is in its deoxygenated state. Methane, ethane or propane cure this condition by binding to the valyl groups and blocking the inter-chain interaction (Murayama, 1973).

Certain apolar gases and anaesthetics have the property of forming an ice-like crystalline hydrate (a clathrate) which remains stable at room temperatures. A theory that hydrate formation is the molecular basis of clinical anaesthesia was put forward separately by Miller and by Pauling in 1961. Despite the widespread interest which the theory generated at the time there is remarkably little known of how anaesthetic-induced hydrates can affect physiologically important molecules or such structures as cell membranes. However, one study on the enzyme invertase has been carried out (Lund *et al.*, 1969). Invertase and its substrate sucrose were incubated in buffered solutions in conditions which favoured the formation of the hydrates $CCl_3F$ and of propane. Enzyme activity was measured over the temperature range $+3°C$ to $-10°C$ in the presence of the hydrate. (At the lower temperatures the reaction proceeded in a frozen state but the enzyme is still to be regarded as in solution.) The activity of the enzyme was reduced in the presence of $CCl_3F$ hydrate at all temperatures, but propane hydrate appeared to exert less effect. The reduced activity in the presence of the $CCl_3F$ hydrate could hardly be attributed to the presence of the crystalline hydrate itself, since propane's hydrate is very similar. Nor could any changes in solute concentration arising from hydrate formation account for the effect. The most likely explanation for the inhibition appeared to be an artefact, a non-hydrate effect, very marked with $CCl_3F$ and only mildly present with propane. It occurs when the hydrates decompose at elevated temperatures and the enzyme molecule is in intimate contact with the hydrate-forming substance. Thus invertase activity is actually protected from the anaesthetic by the crystalline lattice of water molecules.

In general, hydrate formation is an unlikely mechanism of anaesthetic action in any physiological preparation which has been investigated.

## 2.2 EFFECTS OF SOME CLINICAL ANAESTHETICS ON SOLUBLE ENZYMES

Glutamate dehydrogenase is a mitochondrial enzyme which may be used experimentally in the soluble state. Its activity is reversibly and non-competitively reduced by a high dose of halothane (Brammall *et al.*, 1973; Hulands *et al.*, 1975). Mitochondria themselves are more sensitive. Other enzymes, pyruvate kinase, phosphoglycerate dehydrogenase and others active in the glycolytic pathway, are unaffected by halothane. Lactate dehydrogenase has been shown to be only slightly inhibited by a very high dose of diethyl ether (Greene and Spencer, 1963).

## 3 Membrane-bound enzymes

Enzymes which are arranged in membranes are part of the integrated state which is characteristic of intracellular organisation. Mitochondria and endoplasmic and sarcoplasmic reticula all contain enzymes which exhibit a number of interesting responses to anaesthetics and are, of course, supremely important in the cells' metabolism. The effect of anaesthetics on the sarcoplasmic reticulum is dealt with in Chapter IV.

### 3.1 CELLULAR RESPIRATION AND ANAESTHETICS

Mitochondria support the organised sequence of enzymes which carry out the citric acid cycle, electron transport and oxidative phosphorylation, and whose normal functioning is intimately bound up with the distribution of cations within the cell. The mitochondrion is an important target for anaesthetics.

#### 3.1.1 *Oxygen consumption of isolated mitochondria*

It is well known that suspensions of mitochondria reduce oxygen to water in the presence of suitable substrates. The rate at which oxygen consumption proceeds may be determined by a number of factors and it is important, when considering the action of anaesthetics or other inhibitors on mitochondrial function, to define the reaction conditions. Consider the experiment in which rat liver mitochondria are respiring glutamate in the presence of excess oxygen and the phosphate acceptor ADP. Under these conditions the rate of oxygen consumption is determined by the functioning of the enzymes in the respiratory chain, the respiratory state 3 of Chance and Williams (1962). Clinical doses of halothane reversibly inhibit the rate of oxygen consumption in this preparation, but have less effect when succinate is the substrate (Miller and Hunter, 1970; Cohen, 1973). Methoxyfluorane, diethyl ether and other general anaesthetics are broadly similar in this respect. So a general finding is that clinical anaesthetics inhibit the respiratory chain in the region of NADH dehydrogenase (Fig. III.1). This explains the earlier results of Jowett and Quastel (1937) and of Greig (1946) who found that the oxygen consumption of rat brain homogenates respiring glucose was depressed in the presence of anaesthetics but was unaffected when respiring succinate.

The rate of oxygen consumption by mitochondria in the presence of limited ADP (respiratory state 4, Chance and Williams, 1962) is much slower than in the preceding case and exhibits a different response to anaesthetics. With glutamate as substrate, anaesthetics exert no effect. With succinate as substrate, anaesthetics cause an increased rate of oxygen

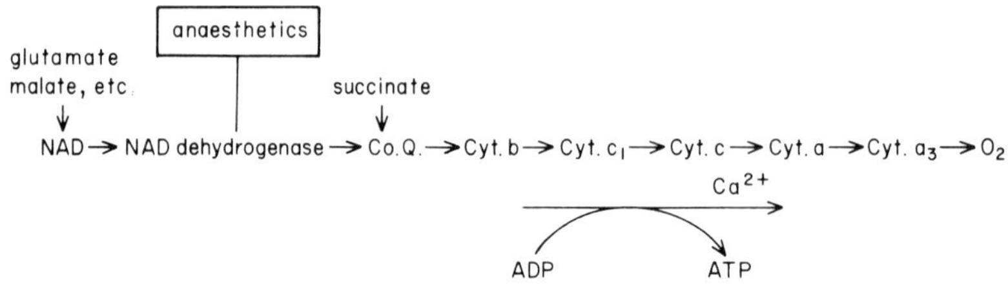


FIG. III.1. Electron transfer, oxidative phosphorylation and $Ca^{2+}$ transport in mitochondria (see text)

consumption, which is important because it implies damage. In the terminology used in mitochondria biochemistry, "respiratory control" is impaired (Cohen and McIntyre, 1972; Cohen, 1973). The differential effects of halothane on the respiratory chain (electron transfer) and respiratory control in mitochondria have been elegantly demonstrated by Cohen and Marshall (1968). The experiment uses mitochondria which are oxidising glutamate. They are exposed to a moderate dose of halothane, which exerts no apparent effect, and are then returned to air free of halothane. There then occurs an *increase* in oxygen consumption. The two stages of the experiment are interpreted as follows. In the first stage the presence of the anaesthetic reduces electron flow and more or less cancels out the impaired respiratory control which would otherwise have manifested itself as an increase in oxygen consumption. When the anaesthetic is removed, the inhibition in the electron flow pathway is rapidly reversed, thereby unmasking the more slowly recovering inhibition of respiratory control. When succinate is used as substrate instead of glutamate, electron flow is not inhibited in the first stage and the increase in oxygen consumption is immediately apparent.

Plant mitochondria exhibit similar effects when exposed to suitable concentrations of the hydrocarbon gases ethylene and butane (Ku and Leopold, 1970; Mehard and Lyons, 1970).

There are substances which are sufficiently lipid soluble to suggest they might have a narcotic action but which in fact exert either no effect or induce convulsions. The volatile fluorocarbon Caroxin-D is an example of the former and flurothyl and *n*-pentane examples of convulsants. When these agents are applied to rat liver mitochondria oxidising glutamate in respiratory state 3 the convulsants exert inhibitory effects like anaesthetics but Caroxin–D does not (Nahrwold *et al.*, 1974). Not surprisingly, perhaps, mitochondria cannot distinguish between a convulsant and an anaesthetic, but it is particularly interesting that they distinguish between Caroxin–D and anaesthetics.

Barbiturates are worth noting because they resemble volatile anaesthetics in blocking the respiratory chain at the NADH–flavoprotein site, but they only do so at high concentrations (Chance *et al.*, 1962).

### 3.1.2 *Uncoupling of phosphorylation*

The impaired respiratory control seen in anaesthetised mitochondria comes close to the condition known as uncoupling, in which oxidative phosphorylation is separated from electron transport and oxygen consumption. Indeed, the suggestion has been made that an anaesthetic such as halothane is an uncoupling agent. Hulme and Krantz (1955), for instance, found that clinical concentrations of diethyl ether caused a 20 per cent uncoupling in the oxidative phorphorylation carried out by rat brain mitochondria. Investigations by Miller and Hunter (1971) and Rosenburg and Hauggard (1973) showed that doses of up to fourfold the clinical level of halothane does not uncouple oxidative phosphorylation in rat liver or brain mitochondria to any serious extent. The design of some of the experiments is particularly interesting because the concentration of $Ca^{2+}$ inside the mitochondria was used as a measure of the degree of uncoupling. The normal steady state level of $Ca^{2+}$ is a reflection of the rate at which ATP is synthesised (Fig. III.1).

As we have seen, halothane inhibits electron transport at NADH-dehydrogenase and so Miller and Hunter (1971) used succinate as substrate and studied the effect of halothane on the downstream coupling sites. The role of NADH was further eliminated by addition of a specific inhibitor. When ATP is being synthesised, that is when electron flow is coupled, there proceeds an energy-dependent transport of $Ca^{2+}$ ions into the mitochondria. This may be followed as a colour change in the dye murexide, to which the mitochondrial membranes are impermeable. Murexide–$Ca^{2+}$ is colourless and so the rate at which the dye assumes its normal colour in a suspension of mitochondria is a measure of the rate at which $Ca^{2+}$ is being removed from solution. The changes in $Ca^{2+}$ concentration take place very rapidly but can be followed accurately by a special form of spectrophotometry. In a typical experiment the reactants, mitochondria, NADH-dehydrogenase-inhibitor, buffer and murexide, were pre-mixed and the calcium transport initiated by the addition of succinate. The half-time of the calcium uptake was about 10 sec. The experiments compared halothane's effect with that of a well established uncoupler of oxidative phosphorylation. Figure III.2 shows that 1–4 per cent halothane reduces the rate of calcium transport but not the steady state concentration attained. Addition of the known uncoupling agent causes a rapid release of $Ca^{2+}$ from the mitochondria. Miller and Hunter (1971) suggest that a high dose of halothane inhibits the rate of $Ca^{2+}$ transport by partially inhibiting electron flow rather than by an uncoupling reaction. At clinical doses halothane may reduce the coupling efficiency but it would overstate the case to describe it as a true uncoupler.

Non-volatile local anaesthetics are easier to work with than inhalational

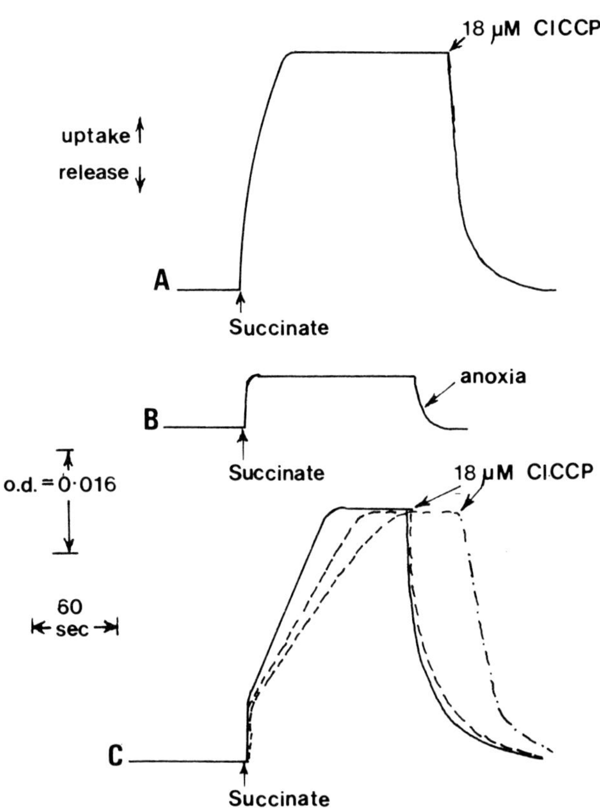


FIG. III.2. Uptake and release of $Ca^{2+}$ from rat liver mitochondria. A: Effect of the uncoupler Cl CCP on the uptake initiated by the addition of succinate. B: Effect of anoxia. C: Effects of 1, 2 and 4 per cent halothane on uptake initiated by the addition of succinate. Note: time scale, see text. (Miller and Hunter, 1971)

anaesthetics and have been shown to exert rather an interesting effect on mitochondrial cation transport. The transport of $Ca^{2+}$ is counter-linked with an efflux of $H^+$. The free concentration of $H^+$ can be followed by the change in colour of a pH indicator which binds to the interior of the mitochondria (Chance *et al.*, 1968). In a normal preparation of isolated mitochondria the addition of $Ca^{2+}$ initiates its inward transport and a simultaneous reduction in $H^+$ within the mitochondria. Mitochondrial suspensions treated with the local anaesthetics butacaine or tetracaine surprisingly show a more rapid reduction in $H^+$ ion concentration following the addition of $Ca^{2+}$. It is considered that the more rapid change in $H^+$ concentration is due to a reduced ability of the inner mitochondrial membrane to buffer cations. In the untreated membrane cations are probably buffered by the phosphate group of the phospholipids, which constitute a large proportion of the membrane. Local anaesthetics bind to the same phosphate groups, thereby diminish their binding capacity and potentiate the observed pH increase (see also p. 134).

Halothane inhibits the uptake of $Ca^{2+}$ by rat brain mitochondria in a variety of conditions (Rosenberg and Haugaard, 1973), but we have no detailed ideas as to how this comes about. It should be stressed that general anaesthetics may not act in the same way as local anaesthetics. The neurological effects of general anaesthetics on intracellular cation transport are dealt with in Chapter VI).

### 3.1.3 *Other effects*

Mitochondria membranes are generally rather impermeable to ions, especially $K^+$. The ionophore valinomycin can be adsorbed by rat liver mitochondria whose membranes subsequently become more permeable to $K^+$. With the mitochondria suspended in a solution of KCl a $K^+$-led osmotic swelling takes place independent of metabolism. Clinical doses of halothane reduce this rate of swelling, which is rather surprising (Miller *et al.*, 1972; Chapter IV). Anaesthetics enhance the valinomycin-mediated $K^+$ permeability of pure phospholipid membranes, probably by fluidising the crystalline structure of the bilayer (Chapter IV). Mitochondria membranes presumably have more complicated properties. The anaesthetic may so alter the mitochondrial membrane structure as to render the ionophore ineffective. Alternatively, it may interfere with an ion or electron flux which is normally linked with $K^+$ entry, and thereby prevent the influx of $K^+$ ions.

Structural changes induced in the constituents of mitochondrial membranes by anaesthetics have special significance in the light of the conformation coupling hypothesis of transduction. According to this hypothesis the "orthodox" configuration of the cristae, as seen with the electron microscope, is formed from the "energised" and morphologically highly convoluted state while simultaneously phosphorylating ADP to ATP. Agents which uncouple oxidative phosphorylation favour the "orthodox" configuration and so does a 4 per cent dose of halothane applied to rat heart mitochondria (Taylor *et al.*, 1972). Lesser doses exert proportionately less effect but it is not clear if clinical doses exert detectable structural effects. The relationship between these structural effects and membrane expansion (Chapter IV) is obscure. In linking these changes to oxidative phosphorylation, or indeed to any other function of mitochondria, it is by no means easy to distinguish between cause and effect.

### 3.1.4 *Oxygen consumption* in vitro *and* in vivo

It is important to remember that the effects of anaesthetics *in vitro* cannot automatically be transferred to the living animal. This point is rather well illustrated by the response of the rat liver to halothane. Let us compare the action of halothane on the isolated and perfused liver with the response of the liver *in vivo* (Biebuyck *et al.*, 1972; Biebuyck and Lund, 1974). One of the effects of a clinical dose of halothane on the isolated perfused rat liver is to inhibit oxygen consumption. In livers taken from recently fed rats the $\underset{\sim}{Q}O_2$ was decreased to about half by halothane whereas livers from starved rats were less affected. Lactate production was measured and in livers from fed rats it was increased some sixteen fold by halothane. In the whole animal a comparable dose of halothane had little effect on liver lactate and a consideration of the unchanged levels of ATP and the oxidation-reduction state of NAD showed that liver oxygen consumption was probably unaffected. A likely explanation is that in the intact animal the liver can utilise substrates which bypass the NADH-dehydrogenase site of halothane inhibition. Fatty acids are a possible substrate and a switch to their synthesis and respiration is consistent with the relatively constant lactate levels and the increase in glucose and insulin in the blood of halothane-anaesthetised animals. The oxygen consumption of perfused livers utilising a fatty acid (oleate) is not inhibited by halothane. Further, it is consistent with the fact that the oxygen consumption of isolated livers from fed animals is more sensitive to halothane than livers from starved animals whose aerobic metabolism utilises fatty acids.

The relationship between the oxygen consumption of nerve cells, nervous activity, consciousness and the inhibition of all three by anaesthetics at one time attracted a great deal of attention. Intuition may have played a significant role in this approach, and perhaps a misguiding role at that. Loss of consciousness and sleep are superficially restful and similar and it was natural to imagine that either condition might be caused by a reduction in the rate of energy-yielding reactions in nerve cells. Now we know that sleep is not neurologically restful and the relationship between neural activity and cellular respiration is not particularly close, although obviously the former ultimately depends on the latter. Here we shall note some of the effects of anaesthetics on cellular, as opposed to mitochondrial, respiration, and then consider the relationship between nerve function, oxygen consumption and the inhibition of both by anaesthetics. Finally, we need to consider measurements of brain metabolic activity carried out with anaesthetised animals.

Anaesthetics do impair the oxygen consumption of cells as we have seen. Mouse heteroploid fibroblasts, for example, show a 25 per cent reduction in oxygen consumption when exposed to a clinical dose of halothane. Glycolysis is enhanced but it fails to compensate for the reduced oxidative phosphorylation, and the cells grow in number more slowly than normal (Fink and Kenny, 1968). Cell division in the sea urchin egg is blocked by a dose of urethane or chloral hydrate which only reduces oxygen consumption by 25–30 per cent. In neither of these cases is it certain that division inhibition is caused by depressed respiration and, as we have seen, anaesthetics can block division by ways other than metabolic.

An analogous situation exists in nerve tissues. The rat sciatic nerve exposed to cyclopropane (0·8 atm) or nitrous oxide (5 atm) shows a reduction in oxygen consumption of 24 per cent and 35 per cent respectively, with no discernible effect on conduction. Higher partial pressures which are capable of blocking conduction inhibit oxygen consumption by twice that amount (Carpenter, 1956). In this situation, the anaesthetics act just like many other inhibitors and demonstrate the extent to which conduction is independent of central metabolism. Experiments with the rat superior cervical ganglion shows a radically different result. Clinical doses of ether or chloroform can block synaptic transmission with no significant effect on oxygen consumption, and specific inhibitors may be used to depress oxygen consumption (including anoxia itself) without immediately impairing transmission (Larrabee *et al.*, 1952). There is no case here for arguing that transmission is blocked by inhibiting cellular respiration. Anaesthetic doses which inhibit transmission also inhibit the extra oxygen consumption which normally accompanies electrical stimulation. The simple interpretation is that the extra oxygen consumption is abolished with transmission.

The respiratory status of the brain in anaesthetised animals has been investigated by several workers. The experiments require the careful maintenance of anaesthetised animals to ensure proper pulmonary and cardiovascular function. The brain tissue must then be rapidly frozen and subsequently analysed for intermediate metabolites. Halothane, nitrous oxide, cyclopropane and ether have all been used with the general conclusion that during anaesthesia brain respiration in general and electron transfer in particular are normal (Nilsson and Siesjo, 1970; Dedrick *et al.*, 1975).

Thus the evidence that anaesthetics do not depress synaptic transmission by a widespread inhibition of respiratory metabolism is strong. But their effects on the regulation of intracellular cations levels by mitochondria are important in influencing synaptic transmission (p. 205). The near prophetic words of Davies and Quastel (1932) may come close to describing a distinctive action of anaesthetics:

> "This indicates that their (i.e. anaesthetics) behaviour is not due to a disturbance of the brain cell whereby its mechanisms for accomplishing oxidation as a whole are affected. Their behaviour appears to be due to an effect, possibly at cell interfaces, whereby the mechanism which result in an activation of the molecules of lactic and pyruvic acids are greatly disturbed."

## 3.2 MICROSOMES

Microsomes are the membranous vesicles derived from the membranes of the endoplasmic reticulum when cells are fractionated for biochemical experiments. In a loose way the terms microsomes and endoplasmic reticulum may be used interchangeably. The interactions between anaesthetics and microsomes have many ramifications. It will be shown in this section that anaesthetics increase the intracellular concentration of microsomal enzymes by a process of induction. Certain microsomal enzymes are directly affected by anaesthetics, while certain anaesthetics are metabolically transformed by the microsomes. Should the reader still have in mind a naïve molecular "lock and key" notion of anaesthetic action, this section will provide a new frame of reference, still at the biochemical level, in which to develop a more realistic and complete concept. The facts point to anaesthetics acting as modulators of enzymes, as substrates and as inhibitors or activators. Additionally some anaesthetics leave behind reactive metabolic end-products to be coped with by the organism.

In vertebrates it is the task of the enzymes located in the endoplasmic reticulum of liver cells to metabolise foreign hydrophobic molecules to a more water-soluble form, which can then be excreted by the kidney. These liver microsomal enzymes are sensitive to anaesthetics, and they are generally referred to as mixed function oxidases. Their overall reaction, which is a hydroxylation, is given below (Boyd, 1972):

$$SH + O_2 + NADPH + H^+ \rightarrow SOH + H_2O + NADP^+$$

The enzyme which causes oxygen to combine with the foreign molecule is called cytochrome P–450, because of its characteristic absorption peak when complexed with carbon monoxide.

As in the case of mitochondria, we are here concerned with electron flow mediated by membrane-bound enzymes, and cytochrome P–450 is the terminal electron donor in the reaction. Two other components work with it. There is phosphatidylcholine, which carries electrons from NADPH to cytochrome P–450, and cytochrome P–450 reductase (Coon *et al.*, 1972).

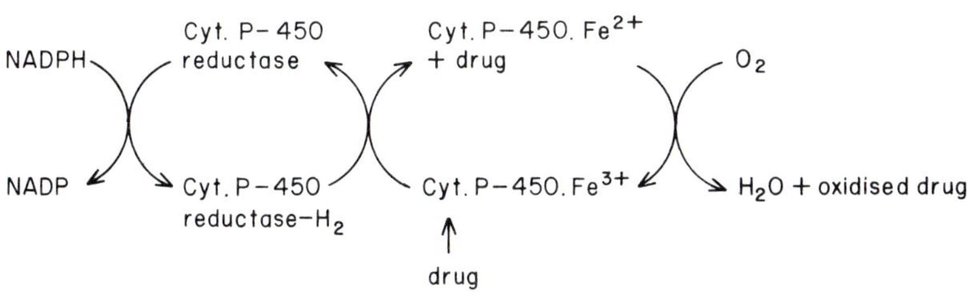


FIG. III.3.

In the above simplified reaction scheme cytochrome P–450 first combines with the drug; it is then reduced before combining with oxygen to subsequently yield oxidised drug and oxidised cytochrome P–450 (Brown, 1973).

### 3.2.1 *Induction of microsomal enzymes*

The activity of enzymes in a cell is controlled by many different mechanisms. In the case of allosteric control, or post-synthetic activation, it is the specific activity of the enzyme which is altered, but a metabolic reaction may also be enhanced by increasing the concentration of the appropriate enzyme, which can be done either by stabilising the enzyme against breakdown processes, or by accelerating its synthesis. The latter process is enzyme induction and it is known to occur by several mechanisms. Of the many foreign substances (xenobiotics) which induce enzymes in mammalian liver cells, phenobarbitone is particularly potent and a convenient one to use experimentally. Phenobarbitone and all other xenobiotic inducing substances share the following properties: they are markedly lipid soluble, they bind with cytochrome P–450 in a characteristic manner and they need to act for a significant length of time to achieve induction. A complete list of substances which accelerate hepatic microsomal enzyme activity *in vivo* would be very long, but a representative selection shown in Table III.1 from Remmer (1972) and Brown (1973) shows clinical anaesthetics to be well represented.

TABLE III.1

Xenobiotics which induce drug-metabolising enzymes in the mammalian liver

| | |
|---|---|
| *Anaesthetics* | *Insecticides* |
| Halothane | DDT |
| Diethyl ether | Dieldrin |
| Chloroform | Chlordane |
| Nitrous oxide | |
| (Methoxyfluorane, see p. 68) | |
| *Tranquillisers* | *Antibiotics* |
| Chlorpromazine | Chloramphenical |
| Chlordiazepoxide | Griseofulvin |
| *Hypnotics* | *CNS stimulants* |
| Phenobarbitone | Nikethamide |
| Ethanol | |
| Chloral hydrate | |
| *Steroids* | *Sulphonamides* |
| Cortisone | Sulphanilamide |
| Methyltestosterone | |

Phenobarbitone induces cytochrome P–450 and other microsomal enzymes by a process involving a binding reaction which, it is thought, achieves some critical conformational change in each of the enzymes, which acts as a trigger. The binding reaction with cytochrome P–450 differs from substrate binding (phenobarbitone is not metabolised by microsomal enzymes), and it can be mimicked by relatively unspecific organic solvents (Imai and Sato, 1967). Other hydrophobic substances, barbituric acid for example, do not achieve induction. The phenobarbitone-cytochrome P–450 complex exhibits a spectrum with a peak at 385 nm and a trough at 420 nm. Considerable protein synthesis follows binding, trebling the amount of enzyme in the liver in about 3 days. Other enzymes are also synthesised, although little is known of them because they are less readily monitored, but NADPH-reductase is an example, whereas cytochrome b.5 is one whose intracellular presence is only slightly increased by phenobarbitone induction. Simultaneous with protein synthesis there occurs marked changes in the cell structure. Induced liver cells grow an extensive smooth endoplasmic reticulum, which in cross-section frequently looks tubular. The whole liver enlarges; this may be due to an increase in cell division in the case of phenobarbitone but in other cases cell enlargement only takes place. Some anaesthetics and related substances (e.g. carbon tetrachloride) are capable of causing changes in the endoplasmic reticulum of liver cells by other, more direct, means (see p. 74). A different pattern of induction is brought about by 3-methylcholanthrene, a polycyclic hydrocarbon. It does not induce cytochrome P–450 but instead a new cytochrome, P–488, appears within 24 hr. The cytochromes P–450 and P–488 are versatile enzymes with substrates in common, but the latter catalyses the hydroxylation of 3,4-benzpyrene more effectively than the former.

In summary, hepatic microsomal enzyme induction is elicited by many hydrophobic substances which exert some kind of conformational change in their target enzymes. The enzymes subsequently, and quite separately, oxidise the hydrophobic substance to a more polar, hydrophilic form, which can then be excreted by the kidneys (Fig. III.4). For a more detailed account the reader should consult Remmer (1972) and Brown (1973). The

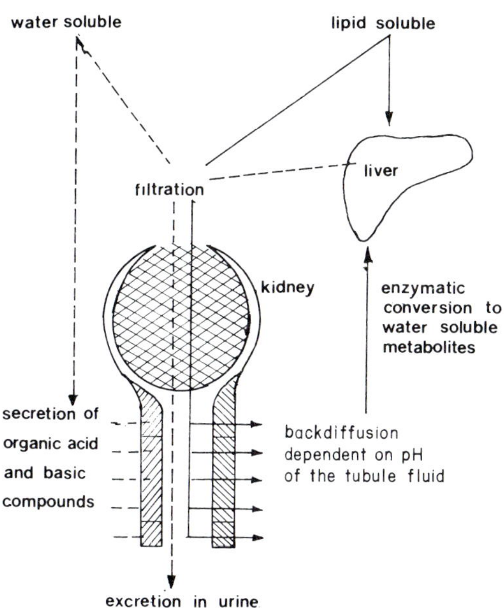


FIG. III.4. Summary of defence mechanism against foreign hydrophobic substances (unbroken arrows) and water soluble substances (broken arrows). After glomelular filtration solutes diffuse back into the extracellular fluids along the concentration gradient generated in the kidney by $Na^+$ reabsorption and osmosis. Lipid-soluble drugs can only be excreted after conversion to water-soluble forms, which do not diffuse back into the extracellular fluids in the kidney so readily, and which may also be actively secreted into the tubule. (Remmer, 1972)

question now to be asked is, how do clinical anaesthetics interact with this comprehensive defence mechanism?

When diethyl ether is applied to rats at half the clinical dose for 7 hr a day over 5 days, there occurs an increase in the cytochrome P–450 concentration in the liver cells (Fig. III.5). There is also an increase in the levels of associated enzymes and an increase in the rate at which xeno-

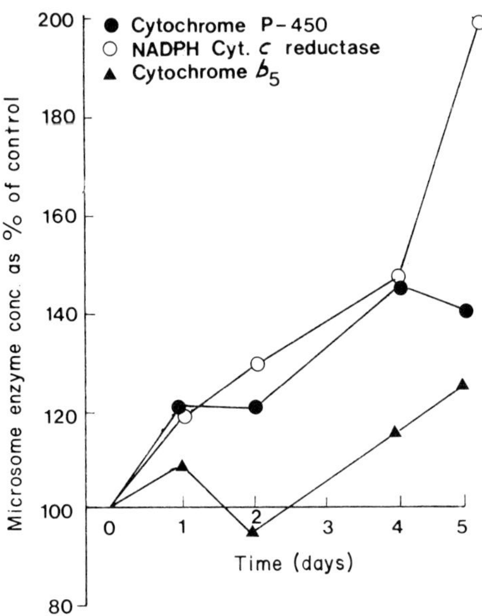


Fig. III.5. The increase in rat liver microsome enzymes during exposure to diethyl ether. The increase in three enzymes is plotted as percentage of control values. The animals were each exposed to 1·6% ether for 7 hr each day (horizontal axis) and examined 16 hr after each exposure. (Brown and Sagalyn, 1974)

biotics such as aniline are metabolised (Brown and Sagalyn, 1974). Nitrous oxide applied to rats in the same way as ether promotes an increase in microsomal enzymes within 2 days (Remmer, 1962). Methoxyfluorane fails to increase cytochrome P–450 levels in rat liver cells and the rate at which aniline and other substances are metabolised is not increased. Ten days' exposure to methoxyfluorane was found to raise the level of NADPH cytochrome *c*-reductase in the liver cells but little else. However, methoxyfluorane causes dechlorinating enzymes to increase their intra-cellular concentration in the rat liver.

The term induction, often loosely used, should not be confused with activation. Induction involves extra synthesis of the enzyme, which in the case of cytochrome P–450 is readily detected by spectrophotometry. Other criteria may be used for other enzymes. Ross and Cardell (1972) were able to show that liver microsomes prepared from rats previously exposed to halothane were capable of dechlorinating a substrate (methoxyfluorane) at a rate faster than control preparations. Simultaneously, the endoplasmic reticulum in the liver cells of treated animals increased considerably, which is good evidence, though not conclusive proof, of a time-induction mechanism at work. An additional interesting finding in this study was the decline in rough endoplasmic reticulum which accompanied the increase in area of the smooth, tubular form. The conversion of rough to smooth endoplasmic reticulum cannot account for the large increase in the latter (Fig. III.6) and accordingly synthesis of membrane is presumed to take place. Concomitant synthesis of enzymes is all the more likely.

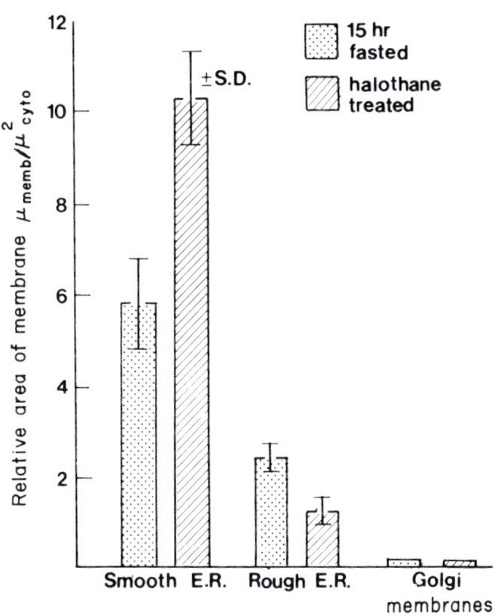


Fig. III.6. Synthesis of intracellular membranes in the rat hepatocyte following halothane anaesthesia. (After Ross and Cardell, 1972)

To a first approximation we can say that diethyl ether, and probably chloroform, nitrous oxide and halothane, resemble phenobarbitone in acting as inducing agents, and methoxyfluorane shows little inducing capability (Table III.1). The potential clinical complications arising from the ability of anaesthetics to induce enzymes are, in general, obvious enough and anaesthetists take them seriously, along with other drug–drug interactions they have to contend with. The broader biological point is, however, that organisms, including humans, respond to anaesthetics much in the same way as they respond to other pollutant substances. The defence mechanisms which come into play cope adequately with some compounds, but with others, including clinical anaesthetics, biochemical reactions create poisonous, water-soluble intermediate compounds (Fig. III.4). Although these are eliminated by the kidneys they can attain a steady state concentration sufficiently high to cause tissue damage. The microsomal mixed-function oxidases are not always capable of coping with the enormous range of xenobiotics being released into the environment or introduced into patients.

### 3.2.2 *Direct effects of anaesthetics on the normal functions of microsomal enzymes*

In this context the normal functions of microsomal enzymes, including cytochrome P–450, are taken to be the hydroxylation of xenobiotics and of endogenous substrates such as cholesterol. A number of endogenous reactions have been studied in liver microsome preparations obtained from anaesthetised rats (Hallén and Johansson, 1975). Nitrous oxide, halothane and chloroform, all at clinical doses for 1 hr, exerted no effects on cholesterol metabolism, which persisted after preparation of the microsomes. This is not surprising, but with animals induced with phenobarbitone, chloroform seems to exert a persistent inhibitory effect, not only on reactions involving cholesterol, but also on other reactions such as the hydrolysis of glucose-6-phosphate. The effect of anaesthetics on the microsomal enzymes *in vivo* is not known, but in uninduced animals it is probably negligible.

The binding of anaesthetics to microsomal enzymes might also impair their hydroxylation of xenobiotics. To investigate this, experiments have been carried out using cytochrome P–450, various drugs as substrates, and clinical anaesthetics as possible inhibitors or activators. An example is Brown's (1971) investigation of halothane's effects on the ability of rat liver microsomes to metabolise several drugs, of which hexobarbitone and aniline are representative. Microsomes which were incubated in a clinical dose of halothane oxidised hexobarbitone at a reduced rate but they hydroxylated aniline at an increased rate. Hexobarbitone, phenobarbitone and many other substances, including volatile anaesthetics, bind to cytochrome P–450 in a manner characterised by a certain spectral pattern whereas aniline binds to cytochrome P–450 in a manner which yields a different spectrum. Compounds represented by hexobarbitone and phenobarbitone are designated Type I, and the aniline-like substances are Type II. The neat correlation found in this work, with the metabolism of Type I compounds being reduced by halothane, a Type I binder with that of the Type II substances being enhanced, prompted Brown to suggest that the binding of Type I substrates blocks essential electron flow and actually promotes the flow of electrons to the site of Type II hydroxylation. A similar symmetry has been noted with other drugs (Korten and Van Dyke, 1973). Thus clinical anaesthetics may *directly* promote or inhibit drug metabolism by microsomal enzymes. It is both consistent and interesting that cytochrome P–450 in microsomes from the human liver, unlike cytochrome P–450 from the rat liver, does not bind with halothane in the Type I mode, nor is aniline hydroxylation stimulated by halothane. The scope for biochemists in this area of direct anaesthetic–enzyme interaction is considerable.

### 3.2.3 *The metabolism of anaesthetics*

Before radioactively labelled inhalational anaesthetics were used it was extremely difficult to measure the amount of anaesthetic which became incorporated into an animal's metabolism. When $^{14}C$ or $^{16}Cl$ labelled anaesthetics became available it was soon apparent that small but significant quantities of labelled metabolites were being formed in treated animals. Labelled carbon dioxide was expired and non-volatile, labelled metabolites of unknown identity were excreted in the urine (Van Dyke *et al.*, 1964). For example, of the total labelled anaesthetic injected into rats the following amounts appeared as $^{14}C$ carbon dioxide: diethyl ether 4 per cent, chloroform 4 per cent, methoxyfluorane 1–2 per cent. After treatment with anaesthetics, rats yielded labelled halogen compounds in the urine made up of the following percentages of the total anaesthetic dose: ether 2 per cent, chloroform 2 per cent, methoxyfluorane 3–5 per cent and halothane 3 per cent. These data show that these anaesthetics enter into the animal's metabolism by reactions which involve removing halogen groups and, in the case of methoxyfluorane, splitting the ether linkage. Neither process was entirely new to biochemistry, but the realisation that clinical anaesthetics were not metabolically inert renewed everyone's awareness of the problem of clinical side-effects.

The metabolism of anaesthetics starts soon after entering the cells. In a mouse breathing 2 per cent halothane, significant chemical change has taken place within 10 min and within 2 hr 9 per cent of the total dose of anaesthetic can be recovered by biochemical methods in the form of $^{14}C$ labelled metabolites. Humans have also been shown to metabolise halothane and other anaesthetics.

By a special low temperature form of autoradiography it has been possible to follow the distribution of inhaled anaesthetics throughout the body of small, experimental animals and thereafter the redistribution of the products of metabolism (Cohen, 1969; Cohen and Hood, 1969). Figure III.7 shows the differential distribution of labelled chloroform and metabolites, the latter being insignificant in the early stage shown in Fig. III.7(a). The concentration of radioactive label may be measured and the changes within specific organs followed at time intervals. Consider the figures in Table III.2. Test animals breathed chloroform for 10 min; some were then sacrificed (time 0); others were allowed to breathe air for a period and then sacrificed at times 15 and 120 min. The concentration of chloroform or its products in the blood is normalised to 1 in the Table.

The figures show an initial high concentration of anaesthetic in fat, which is a direct consequence of the high oil–water partition coefficient of chloroform. After 120 min, however, it is clear that the liver contains a

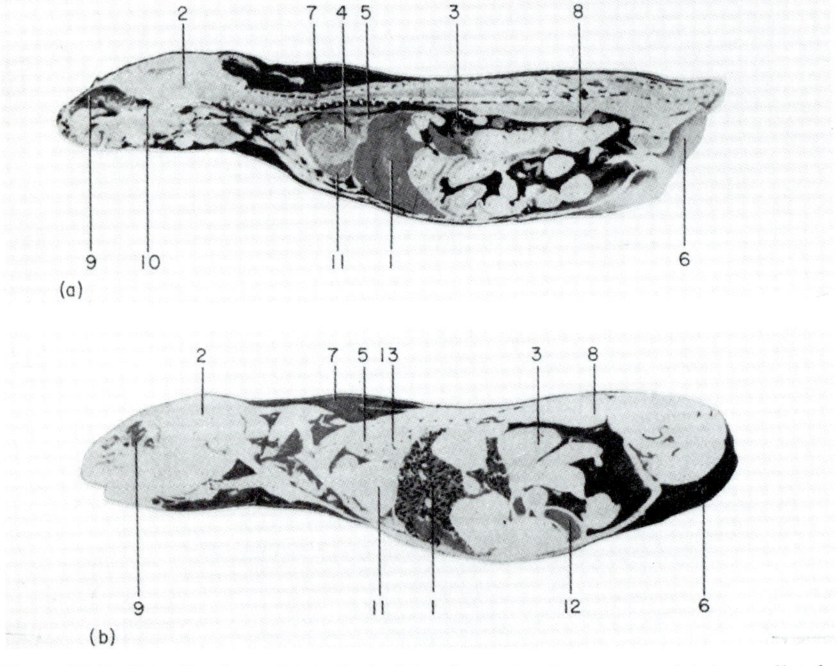


Fig. III.7. Distribution of labelled chloroform in the mouse: (a) immediately after inhalation; (b) 2 hr after a 10 min exposure. The pictures are autoradiographs of longitudinal sections of treated animals: 1, liver; 2, brain; 3, kidney; 4, blood; 5, lung; 6, fat; 7, brown fat; 8, muscle; 9, nasal mucous membrane; 10, Harder's gland; 11, heart muscle; 12, duodenum; 13, bronchi. (Cohen and Hood, 1969)

relatively large proportion of the total radioactivity, of which 85 per cent is non-volatile. Thus fat deposits store the anaesthetic, slowly releasing it by diffusion to be changed chemically in the liver. A significant fraction of the inhaled anaesthetic is metabolised along this route (Table III.2).

TABLE III.2

Distribution of chloroform and its metabolic products in mice, after a 10 min exposure to the anaesthetic (after Cohen and Hood, 1969)

| | 0 min | 15 min | 120 min |
|---|---|---|---|
| Blood | 1 unit concentration | 1 unit concentration | 1 unit concentration |
| Muscle | 0·8 | 1 | 0·7 |
| Lung | 1 | 1·5 | 1·4 |
| Liver | 1·5 | 2 | 6·7 |
| Fat | 6 | 9·25 | 7·2 |
| Brown fat | 12 | 14 | 5·7 |

The solubility of an anaesthetic is therefore relevant to its metabolic processing, because solubility determines its availability (Greenstein *et al.*, 1975). Methoxyfluorane and isofluorane have been compared in this respect in the following experiment. Rats anaesthetised with methoxyfluorane perform a defluorination reaction which is induced by phenobarbitone treatment. Isofluorane defluorination *in vivo* is also detectable, but it fails to respond to similar induction treatment. This is not because of a basic difference in the defluorination mechanism; it is due to the low concentration of the anaesthetic substrate. The inducible defluorination of methoxyfluorane is manifest because the concentration of defluorinating enzymes is rate-limiting before induction and the substrate is present in abundance. Isofluorane has a lower lipid solubility than methoxyfluorane and is not present in abundance, in fact its concentration is probably rate-limiting in the defluorination reaction. Consequently, enzyme concentrations may increase following induction without affecting the overall rate of the defluorination reaction.

Recent work with halothane has indicated that the bromine released from it may exert its own sedative and possibly cytoxic effects (Greene, 1976).

### 3.2.4 *Metabolic products of anaesthetics*

The general picture which has emerged from biochemical studies is that the main toxic effects of anaesthetics are centred in the liver and are caused by intermediates of anaesthetic metabolism and not by the anaesthetics themselves. The idealised view that an anaesthetic is biochemically innocuous has therefore to be modified but not discarded.

The toxic side-effects of an anaesthetic become more severe when the rate at which it is metabolised increases, for example, by enzyme induction. Conversely, inhibiting the rate at which an anaesthetic is metabolised should diminish its toxic effects.

These points can be illustrated by numerous experiments described in the research journals. For example, Fig. III.8 summarises results obtained

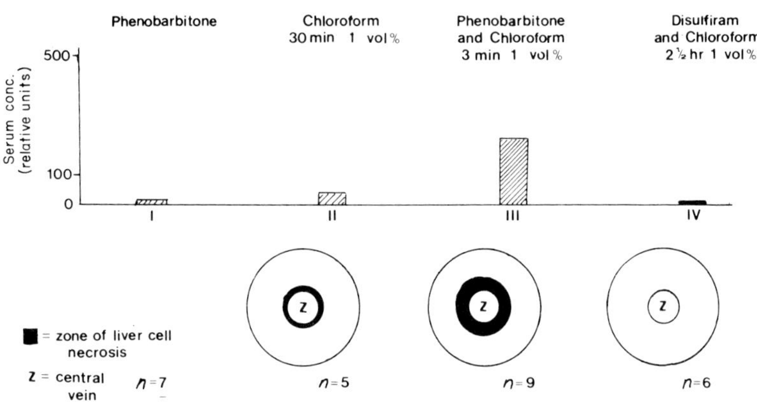


FIG. III.8. Liver damage in rats caused by chloroform and potentiated by phenobarbitone. The histogram shows the serum transamidase activity (GP) 24 hr after chloroform anaesthesia. The circles beneath the histogram depict the necrosis which was measured microscopically. Induction by phenobarbitone potentiates chloroform's effects, whilst the drug tetraethyliuram disulphide, an inhibitor of liver microse enzymes, diminishes its effects. (Scholler, 1970)

from four groups of rats. One group, shown in column 2, was exposed to chloroform at a fairly high dose (1 per cent) for 30 min. Subsequent damage to the liver was measured in two ways. First, the blood level of enzymes which are known to leak from damaged liver cells was determined 24 hr after anaesthesia (see histogram in Fig. III.8). Additionally, liver damage was scored histologically and is summarised diagramatically. The effect of induction with phenobarbitone is dramatic (compare colums 3, 2, 1) and so is the effect of the drug tetraethylthiuram disulphide (column 4) which is known to inhibit liver microsome enzymes.

Experiments of this type point the way to identifying the toxic products of anaesthetic metabolism by identifying agents of greater specificity which protect the liver. One such agent is glutathione, an endogenous compound capable of binding free radical intermediates and in some way suppressing their damaging influence. When the glutathione level in the rat liver is experimentally depleted the radioactive metabolites derived from labelled chloroform were found to be covalently bound to induced microsomal protein to a high degree and damage to the liver was accelerated (Brown *et al.*, 1974). The experiment showed that rats induced with phenobarbitone and then exposed to a clinical dose of chloroform for 2 hr suffered a subsequent destruction of liver cytochrome P–450 and NADPH-cytochrome reductase. Those animals which were not induced contained normal quantities of these enzymes (Table III.3). The metabolism of chloroform seems to lead to the destruction of the metabolising machinery.

TABLE III.3

Rat liver microsomal enzyme content 18 hr after 0·5 per cent chloroform anaesthesia (Brown *et al.*, 1974)

| | Induced | | Non-induced | |
|---|---|---|---|---|
| | Control | $CHCl_3$ | Control | $CHCl_3$ |
| Cytochrome P–450 mμ mol $mg^{-1}$ | 1·26 | 0·28 | 0·48 | 0·45 |
| NADPH-cytochrome reductose mμ mol cyt. red. $min^{-1}$ $mg^{-1}$ | 120 | 40·5 | 87·1 | 84·4 |

Halothane behaves differently, which is slightly surprising. Its metabolic products do not bind to microsomal protein to any great extent and it exerts no histological damage in the rat liver. The slight inhibitory effect it has on microsomal enzymes is not enhanced by induction (Davis *et al.*, 1971). It is important to realise that the toxicity of chloroform is proportional to the rate at which it is turned over, and in particular it is probably determined by the rate at which free radicals become available. Table III.3 actually shows that induction *causes* the damage to the enzymes listed. What applies to chloroform is thought to be true of many general anaesthetics. Another useful generalisation is that the amount of anaesthetic which enters the metabolism of the organism is a product of the duration of the exposure and the dose (Stier, 1968). Typically between 1·5 and 12 per cent of anaesthetic is metabolised (Van Dyke and Chenoweth, 1965). Halogenated anaesthetics are clearly the most toxic and the details of dehalogenation reactions are central to a proper understanding of their toxicity. The steps by which intermediates of anaesthetic metabolism achieve visible histological damage in the liver are equally important (Brown, 1972).

The anaesthetic-like substance carbon tetrachloride can cause severe damage to the mammalian liver, and so to a lesser extent can chloroform and in special cases halothane. Carbon tetrachloride causes the liver endoplasmic reticulum to swell and the ribosomes become freely distributed away from membranes. Both these changes occur within 1 hr of administration to a test animal, which is usually a rat. A typical oral dose is 0·25 ml carbon tetrachloride per 100 g animal. Four hours after treatment dense aggregations of membranes are to be seen in the liver. The membrane bounding the individual cells is not very susceptible to carbon tetrachloride, judging from the persistent electrical potential across it. This potential initially decreases but then recovers its normal value (Fig. III.9).

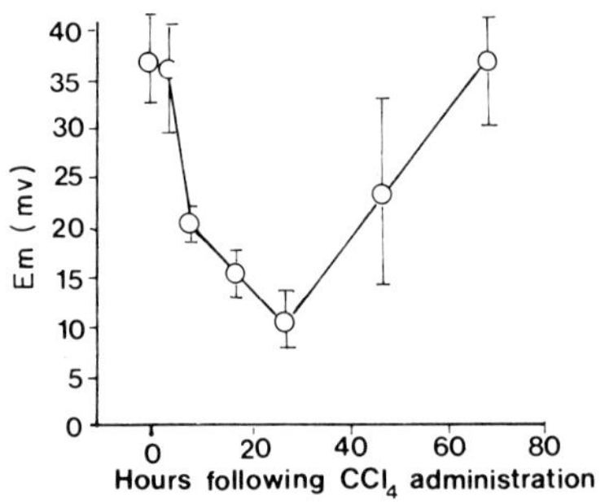


FIG. III.9. The transmembrane potential in rat liver cells following a dose of carbon tetrachloride. (Wands *et al.*, 1970)

The ability of the liver cells to synthesise protein is reduced by the time the gross morphological changes have become established. *In vitro* experiments with ribosomes show they are damaged. Fat accumulates in the liver cells, but the details of this process are not clear. Cell death soon follows these changes.

Halothane's effects on the human liver are very much milder than that of carbon tetrachloride, but nevertheless subject for concern. The overall death rate from hepatic necrosis among patients exposed to halothane is 1 in 10 000 (Cohen, 1969). The irregular occurrence of liver damage in humans has led to the suggestion that some special immunological mechanism may be at work in uniquely sensitive people, but there is no real evidence for this. The widespread view is that free radicals initiate covalent binding reactions and that lipoperoxidation reactions ensue (Brown, 1972). Lipid bilayer organisation is likely to suffer from the latter in particular. Intracellular membranes suffer much more damage than the bounding membrane of the whole cell, probably because they are close to the source of the toxic substances.

## 4 Some curious effects of inert gases

Most physiologists would agree that in human physiology, nitrogen merely serves as a diluent for oxygen and carbon dioxide in the air we breathe, and is metabolically inert. The considerable body of knowledge concerning nitrogen metabolism in plants would not move many physiologists to expect nitrogen to be biochemically active in higher animals. Nevertheless, it has been asserted that nitrogen inspired by humans may not be entirely metabolically inert. Cissick *et al.* (1972) found subjects tend to retain nitrogen gas, with fasting subjects retaining more than fed subjects. Exercising subjects, even in the steady state, expire more nitrogen than they inspire, however. The amount of "excess" nitrogen increases with the protein content of a meal prior to the exercise test. The net production of nitrogen in exercising subjects range from 0·217 l $min^{-1}$ (in a fasting state) to 0·509 l $min^{-1}$ (after a protein-rich meal). The origin of this nitrogen imbalance probably lies in errors in the estimation of gas exchange.

### 4.1 HELIUM

Cellular effects of helium have been described in the previous chapter. It was pointed out they arise at very high partial pressures, consistent with helium acting as an ultra-weak anaesthetic gas, or at a much lower partial pressure, in which case a variety of possibilities, none convincing, exist to account for its action. One might reasonably anticipate a similar distinction when looking at helium's effect at the biochemical level. To date there are no published experiments reporting biochemical effects of very high partial pressures of helium, but there are some intriguing reports of helium apparently acting at the remarkably low partial pressure of around 1 atm.

Helium, at close to atmospheric pressure, appears to stimulate the oxygen consumption of isolated mouse tissues (Cook *et al.*, 1951). Mouse liver slices have been shown to increase their rate of oxygen consumption and carbon dioxide production in Warburg respiration experiments using helium in place of nitrogen in the gas phase. The increases were around 11 per cent and the R.Q. was unaltered. The site of helium's action has not been established, but it appears to be upstream of the enolase reaction. Helium also alleviates cyanide-induced inhibition of oxygen consumption and in anaerobic experiments helium exerted an inhibitory effect upstream of the aldolase reaction (South and Cook, 1953a).

Brain tissue and sarcoma, both from the mouse, have also been studied in the presence of 1 atm of helium. Sarcoma slices showed a 28 per cent increase in oxygen consumption and no change in glycolysis. Brain slices showed a lesser increase in oxygen consumption and a slight depression of glycolysis (Cook and South, 1953).

Other workers have found that rat liver slices in a helium–oxygen mixture show a reduced rate of oxygen consumption which confuses the issue further. Yeast oxygen consumption is unaffected by helium, according to Cook (1950).

It seems highly unlikely that contaminants are responsible for all these effects (Cook and South (1953) deal with the question of gas purity), but it seems equally unlikely that a convincing interpretation will be made without much more systematic work. There exists the possibility that helium does act biochemically at partial pressures of about 0·5 per cent of its "narcotic" dose. The extraordinary effects of ethylene provide the precedent which counters any physical-chemical predictions to the contrary.

The comparison with the effects of more potent inert gases is interesting (South and Cook, 1953b). Xenon at a partial pressure of 1 atm acts very much like helium on mouse liver, brain and sarcoma slices. Argon and hydrogen had much less effect than helium, providing a pattern of potency utterly different to the classical narcotic series. These comparisons strengthen the possibility of a special "low dose" effect of helium. And so do the observations of Cook (1950) on the way helium accelerates metamorphosis in the insects *Drosophila* and *Tenebrio*. It is an entertaining speculation to draw the parallel here between ethylene's effects in plants (Chapter II) and helium's alleged effect in insects!

## 5 Bioluminescence

Bioluminescent reactions are light-emitting, oxidative reactions and occur in many different organisms. Although diverse in mechanism they have several features in common, chief of which is the reduction of oxygen, and the involvement of intermediate compounds ATP or FMN, which also play a central role in the cell's biochemistry. Under appropriate conditions light intensity is a measure of the rate at which the bioluminescent reaction proceeds, and so the reaction rate can be readily monitored. The influence of anaesthetics on two bioluminescent reactions has been studied; the firefly (*Lucida cruciata*) reaction and the bioluminescence of marine bacteria, notably *Photobacterium phosphoreum* and *P. fischeri*.

Anaesthetics exert biphasic effects on these reactions. Low doses actually accelerate, and higher (clinical) doses, of halothane or methoxy-fluorane for example, reversibly inhibit (or dim) the reactions. The correlation between the potency of such agents in inhibiting bacterial luminescence and in anaesthesia is close. It has been argued that luminescent reactions are instructive systems to analyse with a view to learning more about the action of anaesthetics in clinical anaesthesia. From the neurophysiological viewpoint, which surely is the relevant one, this seems a vague line of reasoning, but undoubtedly detailed study of anaesthetic action in any physiological system might indirectly guide thinking into the complexities of anaesthetic action in clinical anaesthesia. More generally, it is historically true that the study of the action of anaesthetics and other agents on luminescent systems has been carried out as an exercise in applying rate process theory to reactions proceeding *in vivo* the value of which the reader will wish to judge for himself.

The firefly luminescent reaction may be carried out *in vitro*, although *in vivo* it is under the nervous control of the insect intent on locating a mate (McElroy and Seliger, 1966). The reaction may be set out schematically, thus (Cormier *et al.*, 1975):

$$E + LH_2 + ATP\text{-}Mg^{2+} \rightleftharpoons E(LH_2 - AMP) + P\text{-}P. \qquad (1)$$

The enzyme, E (luciferase), combines with luciferin, $LH_2$, and ATP to form the luciferyl-adenylate enzyme complex and pyrophosphate, P-P. The complex ionizes and reacts with oxygen becoming excited:

$$E(LH^- - AMP) + O_2 \rightarrow E\ (\text{oxyluciferin dianion AMP})^* + CO_2. \qquad (2)$$

Light is emitted as the molecule descends to the ground state

$$E\ (\text{oxyluciferin-dianion AMP})^* \rightarrow E\ (\text{oxyluciferin-AMP}) + \text{light}. \qquad (3)$$

Although bacterial luminescence is affected by clinical anaesthetics in the same way as the firefly reaction, there are important differences between the two light-emitting reactions. The bacterial reaction involves an aldehyde (RCHO), $FMNH_2$ and luciferase (E). The luciferase enzyme in this case is a two sub-unit protein of molecular weight 80 000. The scheme below (Fig. III.10), shows two intermediate compounds which are synthesised prior to the light-emitting reaction. Molecular oxygen is involved in the reaction converting intermediate I to II.

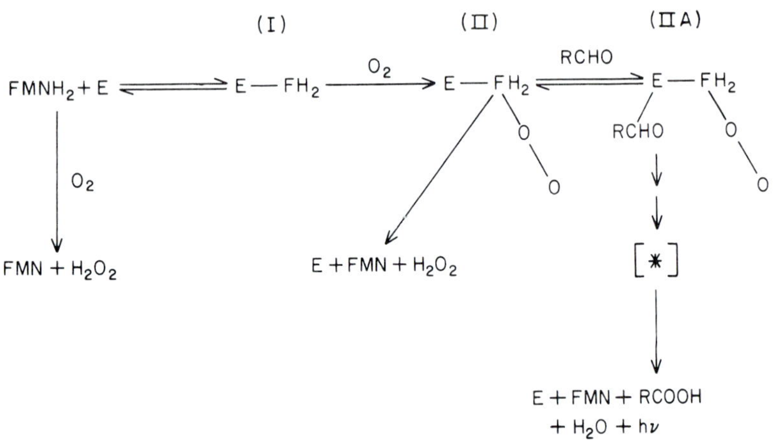


FIG. III.10. Hypothetical reaction pathway in bacterial luminescence. (Shiao-Chun Tu and Hastings, 1976)

The site of action of anaesthetics in these two different luminescent reactions is known only approximately. Taking the bacterial reaction first there are several pieces of evidence which suggest that the critical site is in the light pathway (Halsey and Smith, 1970; White *et al.*, 1973). First, anaesthetics do not dim the emission of light by poisoning the cells' intermediate metabolism. This is shown by the rapidly reversible effect of the agents and by their negligible effect on the cells' aerobic respiration at partial pressures which inhibit the light emission considerably. Secondly, the anaesthetics almost certainly do not interfere with the light-emitting molecule because they do not alter its emission spectrum. A third piece of evidence comes from experiments which utilise the finding that when oxygen is excluded from the cell suspension, intermediate compound I accumulates. The flash of light which occurs when oxygen is subsequently introduced is a measure of the amount of compound I which has built up. Although anaesthetics depress this anoxic flash, they do not affect the time taken for its maximal accumulation, and therefore do not act upstream of the oxidation reaction. Finally, an elegant experiment, using a mutant strain of *P. fischeri* which lacked the aldehyde precursor required for the light reaction, showed that anaesthetics probably act downstream from compound II (White *et al.*, 1973). The mutant normally emits only a dim light in its early stages of culture, but this is intensified when an aldehyde is introduced as a vapour to the cells (see Fig. III.10). Cells exposed to pulses of aldehyde in this way exhibit "slow flashes" of light whose peak is related to the intervals between exposures. Thus, a long delay between pulses of aldehyde is correlated with an intense burst of light. Therefore, these flashes are taken as a measure of the accumulation of the compound II. Anaesthetics do not affect the relationship between the interval separating pulses of aldehyde and the light flashes and, accordingly, they probably have no effect on the accumulation of compound II. Inspection of the scheme in Fig. III.10 shows that anaesthetics therefore act downstream of compound II. Light intensity is thus reduced by anaesthetics because they slow down the rate at which $IIA_a$ and hence light ($h\nu$) is formed.

In the firefly reaction, anaesthetics appear to act by altering the conformation of the luciferase, E, in equation (1). In the following section the arguments leading to this and other conclusions about the anaesthetic–target interaction *in vivo* will be set out.

## 5.1 KINETIC ANALYSIS OF BIOLUMINESCENT REACTIONS AND THE EFFECTS OF ANAESTHETICS

### 5.1.1 *Introduction*

Increasing the temperature of luminescent bacteria increases the intensity of the emitted light up to a certain maximum temperature, after which a further increase in temperature causes a decline in light output (Fig. III.11). From this experimental finding a great deal of work, both experimental and theoretical, has flowed (Brown *et al.*, 1942).

The biphasic temperature effect is accounted for in the following way. It is argued that light intensity is a measure of the reaction velocity and hence a measure of the concentration of rate-limiting enzyme. The enzyme, luciferase, exists in equilibrium with its thermal denaturation product. Using $A_n$ to represent the native enzyme we have:

$$\begin{array}{l} \mathrm{An} \xrightarrow{(2)} \mathrm{An}' \rightarrow \mathrm{An} + \text{light.} \\ {\scriptstyle (1)}\ \upharpoonleft\!\downharpoonright \\ \mathrm{Ad} \end{array}$$

The thermal denaturation is reversible and it may be envisaged as a conformational change in the pure enzyme or the enzyme-substrate

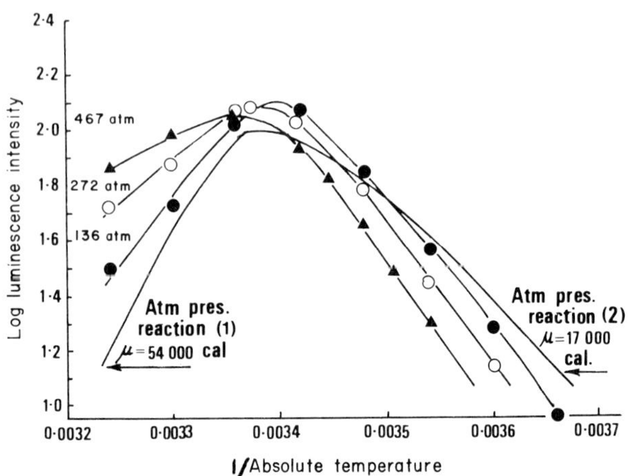


FIG. III.11. The relationship between luminescence intensity in *Photobacterium phosphoreum* and temperature at atmospheric and high hydrostatic pressure. (Brown *et al.*, 1942)

complex. From the data in Fig. III.11 the activation energy for the two reactions at atmospheric pressure can be calculated using the Arrhenius equation. At low temperatures An $\rightarrow$ An′ dominates, with an activation energy of 17 000 cal. The slope of the line at high temperatures yields an apparent activation energy of $-37\,000$ cal mol$^{-1}$, but clearly this is a notional figure only. The measured rate is the net result of two opposing reactions, with the thermal denaturation of the enzyme increasingly influential at high temperatures. The activation energy for the side-reaction An $\rightleftharpoons$ Ad is 17 000 + 37 000 cal, i.e. 54 000 cal mol$^{-1}$, a high value to be expected in the thermal denaturation of a protein.

What other evidence exists to show that the biphasic temperature effect is caused in this way? The main evidence is that the application of hydrostatic pressure at temperatures below the optimum decreases the reaction rate (i.e. dims the light) and at supra-optimum temperatures it accelerates the rate (brightens the light) (Johnson *et al.*, 1942a). Thermal denaturation is likely to proceed with an increase in molar volume which pressure opposes by shifting the equilibrium An $\rightleftharpoons$ Ad to the left, thus counteracting high temperature. Figure III.11 shows that at high hydrostatic pressure the slope of the rate curve remains straight at high temperatures, supporting the idea that the main reaction persists at high temperatures but is merely limited by the concentration of the effective, i.e. native, enzyme. This argument is fairly good, but not conclusive.

For the firefly bioluminescence reaction proceeding *in vitro*, at a rate determined by the concentration of effective enzyme, Ueda and Kamaya (1973) calculated an activation energy of 9 800 cal $mol^{-1}$ for the main reaction at low temperature.

### 5.1.2 *Rate process theory applied to luminescent reactions*

The significance of temperature in understanding anaesthetic action is a long standing theme in the application of rate process theory to bioluminescent reactions. To orientate the reader in unfamiliar territory a brief outline of enzyme kinetics is given before proceeding to the application of rate process theory to particular cases. This is an essential step in the argument leading to an understanding of the thermodynamic and molecular aspects of anaesthetic action on bioluminescence. In all respects the treatment given here is preliminary and it is hoped that the reader will eventually progress to the authoritative physical-chemical texts (Eyring and Magee, 1942; Johnson *et al.*, 1942b, 1945; McElroy, 1943; Johnson *et al.*, 1974).

(i) *Uninhibited reactions.* The rate of the bioluminescent reaction may be defined as

$$I_1 = sk_2(\mathrm{LH_2})(\mathrm{An}) \tag{4}$$

in which $I$ is light intensity under standard conditions, $s$ a proportionality constant which takes into account the units of measurement employed (intensity, concentration), $k_2$ is the rate constant, $LH_2$ and An are as before (McElroy, 1943).

Equation (4) expresses the reaction rate measured at low temperatures. At high temperatures An alters, according to the interpretation of the temperature effects just discussed. Let the total quantity of enzyme, $g$, be composed of An and Ad, native and denatured enzyme, respectively

$$g = \mathrm{An} + \mathrm{Ad}$$
$$\mathrm{An} \rightleftharpoons \mathrm{Ad}, \qquad \mathrm{K_1} = \mathrm{Ad}/\mathrm{An}$$

$$\mathrm{Ad} = \mathrm{K_1An}$$
$$g = \mathrm{An} + K_1\mathrm{An}$$
$$g = \mathrm{An}\,(1 + K_1)$$
$$\therefore \mathrm{An} = g/(1 + K_1). \tag{6}$$

Thus, when temperature alters An we may substitute equation (6) into equation (4) and write

$$I_1 = sk_2(\mathrm{LH}_2)\, g/(1 + K_1). \tag{7}$$

Buried in two of the terms in equation (7) are a number of features which we have to make explicit. First, consider the rate constant, $k_2$. In common with all reaction rates, this is composed of several interacting components, one of which is the concentration of an activated complex which is in equilibrium with reactants and which therefore may be designated an equilibrium constant, $K^{\ddagger}$. The rate at which the activated complex decays to products is determined by temperature, $T$, the transmission coefficient, $k_1$, the Boltzmann constant, $\overline{K}$, and Planck's constant, $h$

$$k_2 = k_1 \frac{\overline{K}T}{h} K^{\ddagger}. \tag{8}$$

Equation (7) also contains the more familiar equilibrium constant $K_1$, for the thermal denaturation of the enzyme An. Both $K_1$ and $k_2$ can be expanded to manifest the experimentally measurable parameters. Having, expanded $k_2$ to its constituents, i.e. substitute equation (8) into equation (7)

$$I_1 = \frac{s[\mathrm{LH}_2]g\, k_1 \dfrac{\overline{K}T}{h} K^{\ddagger}.}{1 + K_1} \tag{9}$$

Now the equilibrium constants $K_1$ and $K^{\ddagger}$ may be expanded, given that in general such constants are made up of enthalpy and entropy terms

$$K = \exp\left(\frac{-\Delta H}{RT}\right) \exp\left(\frac{\Delta S}{R}\right). \tag{10}$$

According to Johnson *et al.* (1974) it is advisable to lump reactants ($g$ and ($\mathrm{LH}_2$)) which in a steady state within the cell are constant but unknown, along with $\Delta S$ and other constants, to give $c$

$$\mathrm{C} = g\, (\mathrm{LH}_2)\, \overline{K}k_1 s \exp \frac{\Delta S^*}{R}$$

Thus,

$$I_1 = \frac{cT \exp\left(\dfrac{-\Delta H^{\ddagger}}{RT}\right)}{1 + \exp\left(\dfrac{-\Delta H_1}{RT}\right) \exp\left(\dfrac{\Delta S_1}{R}\right).} \tag{11}$$

Equation (11) reveals how light intensity is related to temperature and to enthalpy and entropy changes in the reaction.

At this stage it is worth noting some examples of the enthalpy and entropy changes in bioluminescent reactions.

In the case of firefly luminescence proceeding *in vitro* and rate-limited by the concentration of native luciferase, Ueda and Kamaya (1973) calculated from the activation energy already mentioned an activation enthalpy $\Delta H^{\ddagger}$ of 9200 cal mol$^{-1}$ for the main reaction. The thermal denaturation of the limiting enzyme involves a $\Delta H$ or heat of reaction of 76 400 cal mol$^{-1}$, a large value which is consistent with thermal denaturation of a protein. The entropy change of $+257$ eu for the same equilibrium calculated from equation (11) is also consistent with the disorganisation of a protein. These data therefore complete the case for the existence of a thermally denatured enzyme limiting the velocity of the bioluminescent reaction at all but the lowest temperatures. We can now proceed to examine the reaction when it is affected by anaesthetics.

(ii) *Rate process theory applied to inhibited reactions.* Instead of the enzyme, A, being affected by temperature, we will consider ways in which it can be affected by an anaesthetic, X, binding to it, i.e. $X + A \rightleftharpoons XA$. Temperature itself affects the $X + A \rightleftharpoons XA$ equilibrium, and we start by distinguishing two general and hypothetical kinds of inhibition; one in which temperature plays no part, and the other which is sensitive to temperature. Broadly, these idealised categories correspond to competitive and non-competitive enzyme inhibitors, types I and II of Johnson et al (194h).

Consider an inhibitor which interacts with the thermally denatured form of the enzyme. It will shift the equilibrium to the right, like a temperature increase. It reacts with Ad (below) because the appropriate groups are exposed in the Ad form. Following Ueda and Kamaya:

$$\begin{array}{c} U_s + \mathrm{An} \overset{K_1}{\rightleftharpoons} \mathrm{Ad} + U_s \\ \qquad\qquad\quad \upharpoonleft\!\downharpoonright K_3 \\ \qquad\qquad\quad \mathrm{Ad}\, U_s \end{array} \tag{12}$$

in which $s$ is the number of molecules of the inhibitor, $U$, which combines with and inhibits the enzyme, and $K_3$ is the equilibrium constant for the reaction. As before, by arguing that $g$ is the total amount of enzyme, comprising An, Ad, Ad$U_s$, we may write

$$\mathrm{An} = \frac{g}{1 + K_1 + K_1 K_3 U^s}$$

which is analogous to equation (6). For the effect of temperature on the system we have, as in equation (7), by substitution

$$I_2 = sk_2(\mathrm{LH}_2)\frac{g}{1 + K_1 + K_1K_3U^s} \tag{13}$$

in which $I_2$ is the inhibited rate. Dividing the rate of the inhibited reaction into the rate of the uninhibited reaction, i.e. equation (13) into equation (9), and rearranging we obtain

$$\frac{I_1}{I_2} = \frac{1 + K_1 + K_1K_3U^s}{1 + K_1} \tag{14}$$

$$\left(\frac{I_1}{I_2} - 1\right)\left(1 + \frac{1}{K_1}\right) = K_3U^s. \tag{15}$$

As before the equilibrium constant may be expanded, giving

$$\ln\left(\frac{I_1}{I_2} - 1\right)\left(1 + \frac{1}{K_1}\right) = s\ln U - \frac{\Delta H}{RT} + \frac{\Delta S}{R}. \tag{16}$$

in which $\Delta H$ and $\Delta S$ relate to the equilibrium constant $K_3$.
This equation is of practical value, for by plotting data for the left-hand terms against $1/T$ a straight line should be obtained, if the anaesthetic concentration is constant ($s \ln U$ constant) and the assumptions made apply.

In order to plot the left-hand term, values of $K_1$ are required. These are calculated from the equation

$$\ln K = \left(-\frac{\Delta H + T\Delta S}{RT}\right)$$

In the case of inhibitors which react equally with An and Ad, and which therefore do not *bias* the equilibrium, no temperature effect is postulated. That is, the affinity of the enzyme for the inhibitor is negligibly altered by temperature. Kinetic arguments of the preceding type lead to the following relationship in which $X$ is the concentration of the inhibitor and $r$ the number of molecules of it which combine with the enzyme.

$$\ln\left(\frac{I_1}{I_2} - 1\right) = r\ln X - \frac{\Delta H}{RT} + \frac{\Delta S}{R}. \tag{17}$$

with $\Delta H$ and $\Delta S$ terms applying to the particular inhibitor-enzyme interaction

Thus, a plot of ln $[(I_1/I_2) - 1]$ against $1/T$ yield a straight line, if the assumptions apply. Inhibitors of this type are referred to as Type I by Johnson *et al.* (1942b).

Examples of anaesthetics and other inhibitors of bioluminescent reactions which have been investigated in this way fall into either Type I or II categories rather satisfactorily. For instance, urethane acting on bacterial luminescence *in vivo* yield data which clearly conform to a Type II inhibition (Fig. III.12(a)). And clinical anaesthetics (e.g. halothane, methoxyfluorane) applied to firefly luminescence *in vitro* also conforms to the Type II class (Fig. III.12(b)). Urethane and the clinical anaesthetics act to lower the optimum temperature of the light-emitting reaction.

### 5.2 COMPETITIVE AND NON-COMPETITIVE INHIBITORS

If Type II inhibitors are generally expected to be non-competitive, it is worth examining other experimental evidence which bears on this point. In the case of firefly luminescence, Ueda and Kamaya (1973) have clearly shown an absence of competitive inhibition between clinical anaesthetics and ATP, which is a hydrophilic molecule. In bacterial luminescence *in vitro*, White *et al.* (1975) reports some evidence that methoxyfluorane acts on luciferase as if it were competing for the substrate, an aldehyde in this case. It is interesting that the hydrophobic aldehyde is apparently competed for in this way. We will return to the significance of this point shortly.

### 5.3 THERMODYNAMIC DATA FOR ANAESTHETIC INTERACTIONS IN BIOLUMINESCENT REACTIONS

(i) Values for the enthalpy and entropy changes in the anaesthetic–enzyme interaction occurring in firefly luminescence are all large and negative (Table III.4(C)). These are Type II inhibitors.
(ii) Values for the enthalpy and entropy changes in other Type II interactions, obtained with bacterial luminescence under the influence of hydrophobic inhibitors which are not general anaesthetics, are generally quite large and negative (Table III.4(A, B)).
(iii) Enthalpy and entropy changes computed for Type I inhibitors, manifestly not anaesthetic-like, are also negative but smaller (Table III.4(A, B)).
(iv) Type II inhibitors of bacterial luminescence are counteracted by high pressure whereas Type I inhibitors, of which sulphanilamide is one example, are not. The former, therefore, act with a molar volume increase.
(v) The enthalpy change for the clinical agents acting on bacterial luminescence is small and usually negative when calculated by the different "solubility" procedure (Table III.4(D) and below).

In Table III.4 we see a broad correlation between Type II, non-competitive inhibitors and large enthalpy and entropy changes and between Type I or competitive-like inhibitors and small enthalpy and en-

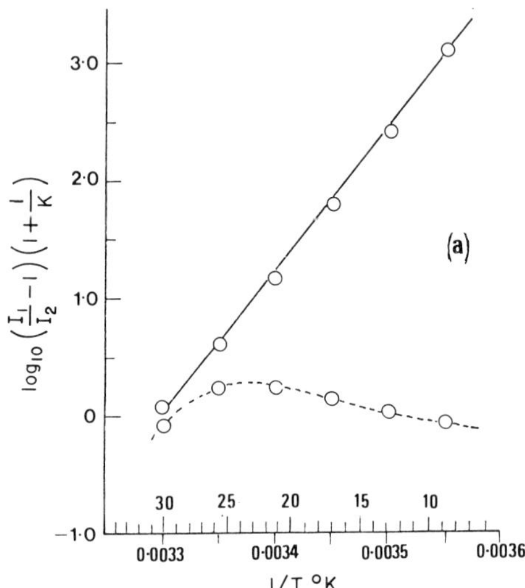


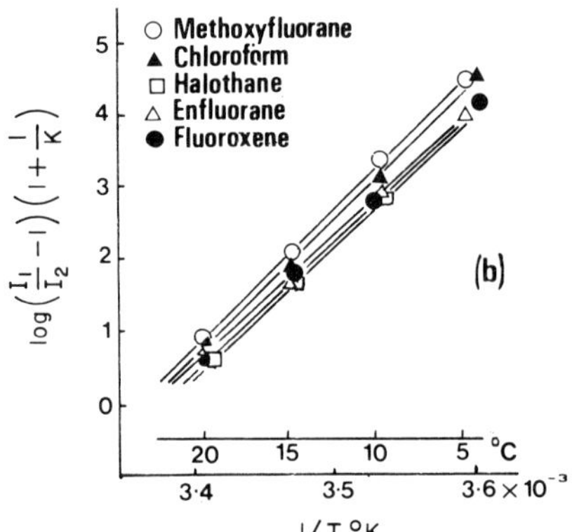


FIG. III.12. Inhibition of bioluminescence by anaesthetics. (a) Urethane acting on luminescence in *Photobacterium phosphoreum*. Rate is plotted against temperature (see text) (after Johnson *et al.*, 1942b). Dotted line shows data according to equation (17). (b) General anaesthetics on the flash of firefly luminescence. Rate is plotted against temperature (see text) (after Ueda and Kamaya, 1973).

TABLE III.4

Apparent enthalpy and entropy changes in anaesthetic–target interactions in bioluminescent reactions

| Substance | Conc. or partial pressure in reaction mixture | $\Delta H$ kcal mol$^{-1}$ | $\Delta S$ eu | $\Delta V$ + or − | | Type of inhibition |
|---|---|---|---|---|---|---|
| (A) *Photobacterium phosphoreum* (*in vivo*) | | | | | | |
| Sulphanilamide | 0·005 M | −12 | −37 | 0/− | Johnson *et al.* (1942b) | I |
| Urethane | 0·075 M | −5 | −165 | + | Johnson *et al.* (1942b) | II |
| Ethanol | 0·400 M | −37 | −128 | + | Johnson *et al.* (1945) | II |
| Chloroform | 0·050 M | — | — | + | Johnson *et al.* (1942) | — |
| (B) *Achromobacter fischeri* (*in vivo*) | | | | | | |
| Phenobarbitone | 0·003 M | −44 | −146 | | McElroy (1943) | II |
| Barbital | 0·015 M | −46 | −152 | | McElroy (1943) | II |
| Benzamide | 0·002 M | −12 | −40 | | McElroy (1943) | I |
| (C) *Firefly* (*in vitro*) | | | | | | |
| Halothane | Clinical range | −83 | −291 | | Ueda and Kamaya (1973) | II |
| Methoxy-fluorane | | −89 | −320 | | Ueda and Kamaya (1973) | II |
| Chloroform | | −89 | −312 | | Ueda and Kamaya (1973) | II |
| Enfluorane | | −80 | −277 | | Ueda and Kamaya (1973) | II |
| Fluoroxene | | −8 | −306 | | Ueda and Kamaya (1973) | II |
| (D) *Photobacterium phosphoreum* (*in vivo*) | | | | | | |
| Halothane | Clinical range | −9 | — | | Flook *et al.* (1974) | — |
| Methoxy-fluorane | | −14 | — | | Flook *et al.* (1974) | — |
| Chloroform | | −5·6 | | see above | Flook *et al.* (1974) | — |
| Diethyl ether | | +2·3 | — | | Flook *et al.* (1974) | — |
| *Photobacterium phosphoreum* (*in vitro*) | | | | | | |
| Methoxy-fluorane | | — | — | | White *et al.* (1975) Apparently competitive with aldehyde substrate | |

tropy changes. Before considering the implications of enthalpy changes, a short digression is required to explain the last set of data in the Table.

The effect of temperature on the anaesthetic action in luminescent bacteria may be treated in a different way from that adopted by Johnson *et al.* (1942b). If anaesthetics "dissolve" in their molecular targets then temperature changes should influence the anaesthetic–target interaction in much the same way as it influences the solubility of anaesthetics in hydrophobic solvents. In other words, temperature should affect anaesthetic potency. By measuring the partial pressure of clinical anaesthetics required to halve the intensity of the light emitted by *Photobacterium phosphoreum* ($ED_{50}$) at different temperatures, it has been shown that apparent potency decreases with increase in temperature (Flook *et al.*, 1974). At least it did so with halothane, methoxyfluorane and chloroform but diethyl ether was exceptional, showing a slight tendency to decrease its potency with a decrease in temperature. Using the change in light intensity as a measure of equipotent doses, the enthalpy of "solution" was calculated using the equation (18) below

$$\ln\left(\frac{P_1}{P_2}\right) = \frac{\Delta H_s}{R}\left(\frac{1}{T_1} - \frac{1}{T_2}\right) \tag{18}$$

in which $P_1$ and $P_2$ are the partial pressures which halve the light intensity at the two temperatures $T_1$ and $T_2$. The resultant values are listed in Table III.4(D).

These values were compared with the values for enthalpy of solution in lipid solvents and are seen to be in agreement except in the case of diethyl ether. The conclusion drawn was that halothane, methoxy and chloroform have changes in apparent potency with temperature which are entirely accounted for by changes in solubility at the site of action, and diethyl ether has a potency change which includes a solubility independent factor.

The significance of a large value for $\Delta H$ is that it implies a drastic conformational change in the target molecule which in some cases also involves an increase in molar volume. It appears as if clinical anaesthetics inhibit enzymes by a non-specific, disorganisation (Type II inhibition), but may also act competitively. It is the target molecule, the enzyme, which determines how the anaesthetic acts. If a hydrophobic active site exists for the enzyme's normal function, then anaesthetics will achieve their disturbance by direct action there. If the only hydrophobic groups in the enzyme are structural and not primarily catalytic in function, then anaesthetics will only perturb the enzyme's function indirectly by way of conformational, structural disturbances (Type II).

## 5.4 STIMULATION OF BIOLUMINESCENCE BY ANAESTHETICS

Figure III.13 shows how low doses of ethanol stimulate the light-emitting process in bacteria at low temperatures. Other clinical agents do the same. Should this effect be regarded as comparable to the way in which amoebae or whole animals are excited by low doses of anaesthetic? The biphasic action of these agents is much overlooked and will have to be accounted for in any comprehensive account of their action. In the case of bioluminescence it is conceivable that the presence of a few anaesthetic molecules at the critical site somehow "strains" the enzyme in a way which lowers the activation energy, i.e. increases the efficiency of catalysis. or perhaps fluidises the otherwise rigid bilayer environment of certain enzymes. Goldacre (1952) speculated along similar lines to account for the "hyper-excitable" effects of anaesthetics in amoebae (Chapter II). Alternatively, low doses of anaesthetics may differentially inhibit competing reactions, in which case this might be analogous to the excitation of a nervous system by the depression of inhibitory pathways.

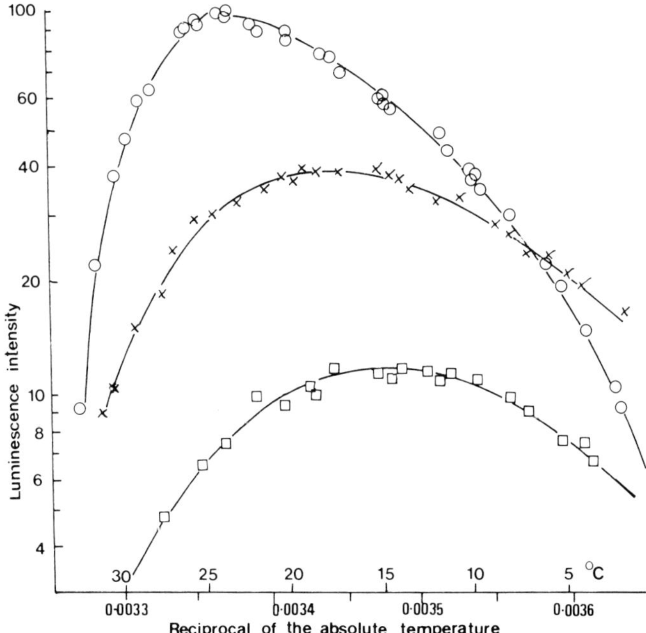


FIG. III.13. Stimulation of bioluminescence in *Photobacterium phosphoreum* by ethanol at low temperature. Upper curve: control; middle curve: 0·4 M; lower curve: 0·8 M ethyl alcohol. (Johnson *et al.*, 1945)

## 6 Conclusions

The biochemical mechanisms by which anaesthetics disturb the normal metabolism of cells and organisms are subtle and unpredictable. Anaesthetics as enzyme inhibitors have a mode of action which is largely dependent on the nature of the enzyme in question, and they are capable of disturbing enzymes organised in membranes by various indirect means. Intact animals possess inbuilt defence mechanisms which handle the clinically important general anaesthetics. The weak inert gases seem to exert certain biochemical effects at partial pressures which are much lower than an orthodox narcotic dose.

## References


Alexander, S. C., Colton, E. T. and Tyler, J. R. (1972). Some effects of $\alpha$ and $\beta$ andrenergic blockade on the metabolic response to ether anesthesia in man. *In* "Cellular Biology and Toxicity of Anaesthetics" (B. R. Fink, ed.). Williams and Wilkins, Baltimore.

Biebuyck, J. F. and Lund, P. (1974). Effects of halothane and other anaesthetic agents on the concentrations of rat liver metabolites *in vivo*. *Molec. Pharmacol.* **10**, 474–483.

Biebuyck, J. F., Lund, P. and Krebs, H. A. (1972). The effects of halothane (2-bromo-2-chloro-1,1,1-trifluoroethane) on glycolysis and biosynthetic processes of the isolated perfused rat liver. *Biochem. J.* **128**, 711–720.

Boyd, G. S. (1972). Biological hydroxylation reactions. *In* "Biological Hydroxylation Mechanisms", Biochem. Soc. Symp. 1971, vol. 34. Academic Press, New York and London.

Brammall, A., Beard, D. and Hulands, G. H. (1973). The effects of inhalational anaesthetic agents on the enzyme glutamate dehydrogenase. *Brit. J. Anaesth.* (Abstract), **45**, 924.

Brown, B. R. (1971). The biphasic action of halothane on the oxidative metabolism of drugs by the liver. *Anesthesiol.* **35**, 241–246.

Brown, B. R. (1972). Hepatic microsomal lipoperoxidation and inhalational anesthetics. *Anesthesiol.* **36**, 458–465.

Brown, B. R. (1973). Hepatic microsomal enzyme induction. *Anesthesiol.* **39**, 178–187.

Brown, B. R. and Sagalyn, A. M. (1974). Hepatic microsomal enzyme induction by inhalational anesthetics. *Anesthesiol.* **40**, 152–161.

Brown, B. R., Sipes, I. G. and Sagalyn, A. M. (1974). Mechanisms of acute hepatic toxicity: chloroform, halothane, and glutathione. *Anesthesiol.* **41**, 554–561.

Brown, D. E., Johnson, F. H. and Marsland, D. A. (1942). The pressure, temperature relations of bacterial luminescence. *J. Cell. Comp. Physiol.* **20**, 151–168.

Carpenter, F. G. (1956). Alteration in mammalian nerve metabolism by soluble and gaseous anesthetics. *Amer. J. Physiol.* **187**, 573–578.

Chance, B. and Williams, G. R. (1962). The respiratory chain and oxidative phophorylation. *Adv. Enzym.* **17**, 65–134.

CHANCE, B., COHEN, P., JOBSIS, F. and SCHOENER, B. (1962). Intracellular oxidation-reduction states *in vivo*. *Science*, **137**, 499–507.

CHANCE, B., MELA, L. and HARRIS, E. J. (1968). Interaction of ion movements and local anaesthetics in mitochondrial membranes. *Fed. Proc.* **27**, 902–906.

CISSICK, J. H., JOHNSON, R. E. and HERTIG, B. A. (1972). Production of gaseous nitrogen during human steady state exercise. *Aerospace Med.* **43**, 1245–1250.

COHEN, E. N. (1969). Metabolism of halothane. 2 $^{14}$C in the mouse. *Anesthesiol.* **31**, 560–565.

COHEN, E. N. and HOOD, N. (1969). Application of low temperature autoradiography to studies of the uptake and metabolism of volatile anaesthetics in the mouse. I. Chloroform. *Anesthesiol.* **30**, 306–314.

COHEN, P. J. (1973). Effect of anaesthetics on mitochondrial function. *Anesthesiol.* **39**, 153–164.

COHEN, P. J. and MARSHALL, B. E. (1968). Effects of halothane on respiratory control and oxygen consumption of rat liver mitochondria. *In* "Toxicity of Anaesthetics" (B. R. Fink, ed.), p. 332. Williams and Wilkins, Baltimore.

COHEN, P. J. and McINTYRE, R. (1972). The effects of general anesthesia on respiratory control and oxygen consumption of rat liver mitochondria. *In* "Cellular Biology and Toxicity of Anesthetics" (B. R. Fink, ed.), p. 327. Williams and Wilkins, Baltimore.

CORMIER, M. J. LEE, J. and WAMPLER, J. E. (1975). Bioluminescence: recent advances. *Ann. Rev. Biochem.* **44**, 255–272.

COOK, S. F. (1950). The effect of helium and argon on metabolism and metamorphosis. *J. Cell. Comp. Physiol.* **36**, 115–127.

COOK, S. F. and SOUTH, F. E. (1953). Helium and comparative *in vitro* metabolism of mouse tissue slices. *Amer. J. Physiol.* **173**, 542–544.

COOK, S. F., SOUTH, F. E. and YOUNG, D. R. (1951). Effect of helium on gas exchange of mice. *Amer. J. Physiol.* **164**, 248–250.

COON, M. J., STROBEL, H. W., AUTOR, A. P., HEIDEMA, J. and DUPPEL, W. (1972). Functional components and mechanism of action of the resolved liver microsomal enzyme systems catalyzing fatty acid, hydrocarbon and drug hydroxylation. *In* "Biological Hydroxylation Mechanisms", Biochem. Soc. Symp. 1971, vol. 34. Academic Press, New York and London.

DAVIS, D. C., SCHROEDER, D. H., GRAM, T. E., REAGAN, R. L. and GILLETTE, J. R. (1971). A comparison of the effects of halothane and $CCl_4$ on the hepatic drug metabolizing system. *J. Pharm. Exp. Therap.* **177**, 556–566.

DAVIES, D. R. and QUASTEL, J. H. (1932). Dehydrogenations by brain tissue. The effects of narcotics. *Biochem. J.* **26**, 1672.

DEDRICK, D. F., SCHERER, Y. D. and BIEBUYCK, J. F. (1975). Use of a rapid brain sampling technique in a physiologic preparation. *Anesthesiol.* **42**, 651–657.

EYRING, H. and MAGEE, J. L. (1942). Application of the theory of absolute reaction rates to bacterial luminescence. *J. Cell. Comp. Physiol.* **20**, 169–177.

FINK, B. R. and KENNY, G. F. (1968). Effects of halothane on cell culture metabolism. *In* "Toxicity of Anaesthetics" (B. R. Fink, ed.). Williams and Wilkins, Baltimore.

FLOOK, V., ADEY, G. D., DUNDAS, C. R. and WHITE, D. C. (1974). Effects of temperature on potency of anaesthetic agents. *J. Appl. Physiol.* **37**, 552–555.

GREENE, N. M. (1976). A new aspect of the metabolism of halothane. *Anesthesiol.* **44**, 191–193.

GREENE, N. M. and SPENCER, E. L. (1963). Diethyl ether and lactate dehydrogenase activity *in vitro*. *Anesthesiol.* **24**, 23–28.

GREENSTEIN, L. R., HITT, B. A. and MAZZE, R. I. (1975). Metabolism *in vitro* of enflurane, isoflurane and methoxyflurane. *Anesthesiol.* **42**, 420–424.

GREIG, M. E. (1946). The site of action of narcotics on brain metabolism. *J. Pharmacol.* **87**, 185–192.

HALLÉN, B. A. and JOHANSSON, G. (1975). Inhalational anesthetics and cytochrome P–450-dependent reactions in rat liver microsomes. *Anesthesiol.* **43**, 34–40.

HALSEY, M. J. and SMITH, E. B. (1970). Effects of anaesthetics on luminous bacteria. *Nature*, **227**, 1363–1365.

HULANDS, G. H., BEARD, D. J. and BRAMMALL, A. (1975). The *in vitro* interaction of inhalational anaesthetics with glutamate dehydrogenase and other enzymes. *In* "Progress in Anesthesia I" (B. R. Fink, ed.). Raven Press, New York.

HULME, N. A. and KRANTZ, J. C. (1955). Anesthesia XLV: Effect of ethyl ether on oxidative-phosphorylation in the brain. *Anesthesiol.* **16**, 627–631.

IMAI, Y. and SATO, R. (1967). Studies on the substrate interactions with P–450 in drug hydroxylation by liver microsomes. *J. Biochem, Japan*, **62**, 239–249.

JOHNSON, F. H., BROWN, D. E. S. and MARSLAND, D. A. (1942a). Pressure reversal of the action of certain narcotics. *J. Cell. Comp. Physiol.* **20**, 269–276.

JOHNSON, F. H., EYRING, H., STEBLAY, R., CHAPLIN, H., HUBER, C. and GHERARDI, G. (1945). The nature and control of reactions in bioluminescence. With special reference to the mechanism of reversible and irreversible inhibitions by hydrogen and hydroxyl ions, temperature, pressure, urethane, and sulfanilamide. *J. Gen. Physiol.* **28**, 463–537.

JOHNSON, F. H., EYRING, H. and STOVER, B. J. (1974). "The Theory of Rate Processes in Biology and Medicine". Wiley, New York.

JOHNSON, F. H., EYRING, H. and WILLIAMS, R. W. (1942b). The nature of enzyme inhibitions in bacterial luminescence: sulfanilamide, urethane, temperature and pressure. *J. Cell. Comp. Physiol.* **20**, 247–268.

JOWETT, M. and QUASTEL, J. H. (1937). The effects of ether on brain oxidations. *Biochem. J.* **31**, 1101–1112.

KORTEN, K. and VAN DYKE, R. A. (1973). Acute interactions of drugs. 1. The effect of volatile anesthetics on the kinetics of aniline hydroxylase and aminopyrine demethylase in rat hepatic microsomes. *Biochem. Pharmacol.* **22**, 2105–2112.

KU, H. S. and LEOPOLD, A. C. (1970). Mitochondrial responses to ethylene and other hydrocarbons. *Plant Physiol.* **46**, 842–844.

LARRABEE, M. G., RAMOS, J. G. and BULBRING, E. (1952). Effects of anesthetics on oxygen consumption and on synaptic transmission in sympathetic ganglia. *J. Cell. Comp. Physiol.* **40**, 461–494.

LUND, D. B., FENNEMA, O. and POWRIE, W. D. (1969). Effects of gas hydrates and hydrate formers on invertase activity. *Arch. Biochem. Biophys.* **129**, 181–188.

MCELROY, W. D. (1943). The application of the theory of absolute reaction rates to the action of narcotics. *J. Cell. Comp. Physiol.* **21**, 95–116.

MCELROY, W. D. and SELIGER, H. H. (1966). Firefly bioluminescence. *In* "Bioluminescence in Progress" (F. H. Johnson and Y. Haneda, eds), pp. 427–458. Princeton University Press, Princeton.

MEHARD, C. W. and LYONS, J. M. (1970). A lack of specificity for ethylene-induced mitochondrial changes. *Plant Physiol.* **46**, 36–39.

MILES, J. L., ROBINSON, D. A. and CANADY, W. J. (1963). Thermodynamics of the solution process. Some applications to enzyme catalysis and inhibition of α-chymotrypsin by aromatic hydrocarbons. *J. Biol. Chem.* **238**, 2932-2937.

MILLER, R. N. and HUNTER, F. E. (1970). The effect of halothane on electron transport, oxidative phosphorylation, and swelling in rat liver mitochondria. *Molec. Pharmacol.* **6**, 67–77.

MILLER, R. N. and HUNTER, F. E. (1971). Is halothane a true uncoupler of oxidative phosphorylation? *Anesthesiol.* **35**, 256–261.

MILLER, R. N., SMITH, E. E. and HUNTER, F. E. (1972). Halothane-induced alterations in energy dependent and energy independent membrane carrier function in isolated rat liver mitochondria, with some electron microscopic observations. *In* "Cellular Biology and Toxicity of Anaesthetics" (B. R. Fink, ed.). Williams and Wilkins, Baltimore.

MILLER, S. L. (1961). A theory of gaseous anaesthetics. *Proc. Nat. Acad. Sci. U.S.A.* **47**, 1515–1524.

MURAYAMA, M. (1973). Sickle cell hemoglobin. *Crit. Essays. Biochem.*, **1**, 461–492.

NAHRWOLD, M. L., CLARK, R. C. and COHEN, P. J. (1974). Is depression of mitochondrial respiration a predictor of *in-vivo* anesthetic activity? *Anesthesiol.* **40**, 566–570.

NILSSON, L. and SEISJÖ, B. K. (1970). The effect of anesthetics upon labile phosphates and upon extra- and intra-cellular lactate, pyruvate and bicarbonate concentrations in the rat brain. *Acta. Physiol. Scand.* **80**, 235–248.

PAULING, L. (1961). A molecular theory of general anaesthesia. *Science*, **134**, 15–21.

REMMER, H. (1962). Drugs as activators of drug enzymes. *Proc. 1st Internat. Pharm. Meeting*, **VI**, 235.

REMMER, H. (1972). Induction of drug metabolizing enzyme systems in the liver. *Eur. J. Clin. Pharm.* **5**, 116–136.

ROSENBERG, H. and HAUGAARD, N. (1973). The effects of halothane on metabolism and calcium uptake in mitochondria of the rat liver and brain. *Anesthesiol.* **39**, 44–53.

ROSS, W. T. and CARDELL, R. R. (1972). Effects of halothane on the ultrastructure of rat liver cells. *Am. J. Anat.* **135**, 5–21.

SCHOENBORN, B. P. and NOBBS, C. L. (1966). The binding of xenon to sperm whale myoglobin. *Molec. Pharmacol.* **2**, 495–498.

SCHOLLER, K. L. (1970). Modification of the effects of chloroform on the rat liver. *Brit. J. Anaesth.* **42**, 603–605.

SCHRIER, E. E., INGWALL, R. T. and SCHERAGA, H. A. (1962). The effect of aqueous alcohol solutions on the thermal transition of ribonuclease. *J. Phys. Chem.* **69**, 298–303.

SHIAO-CHUN TU and HASTINGS, J, W. (1976). "The Flavin Chromophore of Photoexcitable Bacterial Luciferase in Flavins and Flavoproteins" (T. P. Singer, ed.). Elsevier, Amsterdam.

SOUTH, F. E. and COOK, S. F. (1953a). Effect of helium on the respiration and glycolysis of mouse liver slices. *J. Gen. Physiol.* **36**, 513–528.

SOUTH, F. E. and COOK, S. F. (1953b). Argon, xenon, hydrogen and the oxygen consumption and glycolysis of mouse tissues. *J. Gen. Physiol.* **37**, 335–341.

STIER, A. (1968). The biotransformation of halothane. *Anesthesiol.* **29**, 388–389.

TAYLOR, C. A., WILLIAM, C. H., TAKASHI, W., VALDIVIA, E., HANIS, R. A. and GREEN, D. E. (1972). The effect of halothane on energized configurational changes in heart mitochondria *in situ*. *In* "Cellular Biology and Toxicity of Anaesthetics" (B. R. Fink, ed.). Williams and Wilkins, Baltimore.

UEDA, I. and KAMAYA, H. (1973). Kinetic and thermodynamic aspects of the mechanism of general anaesthesia in a model system of firefly luminescence *in vitro*. *Anesthesiol.* **38**, 425–436.

VAN DYKE, R. A. and CHENOWETH, M. B. (1965). The metabolism of volatile anesthetics. II. *In vitro* metabolism of methoxyflurane and halothane in rat liver slices and cell fractions. *Biochem. Pharmacol.* **14**, 603–609.

VAN DYKE, R. A., CHENOWETH, M. B. and POZNAK, A. V. (1964). Metabolism of volatile anaesthetics. I. Conversion *in vivo* of several anaesthetics to $^{14}CO_2$ and chloride. *Biochem. Pharmacol.* **13**, 1239–1247.

WANDS, J. R., SMUCKLER, E. A. and WOODBURY, W. J. (1970). Transmembrane potential changes in liver cells following CCl4 intoxication. *Am. J. Path.* **58**, 499–508.

WETLAUFFER, D. B. and LOVREIN, R. (1964). Induction of reversible structural changes in proteins by non-polar substances. *J. Biol. Chem.* **239**, 596–606.

WHITE, D. C., WARDLEY-SMITH, B. and ADEY, G. (1973). The site of action of anaesthetics on bacterial luminescence. *Life Sci.* **12**, 453–461.

WHITE, D. C., WARDLEY SMITH, B. and ADEY, G. (1975). Anaesthetics and bioluminescence. In "Progress in Anaesthesiology" (B. R. Fink, ed.), vol. 1. Raven Press, New York.

CHAPTER IV

# Inexcitable membranes



## 1 Introduction

Anaesthetics comprise a large variety of structural types, and these compounds act on membranes other than the plasma membrane of an excitable nerve or muscle cell. Indeed, we have speculated in Chapter II that perhaps all important anaesthetic effects originate in membranes. These agents produce a range of effects in diverse membranes, but this does not imply a common mechanism is at work. It should be noted that although some membrane effects (e.g. lysis) may be mediated via general membrane perturbations, other actions (e.g. on transport processes) could be produced by quite specific molecular mechanisms.

Many studies are undertaken on "simple" inexcitable membranes in order that unambiguous results may be obtained, and all too often on the assumption that such data may be applicable directly to excitable membranes. It is dangerous to assume that *all* data available for inexcitable membranes are strictly relevant to excitable membranes, and it is probably of more value to examine such data with a view to learning about the interaction of such agents with the membrane in question. After all, anaesthetics do produce profoundly interesting effects on inexcitable membranes, which suggest to us either something about the behaviour of the anaesthetic molecule in the membrane phase or, alternatively, how the membrane reacts to the presence of "foreign" molecules. Just as the organic solvent olive oil fails to simulate the nature of the biological membrane in many respects, so too the inexcitable membrane may turn out to be an inappropriate model for the excitable system.

In this chapter, it is our intention to focus on the action of anaesthetic agents on four select systems. Consideration will be given to the artificial bilayer membranes, erythrocyte membranes, some epithelial cell membranes and the muscle sarcoplasmic reticulum. These systems have been chosen because each has a particular attraction which will be brought out below and, at least in the first two cases, the actions of anaesthetics have generated a great deal of interest and consequently a wealth of experimental data. The effects of anaesthetics on mitochondrial membranes have already been mentioned in Chapter III.

Before embarking on a discussion of the action of anaesthetics on these systems, it is worth drawing attention to the finding that at least in the case of erythrocyte membranes, the membranes of the sarcoplasmic reticulum and synaptosomal membrane, the membrane–buffer partition coefficients are similar for certain agents such as the tranquilliser chlorpromazine, the tertiary amine local anaesthetic RAC 109, and pentobarbitone. This is shown in Fig. IV.1. Although we do not suggest that too much significance be attached to this finding when considering molecular mechanisms, these data do serve to point out that uptake into functionally different membranes involves a similar process. Membrane specialisation, therefore, does not alter markedly the association of the anaesthetic with the membrane, and it is perhaps this fact more than any other which has led to the intensive study of the action of anaesthetics with inexcitable membranes.

## 2 Model bilayer membranes

Lipids are amphipathic, i.e. one part of the molecule is water-soluble (hydrophilic), the other part is oil-soluble (hydrophobic). Consequently, lipid molecules at an oil–water interface align themselves with the long-chain hydrocarbon tail immersed in the oil phase, and the polar groups in the aqueous phase. Lipids also form characteristic monolayers at an air water interface and the action of anaesthetics on such monolayers has been intensively studied (e.g. Skou, 1958). In addition, lipids form bilayers in an

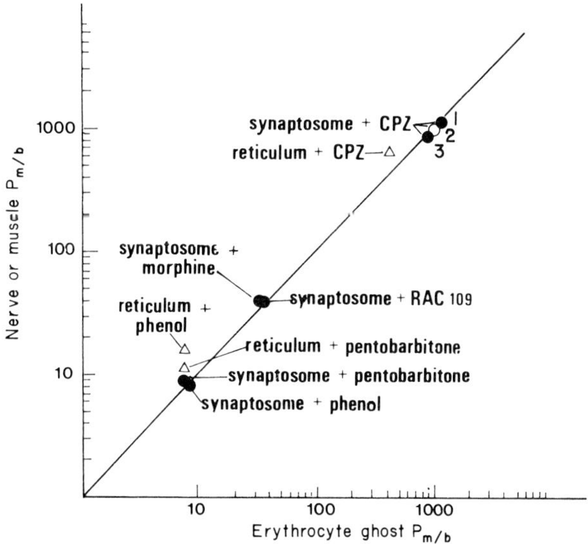


FIG. IV.1. The relationship between the membrane–buffer partition coefficients of anaesthetics in nerve or muscle membranes, and the membrane–buffer partition coefficients in erythrocyte membranes. The chlorpromazine (CPZ) concentrations used with synaptosomes were 0·076, 0·09 and 0·10 mM for the points 1, 2 and 3, respectively; for sarcoplasmic reticulum the CPZ concentration was 0·18 mM. The phenol and pentobarbitone concentrations were 6·5 and 0·18 mM, respectively. The morphine and the quaternary local anaesthetic (RAC 109) concentrations were 5·2 and 2·7 mM, respectively. The line joins the points of equal partition coefficient in erythrocyte and nerve or muscle membranes. (Roth and Seeman, 1972)

aqueous environment, and it is the action of anaesthetics on such bilayer membranes that is the theme of this section.

## 2.1 FORMS

Membranes formed from bimolecular lamellar aggregates of lipids were first made some 15 years ago (Mueller *et al.*, 1962). Lack of space precludes discussion of the techniques of formation of these bimolecular lipid membranes (BLM), and the reader is referred elsewhere for such preparative procedures (e.g. Jain, 1972). A variety of anaesthetics have been tested on bilayer membranes made from phospholipid, which is fortunate in view of the fact that detailed experimental data is available with respect to the interaction of anaesthetic agents with phospholipid monolayers (Feinstein, 1964; Blaustein, 1967; Hauser and Dawson, 1968).

Two types of bilayer lipid membrane have been studied. In some studies a solution of lipid in a solvent such as *n*-decane is applied to an aperture in a partition which separates two compartments of an aqueous solution. The deposited lipid thins out to a bilayer lipid membrane, separating therefore the two aqueous phases. We may term this membrane preparation the planar bilayer model. The planar bilayer membrane is often used to study optical properties, to perform conventional electrical measurements of conductance or capacitance, or to examine permeability characteristics using radioactive isotopes.

One difficulty with this form of the bilayer preparation relates to the uncertainty about its exact composition, i.e. how solvent-free it is. An alternative sort of bilayer preparation may be formed by dispersing phospholipids in an aqueous salt solution above the phase-transition temperature of the lipid. The aggregates formed are termed liposomes or smectic (soap-like) mesophases, and are basically small membrane-bound vesicles. The vesicular structure may take the form of an aqueous sac bounded by a single or, alternatively, an array of lamellar bimolecular layers. The geometry of the aggregate depends on many factors, such as the relative concentration of lipid and water, the nature of the lipid, the temperature and pH. In addition, sonication produces a smaller vesicle of approximately 250 Å in diameter. The phase structure assumed by the lipid is influenced by many factors, including the water content of the phospholipid (see Chapman and Wallace, 1968), and it is difficult to be certain which form is the most relevant physiologically. At present, however, the lamellar model does not seem irreconcilable with existing experimental data, although clearly other phases could exist within the biological membrane which were in equilibrium with the lamellar form (e.g. Lucy, 1968).

Experimental work investigating the action of anaesthetic agents on bilayers has been fruitful and has yielded potentially useful data but it is essential to be extremely cautious when extrapolating from these studies to those on biological membranes. First, biological membranes, whether inexcitable or excitable, are undoubtedly more complex than the bilayer models. Secondly, the composition of the bilayer must modify its permeability characteristics. For example, the addition of cholesterol to phosphatidyl liposomes reduces the membrane permeability to ions and neutral solutes, presumably by virtue of its ordering action on fatty acid chains (Schreier-Muccillo *et al.*, 1973). Finally, the composition of the bilayer may influence the net effect of an anaesthetic agent. Thus, alterations of both the cholesterol and the phosphatidic acid content of bilayer membranes modify the fluidising action (see below) of certain anaesthetics (Miller and Pang, 1976). Despite this caveat, we may still gain valuable information from experiments on the planar bimolecular lipid membrane or the liposome bilayer preparation.

## 2.2 SPECTROSCOPIC METHODS

A lucid description of two spectroscopic methods commonly used in bilayer studies by many investigators within the last decade has been given by Metcalfe (1970). In brief, one of the techniques, the electron spin resonance method (ESR), depends on the fact that the unpaired electron of a nitroxide group ($>\overset{+}{N}—\overset{-}{O}$) will align itself in the direction of an applied magnetic field until it absorbs sufficient energy from an exciting field to "flip" into an antiparallel direction. The absorption of energy results in a characteristic resonance spectrum which is sensitive to the molecular motion of the compound carrying the nitroxide group. Nitroxide analogues of biological compounds, which are referred to as spin labels, can be prepared, and when attached to a membrane component provide a spectrum which will be influenced by any structural rearrangement in neighbouring membrane molecules. The spin-labelled molecule could be an anaesthetic or an actual membrane component.

Use of one agent, 2,2,6,6-tetramethylpiperidine-1-oxyl (TEMPO) has been widely exploited since its introduction by Hubbell and McConnell (1968). Such studies reveal that anaesthetics may increase the binding of TEMPO (a nerve-blocking agent) to membranes.

The other technique, nuclear magnetic resonance (NMR) spectroscopy, relies upon the fact that atomic nuclei ($^{1}H$ or $^{13}C$) possess magnetic moments which become aligned in an externally applied magnetic field. A certain amount of excitation energy is required to destroy this alignment, and measurement of this energy is the basis of NMR absorption spectroscopy. The mobility of an anaesthetic agent in a membrane has been studied by using an NMR relaxation method (see Metcalfe *et al.*, 1968), and this too gives the experimenter information concerning the degree of organisation of the membrane.

Local anaesthetics have well-characterised NMR spectra and are very useful probes of bilayer structure. Such studies have suggested that these agents may penetrate deep into the hydrophobic interior of the membrane fabric. Thus, in the case of tetracaine, the terminal $CH_3$ of the butylene residue could reach to the seventh or eighth $CH_2$ of the fatty acid chain of the phospholipid molecule in a bilayer (Cerbón, 1972).

## 2.3 PERMEABILITY MEASUREMENTS

Early studies on spread monolayers of lipid correlated the nerve-blocking potency of anaesthetics or inert gases with their ability to increase the surface pressure of the film at constant area (e.g. Skou, 1961; Bennett *et al.*, 1967). The advent of the bimolecular lipid membrane preparations provided the opportunity to investigate directly the effects of anaesthetics on bilayer membrane permeability. It is an interesting, and instructive, exercise to compare such data with that obtained from other inexcitable membranes, or from excitable membranes.

The efflux of radioactively labelled cations and uncharged solutes from the aqueous compartments within a single-walled liposome has been measured (Johnson and Bangham, 1969; Johnson *et al.*, 1973). In the initial work it was thought that the liposome had a double wall but later studies showed the presence of only one bilayer (Johnson *et al.*, 1971). These experiments have demonstrated that the liposome permeability to $K^+$, and the valinomycin ($K^+$-ionophore) induced increase in $K^+$ permeability are both enhanced by inhalation anaesthetics such as diethyl ether and chloroform and the alcohol *n*-butanol (Johnson and Bangham, 1969). Since a small increase in activation entropy accompanied these anaesthetic effects, it was suggested that the anaesthetics disordered the liposome membrane, and that as a result the valinomycin carrier operated more efficiently. *n*-Butanol was much more effective at increasing the permeability of the liposome than chloroform and ether. Since the alcohol is the only one of the three molecules with a hydrophobic head group and a hydrophilic tail it was argued that its primary site of action was at the membrane–water interface (Johnson and Bangham, 1969). Butanol also increases the glucose permeability of liposomes more effectively than chloroform (Johnson *et al.*, 1973). Does this imply that the enhancement of the permeability of liposomes to an uncharged solute also depends primarily on an action at the membrane–water interface?

It is fair to say that the action of comparatively few types of anaesthetic agents has been tested on the permeability characteristics of lipid membranes. Probably the alcohols (Bangham *et al.*, 1965a; Gutknecht and Tosteson, 1970) and the local anaesthetics (McLaughlin, 1975; Singer, 1975) have been studied most. The sensitivity of the permeability mechanism to change by the alcohols may be illustrated by the following example. The electrical resistance of the planar bilayer is normally of the order of $10^8\ \Omega cm^2$ or greater. Addition of 8 mM heptanol, a low concen-

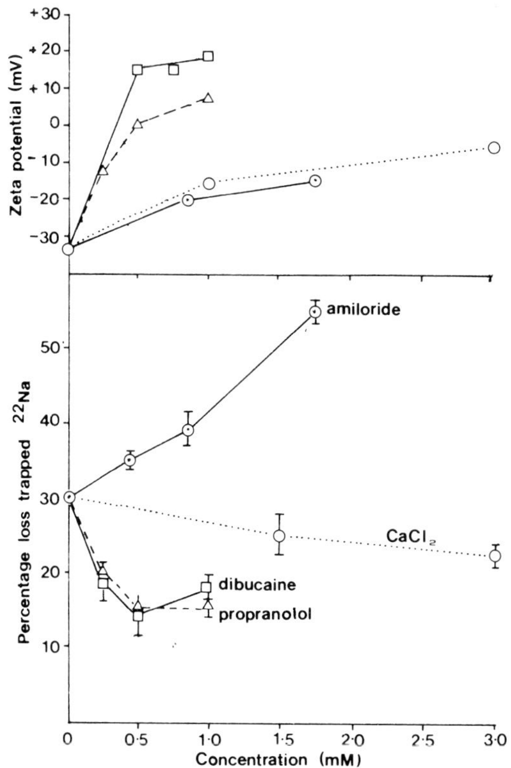


FIG. IV.2. The effects of calcium, dibucaine, propranolol and amiloride on the $^{22}Na^{+}$ efflux and the zeta potential of liposomes composed of 80 per cent phosphatidylcholine and 20 per cent dicetylphosphate. Calcium, dibucaine, propranolol or amiloride were added to the outside aqueous phase at the concentrations indicated on the abscissa. The lower ordinate refers to the percentage of the initial trapped isotope lost during 24 hr (temperature = 25°C). The points represent the means $\pm 1 \times$ SE ($n = 12$) for the radioactive flux data, or the means of twenty measurements for the zeta potential data. (Singer, 1975)

tration in such permeability studies, reduces the resistance a 1000-fold to $10^5$ $\Omega cm^2$ (Gutknecht and Tosteson, 1970)! Not many other membrane systems can exhibit such a remarkable susceptibility to the alcohols.

As pointed out above, the composition of the vesicle is a prime determinant of its permeability characteristics, and also modifies how it responds to anaesthetics. This fact must be born in mind or else the permeability data appear conflicting. For example, vesicles made from egg lecithin and dicetylphosphate are negatively charged and possess a relatively high cation permeability. Local anaesthetics depress the permeability of these vesicles by neutralising the surface charge (Bangham *et al.*, 1965a). Sonicated phosphatidylserine vesicles, on the other hand, exhibit a very low cationic permeability despite their negative surface charge. Local anaesthetics increase the $^{22}Na^+$ efflux from these vesicles in the absence of $Ca^{2+}$, although interestingly enough, in the presence of $Ca^{2+}$ the phosphatidylserine vesicles behave like the egg lecithin/dicetylphosphate variety (Papahadjopoulos, 1970).

It has been proposed that the local anaesthetics act in the charged form to reduce the surface charge potential at the lipid–$H_2O$ interface (McLaughlin, 1975). Let us now consider the evidence in support of this hypothesis, since it has been invoked to explain the action of local anaesthetics on excitable membranes.

The cation permeability of the liposome membrane was found in the early work to be dependent on the negative charge density on the lipid surface (Bangham *et al.*, 1965b). Subsequent studies showed that local anaesthetics could produce a parallel reduction in the $^{22}Na^+$ efflux from negatively charged liposomes, and in the zeta potential (Singer, 1975). The latter parameter is a measure of the surface charge of the vesicles and may be calculated from a knowledge of the electrophoretic mobility of the dispersion. Typical data are shown in Fig. IV.2. Dibucaine and propranolol (a $\beta$ blocker) both caused a reversal of the negative surface charge, which is associated with a marked reduction in $Na^+$ permeability. In this example, the effects of $Ca^{2+}$ and amiloride (a diuretic) are shown for comparison.

Perhaps the strongest piece of evidence in favour of the fixed charge hypothesis derives from experiments with the negative permeant species 5,6-dichloro-2-trifluoromethyl benzimidazole (DTFB), and the positively-charged species, valinomycin–$K^+$ complex. The local anaesthetic tetracaine produced markedly opposite effects on the conductance to these two species (Fig. IV.3). Whereas the conductance of the negative species was increased, that of the positive species was decreased. This supports the notion that the anaesthetic adsorbs to the membrane and reduces the density of surface negative charge. Such a differential action is difficult to

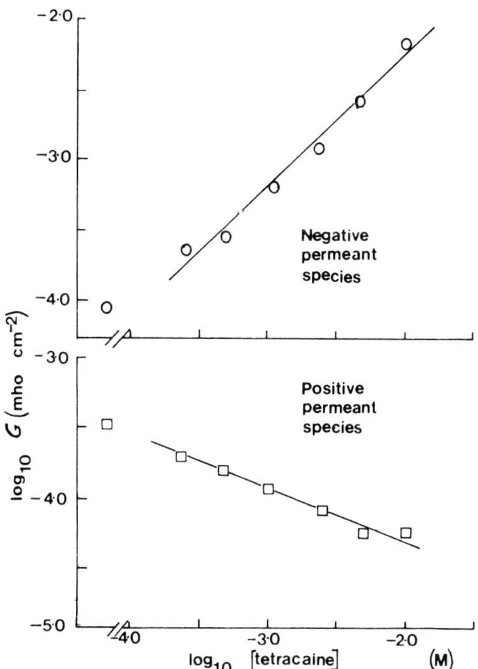


FIG. IV.3. The effects of varying concentrations of the local anaesthetic tetracaine on the conductance of planar bilayer membranes to either the negatively charged permeant species (DTFB), or to the positively charged species (valinomycin–$K^+$ complex). The membranes were formed from phosphatidyl ethanolamine and the aqueous solutions contained $5 \times 10^{-6}$ M DTFB (upper graph) or $5 \times 10^{-7}$ M valinomycin (lower). (McLaughlin, 1975)

account for if it is proposed that the anaesthetic effect is mediated indirectly by a change in the fluidity of the membrane.

Although such an action of the anaesthetic would only be produced by the charged form of the agent, McLaughlin (1975) has shown that the interaction of the anaesthetic with the bilayer is primarily hydrophobic in nature and does not require the presence of polar head groups on the lipid. In the case of tetracaine, NMR studies lend support to this in that the tail of the molecule may penetrate deep into the hydrophobic interior of the membrane phase (Cerbón, 1972).

The above results suggest that the charged local anaesthetic aligns itself in the membrane with the positive charge contributing to the surface charge of the membrane, so reducing its negativity and hence its cation permeability. This is perhaps difficult to reconcile with all the experimental data available. For example, in the presence of valinomycin, dibucaine increases the cation efflux of liposomes (Singer, 1975). Also in the presence of cholesterol the cation permeability of negatively charged vesicles decreases as the concentration of cholesterol is increased, and both dibucaine and propranolol may increase the cation permeability of the vesicles under these conditions. Inspection of these results suggests that such an increase in cation permeability is not wholly consistent with the fixed-charge hypothesis.

In Fig. IV.4 are shown the effects of dibucaine and propranolol on the $^{22}Na$ efflux and zeta potential of negatively charged vesicles containing varying amounts of cholesterol. The action of the diuretic amiloride is also shown again for comparison. It is clear that the addition of cholesterol itself produced a more negative zeta potential. Despite this, the $^{22}Na$ efflux was depressed, since cholesterol probably decreased the bilayer fluidity by restricting the molecular motion of the fatty acid chains. Addition of dibucaine or propranolol (aqueous concentration of both was 1·0 mM) reversed the zeta potential in the absence of cholesterol, and so reduced the $^{22}Na$ efflux.

When cholesterol was added to the vesicles, and the action of the local anaesthetic agents tested, the $^{22}Na$ efflux was enhanced, first, as the zeta potential became more negative and then fell. However, the $^{22}Na$ efflux relative to control was augmented in the presence of the local anaesthetics over much of the cholesterol concentration range tested, despite the reduction in zeta potential relative to control. Dibucaine was more effective than propranolol in opposing the depressant effect of cholesterol on the $^{22}Na$ efflux, possibly because of its increased efficiency in disrupting the ordering influence of cholesterol on the hydrocarbon chains of the lipid. This efficiency could simply relate to the degree of penetration by, and therefore length of, the anaesthetic molecule. In this respect it is interesting to note that both dibucaine and cholesterol have a length of 19 Å, whereas propranolol is 14·5 Å long (Singer, 1975).

The experiment using cholesterol illustrates the pitfalls associated with membrane hypotheses. The composition of the membrane must always be a major consideration. The fixed-charge hypothesis may consequently not have universal application to all membrane models or indeed to real

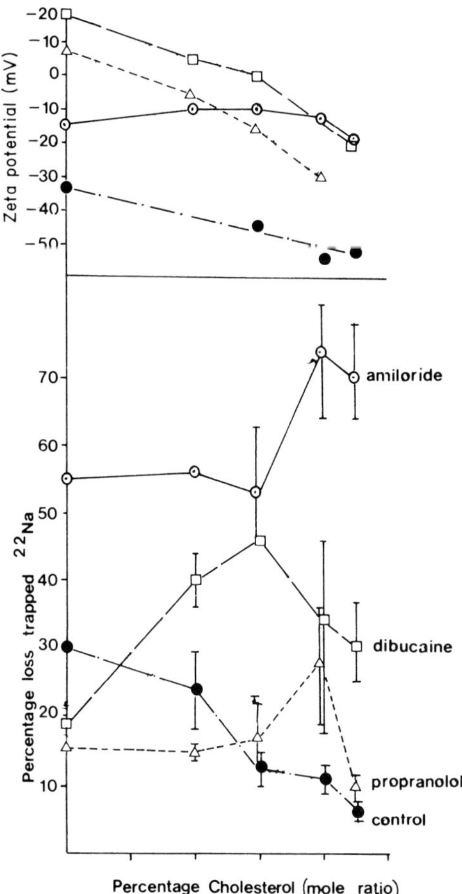


FIG. IV.4. The effects of propranolol, dibucaine and amiloride on the $^{22}Na^{+}$ efflux and the zeta potential in the presence of varying concentrations of cholesterol. The liposomes were composed (by mole ratio) of 20 per cent of dicetylphosphate, cholesterol (concentration given on abscissa), and phosphatidylcholine (to make 100 per cent). Dibucaine, amiloride or propranolol was added to the outside aqueous phase at a concentration of 1·0, 1·73 and 1·0 mM, respectively. The upper ordinate refers to the zeta potential (twenty measurements) and the lower ordinate to the percentage of initial trapped isotope lost during 24 hr (temperature = 25°C). For the radioactive flux data each point represents the mean of four or eight measurements, the bar height representing the range of values obtained. (Singer, 1975)

membranes, and of course it is restricted to the action of charged anaesthetics.

## 2.4 CHANGES IN FLUIDITY

There can be little doubt that the concept of membrane fluidity dominates much of our thinking about the nature of molecular packing in lipid bilayers, inexcitable and excitable membranes. Its importance has been expressed in discussions of the structure of membranes (Singer and Nicolson, 1972), and an exciting development has been the direct observation of such membrane motion using the electron microscope (Hui, 1976).

Interpretation of fluidity changes at a molecular level demand a precise knowledge of the composition of the lipid bilayer. In this respect, spherical liposome preparations are preferable since these are solvent-free, in contrast to the solvent-containing planar bilayers.

It is generally accepted that considerable orientation of the hydrocarbon chains in a bilayer can occur. The configuration and packing of the "liquid" chains may not necessarily be the classical parallel chain format and the terminal $CH_3$ groups may be more uniformly distributed throughout the hydrocarbon region. Factors such as the degree of hydration alter the molecular organisation of the chains, an increase in hydration tending to disorder the packing (Levine and Wilkins, 1971). Also, the influence of cholesterol on the fluidity or mobility of the phospholipid chains is important, and depends on the type of phospholipid, but in egg lecithin bilayers, cholesterol reduces the molecular motion of the chains (Levine and Wilkins, 1971; Schreier-Muccillo *et al.*, 1973).

We may ask, therefore, what anaesthetic agents do, if anything, to the fluidity of the bilayer membrane, and can changes in fluidity explain the observed permeability effects? A small digression might help predict the type of effect to be observed.

NMR studies provide some clue to the position of the anaesthetic agents in bilayer membranes and, as stated above, it has been proposed that a local anaesthetic such as tetracaine may push its butylene tail deep into the hydrophobic core of the membrane. It is likely that certain local anaesthetics may be at least half as long as many of the lipid molecules in a bilayer. Dibucaine, for example, is 19 Å long. On this basis, then, we might expect that insertion of a local anaesthetic molecule into the membrane phase might disrupt hydrophobic interactions between, say, cholesterol and the lipid hydrocarbon chain, and hence cause disordering of the membrane interior. Similarly, the marked hydrophobicity of some of the general anaesthetics may result in these agents producing considerable changes in the bulk fluidity of the membrane fabric.

The order of spin-labelled phosphatidylcholine vesicles is decreased by halothane and methoxyfluorane in a concentration-dependent way (Trudell *et al.*, 1973a), and the authors concluded that these agents produced not a local disordering effect but a general fluidisation of the hydrophobic region of the membrane. Such an effect may be the basis of anaesthetic-induced permeability increases in all membranes, although in the case of biological membranes it is difficult at present to ascribe the increase directly to a change in the lipid environment, or to an induced configurational change of a neighbouring protein which was solvated by the hydrocarbon chains of the lipid.

Experiments with hydrostatic pressure have confirmed that the disorder caused by anaesthetics is secondary to the presence in the membrane of the anaesthetic molecule itself, since application of pressure could restore order in phospholipid vesicles without displacing the spin-labelled molecule TEMPO (Trudell *et al.*, 1973b).

The fluidising efficiency of many anaesthetic agents is now well documented, but as noted above is concentration-dependent. We may well ask whether such an effect occurs at concentrations of anaesthetic agents producing anaesthesia. Experiments with a steroid spin probe suggest that the aliphatic alcohols may disorder at anaesthetic concentrations (Paterson *et al.*, 1972), but in contrast, studies with fatty acid spin labels have shown that no change in the order of bilayer vesicles accompanied general anaesthetic levels of halothane, chloroform, diethyl ether, butanol and benzyl alcohol (Boggs *et al.*, 1976). Figure IV.5 shows the action of one of these

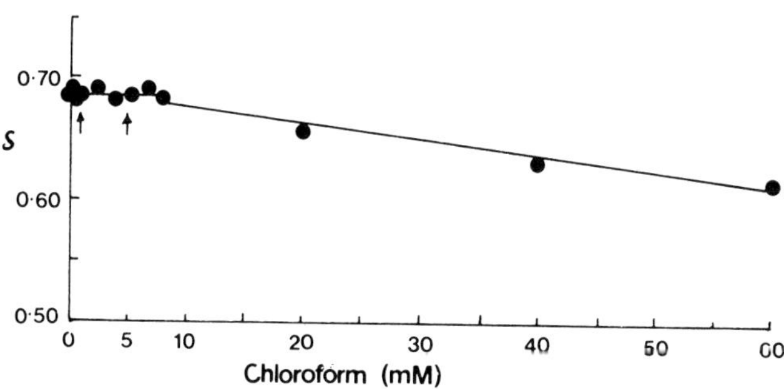


FIG. IV.5. The effect of varying aqueous concentrations of chloroform on the order parameter ($S$) of 8-doxylpalmitate in phosphatidylcholine–cholesterol vesicles. A reduction in $S$ represents an increase in fluidity. The arrows indicate the aqueous concentrations of chloroform which produce general (left-hand arrow) or local anaesthesia (right-hand arrow). (Boggs *et al.*, 1976)

agents on the spin order parameter, $S$, of a labelled fatty acid. A reduction in $S$ value indicates an increase in fluidity, and it can be seen that only small changes in this parameter accompany concentrations of anaesthetic some tenfold greater than that producing general anaesthesia.

This finding does not preclude the possibility, however, that a small local change in fluidity contributes to the depression of ion channel conductance in an excitable membrane. What these data simply suggest is that the bulk fluidity of the membrane of phospholipid vesicles is not altered.

Perhaps an equally important, and certainly as interesting, discovery was that vesicles low in cholesterol content ($\leqslant 5$ per cent) exhibit a low degree of order, and that anaesthetics may order these structures (Neal *et al.*, 1976). Does this mean that anaesthetics are more selective in their fluidising action than was initially presumed? Miller and Pang (1976) have posed and partially answered this question. Spin label studies demonstrated that all anaesthetics fluidised phosphatidylcholine bilayers low in phosphatidic acid (4 per cent) and high in cholesterol (33 per cent). This was indicated by the decrease in order parameter observed (see Table IV.1).

TABLE IV.1

The change in order parameter, $\Delta S$, measured by 5-doxylstearic acid in lipid bilayer membranes exposed to anaesthetics (Miller and Pang, 1976)

| | | $\Delta S$ | | |
|---|---|---|---|---|
| | | Bilayer composition, balance PC | | |
| Anaesthetic | Concentration (aqueous) | 4% PA | 20% PA | 4% PA/ 33% Chol |
| Halothane | 11 mM | −0·03 | −0·03 | −0·06 |
| Urethane | 90 mM | −0·08 | −0·03 | −0·03 |
| *n*-Octanol | 46 mM | −0·02 | −0·06 | −0·08 |
| Ketamine | 25 mM | +0·03 | 0 | −0·03 |
| Alphaxalone (steroid) | 18 mM | +0·02 | +0·03 | −0·02 |
| Pentobarbitone | 16 mM | +0·02 | +0·02 | −0·03 |

PC = egg yolk phosphatidylcholine; PA = egg yolk phosphatidic acid; Chol = cholesterol. Anaesthetics were equilibrated with the bilayers for 24 hr.

A negative $\Delta S$ (order parameter) reflects fluidisation, a positive value, an ordering effect. Whereas halothane, urethane and octanol fluidised all the compositions tested. Alphaxalone and pentobarbitone fluidised only the high-cholesterol bilayer membrane. The action of pentobarbitone could be changed from a fluidising effect to an ordering action by lowering the cholesterol to 5–10 per cent or raising the phosphatidic acid from 4 per cent to 10 per cent.

It must be concluded, therefore, that the fluidising action of an anaesthetic agent could be restricted to a prescribed region of a bilayer of a particular composition and perhaps high order. Other regions of a bilayer membrane, of different composition and lower order, may actually be ordered by the presence of an anaesthetic agent. If the permeability of such lipid membranes is modulated by the fluid state of the lipid hydrocarbon chains, then it is to be anticipated that permeability could increase, or decrease, as is observed experimentally, depending on which action prevails.

## 2.5 EFFECTS ON THE PHASE-TRANSITION TEMPERATURE

Perhaps a less exploited action of anaesthetics is their effect on the phase-transition temperature of lipid bilayers. The physical state of bilayer-lipids depends on the temperature, and if the latter is increased, lipid bilayers undergo an endothermic transition in which the lipid changes from a (gel) crystalline state to a liquid crystalline form. Such a change is associated with an increased disorder in the fatty acid chains. The phase transition is therefore accompanied by a positive enthalpy, and entropy change (e.g. Hill, 1974) and occurs at a temperature determined by the lipid composition of the bilayer (e.g. at 41–42°C for dipalmitoyllecithin). The transition temperature of a given lipid in turn depends on several factors, including the length and the degree of saturation of the fatty acid chain, the nature of its polar head group and, very important, whether the lipid dispersion has been sonicated. Such sonicated vesicles (diameter 250 Å) can have wide transitions occurring at a much lower temperature, about 22°C (see Lee, 1976), or alternatively may fail to show good transitions.

At the phase transition the thickness of the bilayer decreases slightly and an increase in the volume of the lipid occurs. A variety of methods are currently available for the measurement of this transition. It can be monitored using spin labels, or X-ray techniques, by a calorimetric method, or by measuring the volume change. One of the simplest methods employed measures the change in turbidity of the lipid dispersion using a spectrophotometer.

Such studies show that the phase transition occurs very sharply over a small temperature range (1°C). This is suggestive of a co-operative effect of lipid molecules, i.e. the molecules change from one phase to another in unison and the tendency of one molecule to change phase is not independent of that of its neighbours. The "width" of the transition reflects the magnitude of such co-operative effects, so that a sharp or narrow transition implies that a large number of molecules are involved in such a co-operative action, and a wider transition suggests that a smaller number of molecules are involved. Sonicated vesicles, for example, exhibit a comparatively wide transition, indicating that the co-operative unit is small. Aqueous dispersions of binary mixtures also possess less sharp transition temperatures compared with that of the pure (one-component) lipid bilayers (Trudell *et al.*, 1975).

In view of the actions of anaesthetic agents discussed above, it is perhaps possible to predict what effect these compounds may have on the phase-transition temperature. If the agent penetrates deep into the hydrophobic core of the membrane, it might be expected that such an invasion would alter the transition temperature of the lipid by fluidising partially the lipid chains. An anaesthetic agent with a predominantly surface action may, however, produce a smaller change in the transition temperature.

All anaesthetic agents tested thus far reduce the phase transition of lipid bilayer preparations (Hill, 1974; Trudell *et al.*, 1975; Lee, 1976). The action of dibucaine on the phase-transition temperature of two types of bilayer is shown in Fig. IV.6. The study demonstrated that this local amine anaesthetic reduced the transition temperature without altering the transition width. The effect was non-linear with respect to concentration, and it was proposed that an alteration in surface charge of the liposome accompanied the binding of the positively charged anaesthetic agent, so reducing

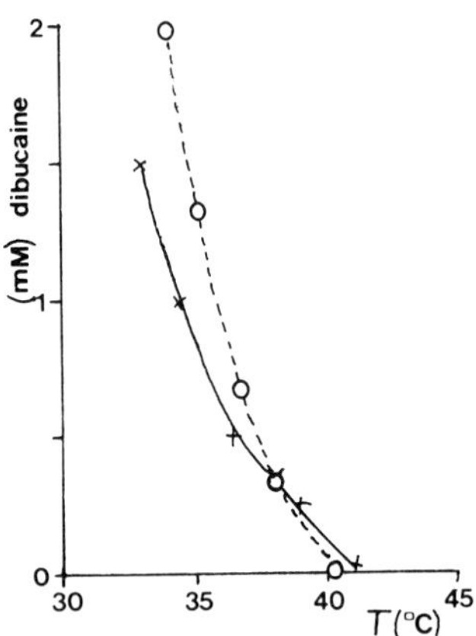


FIG. IV.6. The effect of dibucaine on the phase-transition temperature (°C) of dipalmitoyl phosphatidylcholine (O— — —O) or dipalmitoyl phosphatidylcholine containing 11 mol% myristic acid (+——+). (Lee, 1976)

further binding and therefore efficiency. If 11 mol% of myristic acid were added to the liposome to make it negatively charged, then the efficiency of dibucaine was augmented (+——+). This enhancement of effect could be reversed by addition of divalent ions. This suggests that the action of dibucaine may be dependent on a surface effect.

Figure IV.7 shows the abrupt increase in volume which occurs when multilamellar liposomes of phosphatidylcholine undergo their transition to the liquid crystalline state. The addition of anaesthetics to the liposomes lowers the transition temperature and also modifies the melting dilation in a characteristic way. *n*-Butanol has little effect on the width of the transition (IV. 7a) and its dose-response curve exhibits saturation rather like dibucaine in Fig. IV.6. *n*-Hexanol, however, increases the width of the transition (IV. 7b) (i.e. reduces the co-operative unit), and so do the extremely hydrophobic general anaesthetics methoxyfluorane and trichloroethylene. These substances probably partition preferentially into the interior of the bilayer and consequently affect the kinetics of the melting process in a manner rather different to that seen with butanol, which probably partitions at the bilayer–aqueous interface. Thus the lowering of the transition temperature is not to be viewed in simple colligative terms. These differences relate to the differential effects these substances have on bilayer permeability.

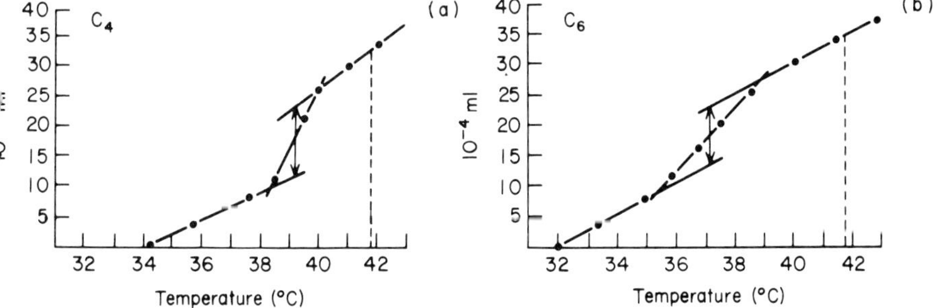


Fig. IV.7. The effect of *n*-alcohols on the transition temperature and transition width of dipalmitoylphosphatidylcholine liposomes: (a) *n*-butanol, 78 mM, (b) *n*-hexanol, 23 mM. The vertical dashed line indicates the transition temperature for untreated liposomes (Macdonald, 1978).

Just as hydrostatic pressure reduces bilayer permeability, so it increases the transition temperature of a bilayer without affecting $\Delta H$ or $\Delta V$. $\Delta H$ and $\Delta V$ and the extent to which pressure raises the transition temperature are all interrelated by the Clausius–Claperyon equation.

The presence of anaesthetics in a bilayer has no effect on $dT/dP$, hence in this system pressure simply counteracts the effect of anaesthetics without affecting the nature of the anaesthetic–target interaction. Strictly speaking this is chemical *reversal*, rather than *antagonism*, of an anaesthetic action.

The precise action of these anaesthetics on the phase-transition temperature may therefore provide information about the location of the anaesthetic in the bilayer membrane. This must certainly be one of the interpretative values of such experiments.

## 3 The erythrocyte membrane

Much of our detailed information on the action of anaesthetics on "real" inexcitable membranes derives from experiments on erythrocyte membranes. In the experiments conducted by Seeman and colleagues (see Seeman, 1972) the haemolysis of human erythrocytes in a hypotonic medium has been studied, and as shown below anaesthetics inhibit such haemolysis (the antihaemolytic effect) and consequently are often said to "protect" the erythrocyte from hypotonic haemolysis.

### 3.1 MEMBRANE CONCENTRATION

One of the principal advantages of studying erythrocyte membranes lies in the ease with which these membranes may be isolated and the uptake of labelled anaesthetics monitored. Determination of the membrane concentrations of these agents is obviously useful when considering their mechanism of action. The action on the erythrocyte membrane of a wide spectrum of anaesthetic compounds has been investigated in detail by Seeman and colleagues (Seeman, 1966a; Seeman, 1972), and the uptake of many anaesthetic agents and inert gases has been measured (Seeman, 1972; Power and Stegall, 1970).

The membrane concentrations of some agents, corresponding to the concentration required to produce a 50 per cent antihaemolytic effect, are presented in Table IV,2. The equieffective membrane concentrations obviously differ by a factor of 10 or so. For the alcohols, it is apparent that the efficacy of the agent increases as the chain length is increased, implying that perhaps some size- or volume-dependent parameter is important in determining potency. Mullins (1954) first suggested that the volume of occupation of the agent in the membrane, i.e. the membrane concentration × molecular volume, might describe its effectiveness more accurately.

TABLE IV.2

The aqueous and membrane concentration of a range of anaesthetics which produces 50 per cent antihaemolysis in human erythrocytes (Adapted from Seeman, 1972)

| | Concentration for 50% anti-haemolysis (mol (l $H_2O)^{-1}$) | Anaesthetic concentration in membrane (mol (kg dry membrane)$^{-1}$) |
|---|---|---|
| Methanol | 2·7 | 0·108 |
| Ethanol | $6 \times 10^{-1}$ | 0·070 |
| Propanol | $1{\cdot}9 \times 10^{-1}$ | 0·098 |
| Butanol | $4{\cdot}2 \times 10^{-2}$ | 0·102 |
| Pentanol | $1{\cdot}45 \times 10^{-2}$ | 0·075 |
| Hexanol | $3 \times 10^{-3}$ | 0·039 |
| Heptanol | $7{\cdot}5 \times 10^{-4}$ | 0·030 |
| Octanol | $2{\cdot}35 \times 10^{-4}$ | 0·036 |
| Nonanol | $4{\cdot}1 \times 10^{-5}$ | 0·024 |
| Decanol | $1 \times 10^{-5}$ | 0·012 |
| Chloroform | $9 \times 10^{-3}$ | 0·112 |
| Diethyl ether | $9 \times 10^{-3}$ | 0·059 |
| Halothane | $8 \times 10^{-3}$ | 0·065 |
| Chlorpromazine | $1 \times 10^{-5}$ | 0·016 |
| Pentobarbitone | $1{\cdot}6 \times 10^{-3}$ | 0·016 |

However, this too is likely to be simplistic since, at least for the antihaemolytic effect, the negatively charged agents such as the barbiturates do not conform to this prediction (Roth and Seeman, 1972). It appears that these agents exert an equieffective action at membrane concentrations (or volumes) up to fortyfold lower than expected. At present, no satisfactory explanation has been put forward for this finding.

It has been proposed that the membrane concentrations should be multiplied by the free energy of binding value ($\Delta F$) for each agent, since this product turns out constant. The $\Delta F$ value has been calculated for the alcohols (Seeman *et al.*, 1971c) and chlorpromazine (Kwant and Seeman, 1969) and is estimated from the following relation

$$\Delta F = -2{\cdot}303\, RT \log K$$

where $K$ is the alcohol-receptor affinity constant, $R$ the general gas constant and $T$ the absolute temperature.

Such a calculation indicates that the free energy of adsorption of the alcohols in experiments investigating this antihaemolytic action is about $-800$ cal mol$^{-1}$, which compares favourably with the free energy of transfer of methylene groups from a water to a non-polar phase (Schneider, 1968). The inference is therefore that the membrane–alcohol interaction is hydrophobic, and it appears that membrane concentration (for 50 per cent antihaemolysis) $\times$ the free energy of adsorption is a constant value for the alcohols at least (Seeman *et al.*, 1971c). Insufficient data are available at present, however, to extend this hypothesis to the action of other anaesthetics.

At low concentrations, then, anaesthetics appear to "protect" the erythrocyte against hypotonic haemolysis. Interestingly it seems, although not much data are available, that the maximum membrane concentration of the agent producing this stabilisation is not affected by temperatures within the range 5–37°C. This has been shown to be the case, for example, for chlorpromazine (Kwant and Seeman, 1969). This must preclude any notion that the state of membrane lipid (i.e. liquid versus gel) determines the extent of the anaesthetic–membrane interaction.

At higher concentrations many lipid-soluble agents, including anaesthetics, produce membrane lysis (breakdown). For chlorpromazine, this action occurs at a membrane concentration about five times that producing a 50 per cent antihaemolytic effect (see Table IV.2) and corresponds to about 1 anaesthetic molecule every 200 Å$^2$, or 1 anaesthetic molecule every 4–5 phospholipid molecules. This lytic action has been described for many anaesthetics and it presumably reflects the inability of the membrane structure to withstand such an extensive penetration by hydrophobic molecules. Although such concentrations of anaesthetics do not appear to alter the membrane thickness markedly, they do produce obvious undulations or invaginations of the erythrocyte membrane which are visible under the electron microscope (Seeman, 1966b). By analogy with the detergents, it is likely that the membrane lysis involves actual breakdown of a lipoprotein complex in the cell membrane (Rideal and Taylor, 1967).

### 3.2 ANTIHAEMOLYTIC ACTION

The ability of hydrophobic agents to reduce the susceptibility of the erythrocyte to haemolysis is well known, although, note that at higher concentration lysis occurs. It has been argued that only lipid-soluble agents such as the anaesthetics exhibit this property (Roth and Seeman, 1971), but it has been reported that water-soluble agents such as propranolol, adrenaline, noradrenaline and many others do protect erythro-

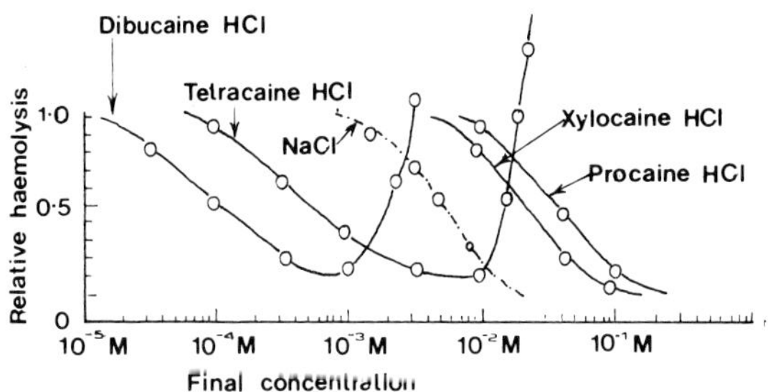


FIG. IV.8. The stabilising and lytic effects of various local anaesthetics on human erythrocytes. A relative haemolysis of 1·0 indicates an absolute degree of 60 per cent haemolysis. Note that the stabilisation by xylocaine or procaine occurred at concentrations higher than equiosmolar concentrations of NaCl or sucrose and thus represents an osmotic protection of the erythrocyte. (Seeman, 1966b)

cytes from haemolysis (Mikikits *et al.*, 1970). This action is not confined, therefore, to anaesthetic agents alone, nor may it only be a property of hydrophobic compounds.

The biphasic effect of local anaesthetic agents on human erythrocyte haemolysis is shown in Fig. IV.8. The erythrocytes were haemolysed in 0·3 per cent NaCl, and it can be seen that the haemolysis relative to the control was reduced as the concentrations of dibucaine or tetracaine were increased up to a certain value, beyond which greater haemolysis occurred due to the lytic action of these agents. The apparent protection afforded by xylocaine or procaine was osmotically based, since it occurred at higher concentrations than those required for protection by NaCl or sucrose.

Such an antihaemolytic effect can be explained if it is assumed that an increase of the erythrocyte membrane area occurs during exposure to the anaesthetic. This would seem to be borne out experimentally. The two principal methods which have been used to study the membrane expansion induced by these agents are shown in Fig. IV.9.

In the ghost formation method (a), the haemolysis induced by 0·3 per cent NaCl is followed in the presence of the anaesthetic agent; the cell then seals and reswells. The mean cell volume is then monitored by a Coulter counter and can be compared with the control volume measured in the absence of anaesthetics. In the alternative method (b), the anaesthetics are added to the prehaemolysed, spherical ghosts. Initially the ghost becomes misshapen and reduced in volume, but after 2 hr or so the cell once again becomes spherical and the mean cell volume may be measured in the same way as above.

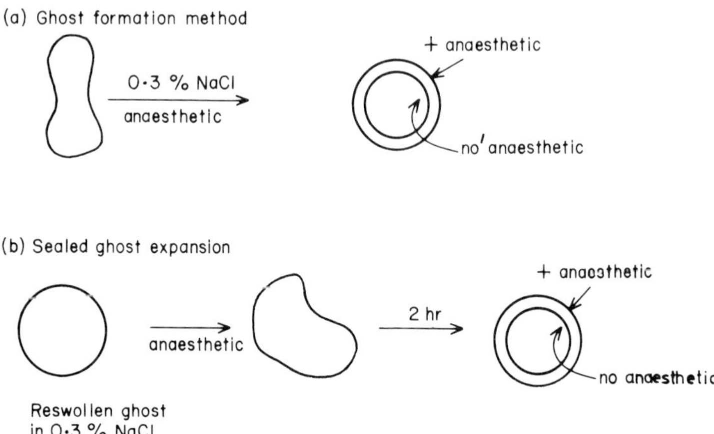


FIG. IV.9. The methods for studying ghost membrane expansion by anaesthetics. In the ghost formation method (a), the intact erythrocytes are haemolysed in hypotonic solution in the presence of the anaesthetic. In the sealed ghost expansion method (b) the anaesthetic is added to preformed, spherical ghosts. In the second method 2 hr after the addition of the anaesthetic the cells are again spherical, and they are then monitored by the Coulter counter and mean cell volume computer. (Seeman *et al.*, 1969a)

Since in both cases the cells are spherical when the cell volume is measured, it is a simple task to calculate the cell membrane area. These two methods give similar results and suggest that at concentrations known to inhibit hypotonic haemolysis by 50 per cent the membrane expansion produced by the alcohols is about 1·5 per cent, and that produced by chlorpromazine is about 2 per cent (Seeman *et al.*, 1969a). The maximum expansion the membrane can accommodate prior to lysis is about 5–7 per cent (Seeman *et al.*, 1969a, 1969b).

As expected, the critical haemolytic volume (maximum volume before haemolysis) is also increased by anaesthetics (Seeman *et al.*, 1969b). More recent data suggest that the actual membrane volume expansion is of the same order of magnitude as the area increase, but it is difficult to derive any information about changes in membrane thickness from these data (Seeman, 1975). An intriguing possibility is that a small change in erythrocyte membrane area accompanies surgical anaesthesia in man. One study suggests that volatile anaesthetics at concentrations equivalent to the minimum alveolar concentration (MAC) value in humans produce about an 8 per cent antihaemolytic effect, which can be accounted for if it is assumed that the erythrocyte membrane area increases by 0·4 per cent (Seeman and Roth, 1972).

A further interesting feature of the protection of erythrocytes by some anaesthetics is the pH dependence. Thus, an increase in extracellular pH increases the protection against haemolysis conferred by dibucaine (Seeman, 1966b) or chlorpromazine (Seeman and Kwant, 1969). A puzzling observation is that although the antihaemolytic effect of chlorpromazine is dependent on extracellular pH and a charged form of this agent is the less potent, the membrane expansion of cell ghosts produced by chlorpromazine is not altered when both intra- and extracellular pH are varied (Seeman and Kwant, 1969). Since this study concludes that both the free base and the charged form of chlorpromazine may expand the erythrocyte membrane by an equal amount, it seems curious that the erythrocyte protection by chlorpromazine should be pH-dependent at all!

A change in extracellular pH from 8 to 6 can produce a 100–1000-fold reduction in the antihaemolytic potency of anaesthetics such as the tertiary amines. This may imply that either the uncharged form is the active species, or else the uncharged form penetrates the membrane more readily and acts in the charged form on the inner aspect of the membrane. The latter possibility is suggested, at least for dibucaine, in experiments where the pH is varied. Such intracellular pH changes were monitored by exposing the erythrocytes to sodium nitrite in order to convert the cells to methaemoglobin erythrocytes (Seeman, 1966b). The spectral characteristics of methaemoglobin within the erythrocyte varies with pH and even to the naked eye the cells appear brown at pH 6·0 and red at pH 10·0. This simple technique thus allowed the potency of agents to be tested as the intracellular pH was varied, demonstrating that intracellular acidification enhances the potency of dibucaine. The notion therefore at present is that the charged form of certain compounds such as dibucaine and chlorpromazine may act on the inside surface of the erythrocyte membrane. This is consistent with the ideas on how local anaesthetics affect excitable membranes, which are dealt with in the next chapter.

## 3.3 NATURE OF THE MEMBRANE EXPANSION

Ponder (1948) was one of the first investigators to suggest that antihaemolytic effects could be accounted for by an increase in critical haemolytic volume. The anaesthetic agents increase the surface area: volume ratio of the erythrocyte, increase the haemolytic critical volume, and hence reduce the osmotic fragility. However, if we now examine the extent of the erythrocyte membrane expansion then one very interesting fact immediately emerges. Although it is undoubtedly difficult to obtain a precise figure for the volume of occupation of anaesthetic in an erythrocyte membrane, and these estimates are therefore probably subject to considerable error, current values suggest that the actual space occupied by these agents is approximately 1/10 of the actual calculated expansion (Seeman, 1972).

Anaesthetic penetration of lipid films produces an expansion (Skou, 1958; Clements and Wilson, 1962), but this volume change probably reflects the addition of a certain volume of anaesthetic molecules. The question therefore arises as to how we account for the allegedly high expansion of the erythrocyte membrane. What is it about the molecular architecture of a biological membrane which makes it so structurally labile?

The very fact that expansion occurs at all suggests that these agents do not simply occupy free space in the membrane. We could ask, first, how the expansion compares in size and in nature with that accompanying an increase in temperature. A rise in temperature does reduce the osmotic fragility of erythrocytes, presumably by increasing their surface area (Seeman, 1969). The membrane expansion is of the order of 0·1 per cent/°C. Thus, anaesthetics at a local concentration may produce an increase equivalent to that observed by about a 20°C elevation in temperature. It is difficult to decide whether the nature of the anaesthetic-induced expansion is similar to that induced by a rise in temperature.

The reason for the large expansion in the presence of anaesthetic agents remains an open question. One obvious and not unreasonable explanation may lie in the known susceptibility of protein structure to anaesthetics. It has been shown that potent anaesthetics such as methoxyfluorane, chloroform and halothane produce marked structural changes in globular protein, whereas less potent agents such as $N_2O$ produce only small effects. Also, in an ascending homologous series the narcotic potency and effects on protein structure increase in a parallel fashion (Balasubramanian and Wetlaufer, 1966). It is possible, therefore, that a protein conformational change such as a rearrangement of a membrane protein side chain is partly responsible for the extra expansion.

Another possibility is that displacement of membrane-bound $Ca^{2+}$ will result in slight expansion, since $Ca^{2+}$ condenses the membrane state. The difficulty with this idea is that some anaesthetics displace $Ca^{2+}$ from phospholipid bilayers, whereas other agents increase the binding (Blaustein and Goldman, 1966). Perhaps it is unwise, however, to rely solely on *in vitro* phospholipids as a model system for the erythrocyte membrane's $Ca^{2+}$ binding characteristics. Other work does, nevertheless, demonstrate that anaesthetics differentially affect erythrocyte membrane-bound $Ca^{2+}$ levels (Seeman *et al.*, 1971a).

One of the most popular theories of expansion suggests that anaesthetic agents fluidise or disorder the membrane. The nature of this fluidising action in liposomes has already been discussed. It is proposed that the anaesthetic agent "loosens up" membrane macromolecules, so that if its own molecular motion is measured then this increases as the concentration of the agent in the membrane rises. At low concentrations, the anaesthetic is restricted by the surrounding semi-rigid membrane structure, but at higher concentrations it becomes more mobile as the surrounding membrane structure disorders (e.g. Metcalfe *et al.*, 1968). However, considerable doubt has been thrown on whether this fluidising action is important physiologically, and perhaps more important is the finding that anaesthetics may also order lipid bilayer membranes, the composition of the membrane governing which effect is the dominant (see p. 110).

Although it is premature to be certain about the nature of the erythrocyte membrane expansion, its physiological significance could be tested further by examining the pressure dependence of the haemolytic protection offered by anaesthetics. Such studies are at a preliminary stage (Brewster *et al.*, 1976, Fenn and Boschen, 1971). The application of hydrostatic pressure to the study of antihaemolytic effects would at least provide an additional estimate for the actual size of the membrane expansion produced by anaesthetics.

## 3.4 EFFECTS ON ION AND NEUTRAL SOLUTE FLUXES

One further advantage in studying erythrocyte membranes is that the action of anaesthetics on ion exchange characteristics and neutral solute exchange may be conveniently investigated with radioactive methods.

The erythrocyte membrane possesses a functional transport system capable of driving ions "uphill" against an electrochemical gradient (active transport), but it also performs facilitated diffusion or carrier-mediated transport. Also, due to the asymmetric distribution of ions and the permeability characteristics of the membrane, passive diffusion of ions down their electrochemical gradients takes place. It has been stated that, generally speaking, anaesthetics depress carrier-mediated transport, but may produce variable effects on passive and active fluxes (Seeman, 1972). There are clearly species variations in the nature of certain fluxes (e.g. Deuticke and Gruber, 1970), and one early report emphasised species variation with respect to the action of alcohol anaesthetics on erythrocyte membrane permeability (Jacobs and Parpart, 1937). The reason underlying these differences is not well defined, thus it is necessary to be cautious in this regard when comparing results from different studies.

### 3.4.1 $Na^+$–$K^+$ *transport*

If we consider the active $Na^+$–$K^+$ transport system first, then depending on the experimental conditions both local and general anaesthetics will depress transport. In the case of the local anaesthetics at pH 6·0, the $K^+$ influx was more sensitive to reduction than the $Na^+$ efflux (Andersen, 1968). This may mean that these agents uncouple the $Na^+$–$K^+$ pump, i.e. alter the stoichiometry of the $Na^+$–$K^+$ exchange. It would be interesting to explore further this possibility. This action may, however, be a secondary one. In the case of general anaesthetics, such as ether and chloroform, this would certainly seem to be so. Near lytic concentrations of these agents depress the pumped $Na^+$ efflux (Halsey *et al.*, 1970), but further investigation of the action of ether and chloroform at lower concentrations revealed that these agents can also potentiate $Na^+$ transport (Hale *et al.*, 1972). This effect is shown in Fig. IV.10.

In this experiment, the intracellular $Na^+$ concentration of erythrocytes, which had been previously cold-stored, was measured after 3 hr in an incubation medium (probably $K^+$-rich) at 37°C. During cold storage, the cells gain $Na^+$ which will be pumped out in the recovery medium at 37°C if ouabain is not present (no ouabain curve), but not if ouabain is added

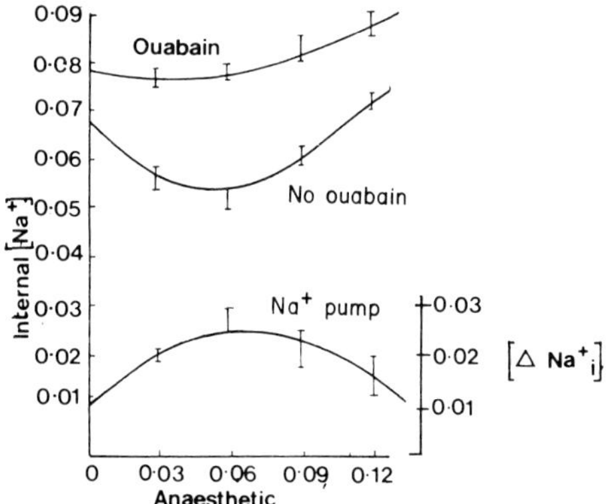


FIG. IV.10. Changes in the intracellular $Na^+$ concentration, μequiv (mg Hb)$^{-1}$, of human erythrocytes after 3 hr incubation at 37°C in the presence of various concentrations of ether (arbitrary units). The incubation was carried out either in a ouabain-containing (top curve) or ouabain-free (middle curve) medium. The bottom curve represents the difference between the other two curves, and thus reflects the Na pump activity for the various ether concentrations. (Hale *et al.*, 1972)

to the recovery medium (ouabain curve). If we examine the ouabain curve first then it seems that the $Na_i$ remains constant at low anaesthetic concentrations but rises at higher concentrations, presumably due to a rise in the erythrocyte membrane permeability to $Na^+$. A similar increase is seen in the ouabain-free situation, although at lower concentrations of anaesthetic the $Na_i$ falls as the $Na^+$ pump extrudes $Na^+$. The lower curve is the difference in intracellular $Na^+$ content of the ouabain, and ouabain-free curves, and if it is assumed that $Na^+$ permeability is the same in these two conditions then this represents the efficiency of the $Na^+$ pump. Thus, at an intermediate concentration of anaesthetic the $Na^+$ efflux is stimulated.

The $Na^+$ pump potentiation induced by chloroform and ether was less if the $Na^+$ efflux was augmented by using external $K^+$, so that when the pump was activated maximally by external $K^+$ then the anaesthetic induced no further stimulation. There is consequently no evidence at present to suggest that the anaesthetics can increase the capacity of the transport system. They simply increase the degree of activation. Reference to these authors' data indicates that even the lower concentrations of ether and chloroform producing potentiation of the $Na^+$ efflux are still five to six times higher than clinical doses.

Despite the considerable interest in the action of anaesthetic agents on the $Na^+$–$K^+$–ATPase enzyme system which supports the $Na^+$–$K^+$ exchange (see Seeman, 1972, Table 9), it is probably true to say that we understand very little about the detailed mechanism of action of anaesthetic agents with membrane-bound enzymes. Consequently, the nature of the $Na^+$–$K^+$ pump potentiation, or depression by anaesthetics, remains obscure.

### 3.4.2 *Carrier-mediated transport*

Many authors have investigated the depressant effects of anaesthetics on carrier-mediated transport of solutes, in particular glucose. It is plausible that the anaesthetic competes with the normal substrate for the membrane carrier site, or alternatively that a modified anaesthetic-carrier complex competes with anaesthetic-free carriers for the normal substrate. In the latter case, the combination of the anaesthetic with the carrier could result in its inactivation, or at least a reduced mobility. Such an interaction of anaesthetics with membrane carriers has been proposed (e.g. Green, 1965).

Since benzyl alcohol is known to disorder erythrocyte membranes, Clayton and Martin (1971) examined the action of this agent on the carrier-mediated transport of choline in human erythrocytes. In these experiments they measured the efflux of choline as a function of external choline concentration in the presence and absence of the alcohol. Their results are shown in Fig. IV.11. Clearly, the efflux of choline into choline-free Ringer is depressed more on a percentage basis than the efflux into

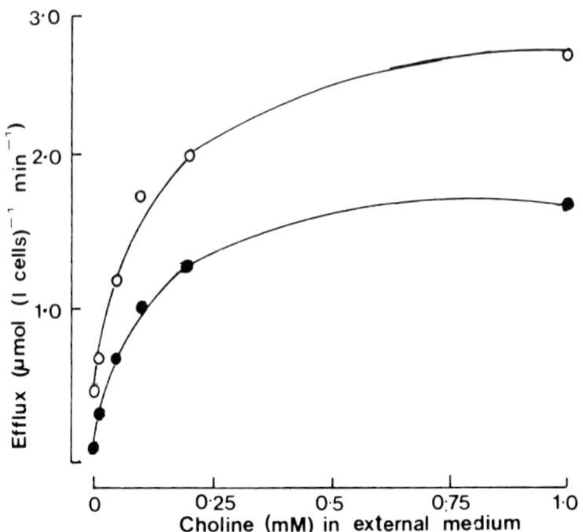


FIG. IV.11. The rate of efflux of choline from human erythrocytes into media containing various concentrations of choline, in the absence (○) and the presence (●) of 20 mM benzyl alcohol. (Clayton and Martin, 1971)

1·0 mM choline-Ringer. This is an interesting finding, since in the case of the transport into choline-free Ringer presumably an empty carrier returns to the inner surface of the membrane, whereas in the experiment performed in 1·0 mM choline-Ringer a choline-carrier complex returns across the membrane. On this assumption, then, one interpretation is that a free carrier is more susceptible to inactivation by benzyl alcohol than an occupied carrier. A more detailed picture of the interaction of membrane carriers with anaesthetics is, however, made all the more difficult to define in view of the scanty knowledge concerning such membrane components.

### 3.4.3 *Passive permeability*

Turning our attention to the effects of anaesthetic agents on the passive permeability characteristics of the erythrocyte membrane, these actions are at first sight a little confusing. Thus it has been shown that for the human erythrocyte, the membrane permeability to water (hydraulic flow) is enhanced by chlorpromazine and ethanol (Seeman *et al.*, 1970), and that the $Na^+$ permeability may be increased by chlorpromazine (Seeman *et al.*, 1971b) and butanol (Burt and Green, 1971). However, if $Ca^{2+}$ is

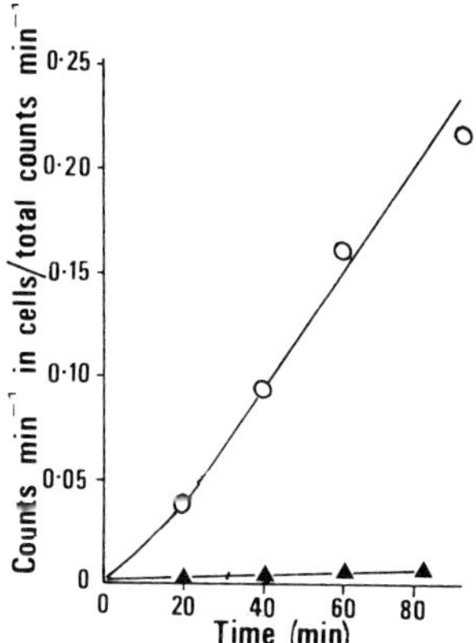


FIG. IV.12. The $^{22}Na^+$ entry into human erythrocytes in the presence and absence of 0·4 M *n*-butanol. Samples of cells were removed from the radioactive medium at the times indicated on the abscissa and assayed for radioactivity. ○ = 0·4 M butanol; ▲ = no butanol. (Burt and Green, 1971)

present in the bathing medium then chlorpromazine decreases $Na^+$ permeability (Seeman *et al.*, 1971b), and the local anaesthetics tetracaine and procaine can reverse the action of butanol (Burt and Green, 1971). Ethanol does not seem to affect the $Na^+$ permeability under any conditions (Seeman *et al.*, 1971b). We may conclude, tentatively, that in the presence of $Ca^{2+}$, charged anaesthetics depress the $Na^+$ permeability whereas neutral anaesthetics may enhance $Na^+$ permeability or have no effect.

One consistent finding is that raising the external $Ca^{2+}$ can oppose any anaesthetic-induced changes in the $Na^+$ permeability. This cannot be explained by assuming that the anaesthetics cause $Ca^{2+}$ displacement, however, since butanol increases the membrane-bound $Ca^{2+}$ in erythrocytes (Seeman *et al.*, 1971a). The most dramatic anaesthetic-induced effect is shown in Fig. IV.12. In this study, 0·4 M butanol produced a reversible twenty to fortyfold increase in $Na^+$ permeability which could be opposed by the divalent cations $Ca^{2+}$, $Sr^{2+}$ or $Ba^{2+}$. This latter effect was, however, dependent on temperature and could not be demonstrated above about 17°C. This finding illustrates the importance of paying due regard to temperature in these permeability studies.

The concentrations required to increase either the permeability to $H_2O$ or $Na^+$ may produce considerable disorder in the erythrocyte membrane. The increase in $H_2O$ permeability is at least consistent with the notion that the water within the membrane is in a more mobile or fluid state.

Anaesthetics can also increase the erythrocyte membrane permeability to $K^+$ (e.g. Andersen, 1968), and in a recent systematic study the action of local anaesthetics on chloride exchange was investigated (Gunn and Cooper, 1975). The principal findings were that local anaesthetics rapidly inhibited the exchange diffusion of chloride and that this effect was not affected by $Ca^{2+}$ ions (Fig. IV.13). Neither the control exchange nor the reduced exchange in the presence of tetracaine was affected significantly by a twentyfold increase in external $Ca^{2+}$. This implies that the anaesthetic effect could not be accounted for by a surface charge density hypothesis (see above).

The inhibition was thought to be non-competitive in that when $Cl^-$ efflux was measured as a function of intracellular $Cl^-$ concentration in the presence and absence of tetracaine, the flux showed the same saturation characteristics. The inhibition is therefore presumably brought about by the anaesthetic acting at a site different from that of the $Cl^-$ binding site. The membrane chemistry of $Cl^-$ exchange has not been characterised completely, but a protein of molecular weight 95 000 daltons has been implicated as the substrate binding site for the anion transport system (Cabantchik and Rothstein, 1974). It is possible, therefore, that a conformational change in this protein may mediate the depressant effects of

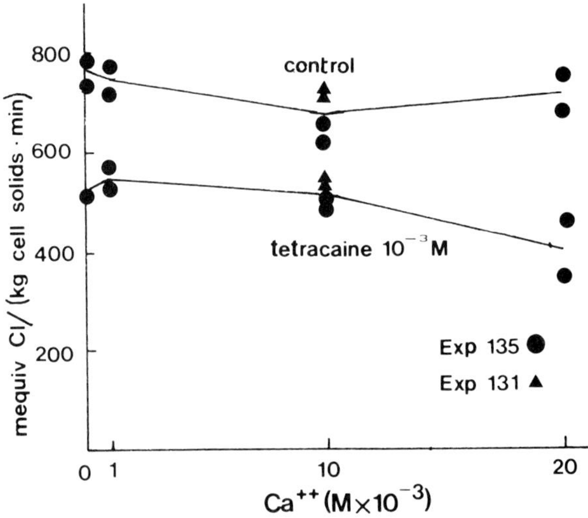


FIG. IV.13. The action of external $Ca^{2+}$ on the chloride self-exchange flux of human erythrocytes measured in the absence (control) or presence of the local anaesthetic tetracaine ($10^{-3}$ M). Data from two experiments are shown. (Gunn and Cooper, 1975)

local anaesthetics, although on the basis of the present evidence it is unlikely that these agents attach to the $Cl^-$ binding site itself.

## 4 Epithelial cell membranes

Studies on the action of anaesthetics on epithelial cell preparations of the amphibian skin or bladder are of interest since these systems also possess an active $Na^+$ transport system. Direct comparison of these systems with that of the erythrocyte membrane is, however, made difficult by their complexity. For example, in the frog skin, movement of the $Na^+$ ion from the outside to the inside surface involves passage across several cell layers. The genesis of transepithelial potential differences is at present not clearly defined and recent discussions have emphasised the need for a new model for epithelial transport (e.g. Finn, 1976).

Nevertheless, it is worth describing a few experiments using these preparations, since the results do again draw attention either to the common or, alternatively, dissimilar features of anaesthetic action. Thus, in the frog skin it has been demonstrated that the local anaesthetics in the charged form can stimulate $Na^+$ transport, but only if added to the outside solution. In the uncharged form these agents inhibit $Na^+$ transport whether added to the outside or inside (Skou and Zerahn, 1959). Hence, if the pump current (or short-circuit current) is measured when procaine at pH 7·8 is present in the outside solution, then an increase in short-circuit current is observed first, which gives way to a decrease, this dual effect being due to the fact that at this particular pH procaine is present in both the charged and uncharged forms.

If, however, procaine at pH 7·8 is added to the inside solution, then only the secondary depression of the short-circuit current is observed, since the charged form is inactive on the inside surface. These authors concluded that the stimulation of the $Na^+$ transport was secondary to an increase of permeability of the epithelial cell membrane to $Na^+$. Thus, by increasing the $Na^+$ permeability of the cellular barrier to diffusion, more $Na^+$ was made available to the transport system and consequently the pump was stimulated. If this interpretation is correct, then this would mean that the local anaesthetic procaine affects the $Na^+$ permeability differently in the epithelial and erythrocyte cell membrane. It would be interesting to examine the $Ca^{2+}$ dependence of the stimulatory effect observed by Skou and Zerahn (1959).

Unfortunately, these authors draw no conclusions about the mechanism of the inhibitory effects of the uncharged local anaesthetics on the $Na^+$ transport process. This depressant effect may be related to the known inhibitory action of local anaesthetics on $Na^+$–$K^+$–ATPase activity (Seeman, 1972, Table 9).

No doubt the prevalent action of an anaesthetic on the $Na^+$–$K^+$ transport process is governed by the concentration of the agent tested. This fact is clearly demonstrated in the results of Andersen and Shim (1971). These authors examined the action of general anaesthetics on the short-circuit ($Na^+$ pump) current of epithelial cells of the toad bladder. In this particular study the view was taken that the transbladder potential is generated by the electrogenic pumping of $Na^+$ ions from the mucosal to serosal side of of the bladder. All the agents tested (cyclopropane, nitrous oxide, halothane, ether, thiopentone) stimulated the $Na^+$ transport at low concentrations and depressed it at higher concentrations. The action of cyclopropane is shown in Fig. IV.14. It was suggested that the stimulatory action of the agents was due to an effect on the serosal surface involving the presence of catecholamines, and that the depressant effect was due to an action on the mucosal surface. However, how these actions were mediated was not discussed further. The relationship between the concentration required for a particular effect and the minimum anaesthetic requirement (MAR) value is shown in Fig. IV.15. Evidently, the concentrations producing

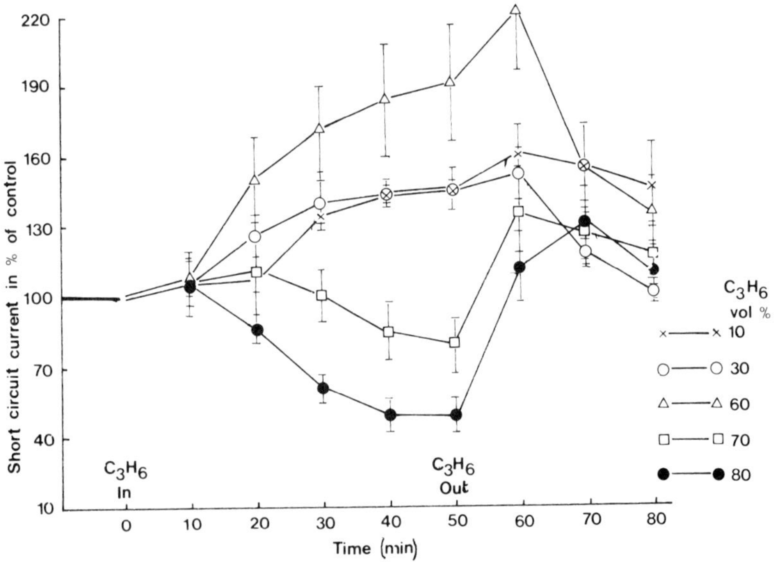


FIG. IV.14. The effect of various concentrations of cyclopropane on the short-circuit current in the toad bladder, as a percentage of untreated control. Points are the means of five experiments $\pm 1 \times$ SE. (Andersen and Shim, 1971)

stimulation can lie below the MAR value, whereas the concentrations producing inhibition may be ten times higher than that value.

The effects of a range of general anaesthetics on the serosa-mucosa permeability to water in the toad bladder have also recently been investigated (Amaranath and Andersen, 1974). Generally, these agents depressed the osmotic flux at low concentrations and increased it at higher concentrations. In Table IV.3 the membrane concentrations which depress, do not change or augment water permeability are compared with the membrane concentrations corresponding to the minimal anaesthetic requirements for the same species,

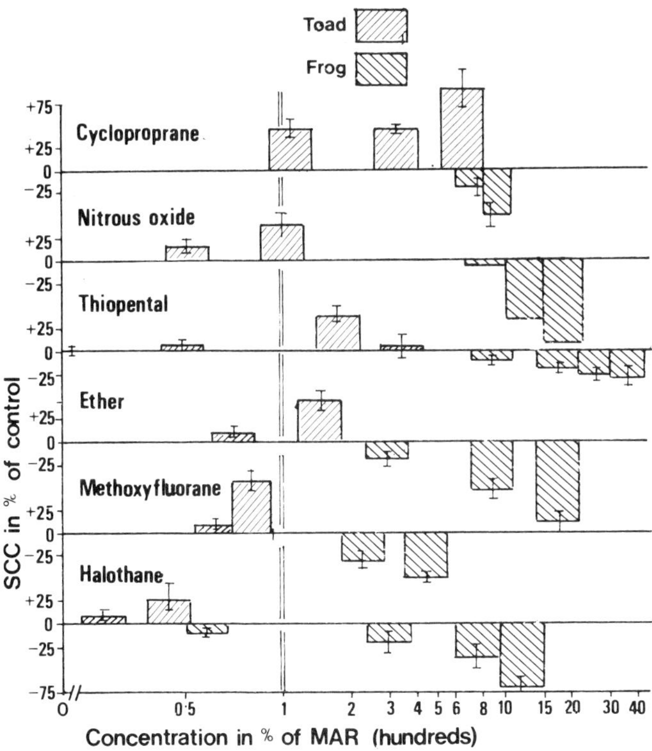


FIG. IV.15. The relationship between the effect on $Na^+$ transport *in vitro*, i.e. short-circuit current (SCC) and the anaesthetic effect *in vivo* in the toad. SCC was measured in the isolated toad bladder. The anaesthetic dose is expressed in multiples of the minimal anaesthetic requirement (MAR). The latter parameter was determined by monitoring the toad's response to a standard stimulus. (Andersen and Shim, 1971)

TABLE IV.3

Estimated membrane concentrations (mM (kg dry membrane)$^{-1}$) of general anaesthetics at minimal anaesthetic requirements (MAR) in the toad, and at levels that decreased, did not change or increased water permeability in the isolated toad bladder (Amaranath and Andersen, 1974)

| | | Anaesthetic concentration per kg dry membrane | | |
|---|---|---|---|---|
| | Minimal anaesthetic requirement (MAR) (mM) | Low concentration and decreased permeability (mM) | Medium concentration and normal permeability (mM) | High concentration and increased permeability (mM) |
| Chloroform | 23·0 | 5·1 | 10·2 | 22·4 |
| Trichloroethylene | 5·1 | 2·5 | | 5·2 |
| Halothane | 5·2 | 15·6 | 23·4 | 31·2 |
| Ether | 21·3 | 10·0 | 15·0 | 20·0 |
| Cyclopropane | 3·2 | 7·2 | 10·8 | 14·4 |
| Thiopentone | 1·0 | 17·6 | 26·4 | 35·2 |

which had been determined previously. This comparison shows that the water permeability of the toad bladder cells may be depressed or increased by clinical concentrations, depending on which agent is considered.

## 5 Sarcoplasmic reticulum

There are many interesting reasons for investigating the actions of anaesthetics on the function of the sarcoplasmic reticulum (SR) membrane. First, this system is capable of sequestering and releasing $Ca^{2+}$ and therefore affords the experimenter the opportunity to study $Ca^{2+}$ movements under the influence of these agents. The SR releases $Ca^{2+}$ in response to depolarisation of the transverse tubular membrane, thereby producing contraction, and subsequently sequesters $Ca^{2+}$, producing resultant relaxation. The SR is therefore intimately involved in the process of excitation-contraction coupling and thus it is not surprising that an SR deficit has been implicated in abnormal contracture states, such as muscle rigidity. Such rigidity may accompany the malignant hyperpyrexia (lethal rise in body temperature) induced, for example, by halothane anaesthesia (Strobel and Bianchi, 1971).

The aetiology of this syndrome is not well understood; however, alleviation of the symptoms of hyperpyrexia may be produced by infusion of local anaesthetics. Since caffeine-induced contractions may be blocked by local anaesthetics, the SR may turn out to be one of the sites of interaction important in the syndrome.

Secondly, anaesthetics undoubtedly produce marked changes in the force of contraction (inotropic effects) of heart muscle (Smith and Smith, 1972), and this action could be mediated by an SR membrane effect. A study of SR function under the influence of anaesthetics is therefore instructive and has clinical application.

## 5.1 $Ca^{2+}$ MOVEMENTS

The results described here are derived from experiments with skeletal muscle preparations. We should realise, however, that what is being measured in these experiments is either the $^{45}Ca^{2+}$ efflux from SR fragments or whole muscle. These data therefore give no information concerning the target for the anaesthetic effect. $Ca^{2+}$ ions are actively pumped into the SR, this process being dependent on ATPase activity. Subsequently the $Ca^{2+}$ ions presumably become bound to some membrane component on the inner surface of the SR. This could be phospholipid or protein. The release of $^{45}Ca^{2+}$ from the SR must be controlled by both the affinity of the membrane site for $Ca^{2+}$ and the permeability of the SR membrane to $Ca^{2+}$.

It is appropriate to begin by examining the well-known interaction between local anaesthetics and the drug caffeine in skeletal muscle. Feinstein (1963) showed that the local anaesthetics procaine and tetracaine block caffeine-induced rigor and the associated $Ca^{2+}$ fluxes from the muscle. Caffeine may cause potentiation of twitch tension, reversible contraction, or irreversible contraction (rigor) depending on the concentration used. Its suggested action is a direct one on the $Ca^{2+}$ release and sequestering mechanisms of the SR, no membrane depolarisation being involved (Bianchi, 1968). Consequently, the proposed site of interaction between caffeine and these local anaesthetics is at the SR.

A simple opposing action of local anaesthetics and caffeine is not always observed, however. Experiments on fragmented SR demonstrated that at pH 7·4 procaine inhibits caffeine-induced release of $^{45}Ca^{2+}$, but at higher pH values it can release up to about 25 per cent of bound $Ca^{2+}$ from the SR, producing contracture (Thorpe and Seeman, 1971). This might suggest that the protonated form of the local anaesthetic inhibits caffeine but the uncharged form is capable of potentiation.

Moreover, the situation is not likely to be as straightforward as this, since lidocaine at pH 6·0, where it is present almost entirely in the protonated form, has no blocking or potentiating action. Also, at pH 7·2, where 82 per cent of lidocaine is present in the protonated form, this agent potentiates the caffeine contracture (Bianchi, 1968). At the same pH, procaine blocks the caffeine-induced contracture and it seems that both agents are taken up equally into the muscle cell, hence another explanation must be sought for the different action of lidocaine. In Fig. IV.16 are shown the structural formulae of benzocaine, tetracaine, procaine, lidocaine and caffeine. The uncharged forms of procaine, tetracaine and lidocaine can induce contracture, although benzocaine will block the caffeine contracture at a pH where it lacks a positively charged tertiary amine group. Thus we may conclude that the structural requirements for producing contracture include a free pair of electrons on the tertiary amine nitrogen of the uncharged form of the local anaesthetics, similar to that existing on the 9 position in the caffeine molecule.

With respect to the inhibition produced by procaine, tetracaine and

Benzocaine $pK_a = 3{\cdot}19$

Tetracaine $pK_a = 8{\cdot}24$

Procaine $pK_a = 8{\cdot}95$

Lidocaine $pK_a = 7{\cdot}85$

Caffeine $pK_a = 0{\cdot}8$

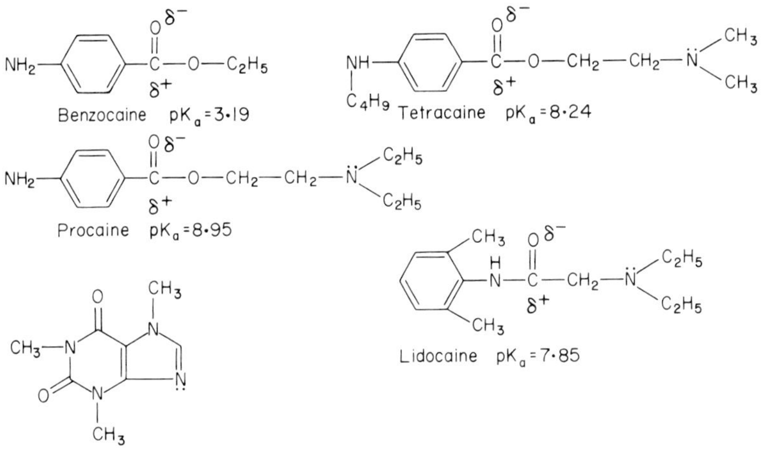

FIG. IV.16. The structural formula and $pK_a$ of benzocaine, tetracaine, procaine, lidocaine and caffeine. Note that since the local anaesthetics procaine, tetracaine and lidocaine contain a tertiary nitrogen atom they can exist in either the uncharged form, viz.

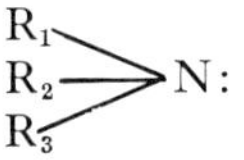

$$R_1, R_2, R_3 > N:$$

or the positively charged substituted ammonium cation, viz.

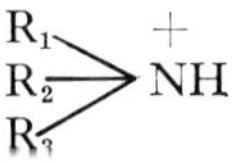

$$R_1, R_2, R_3 > \overset{+}{N}H$$

Since the $pK_a$ of these local anaesthetics lies between 7·85 and 8·95 only about 5–20 per cent will be in the uncharged form at physiological pH. (Bianchi, 1968)

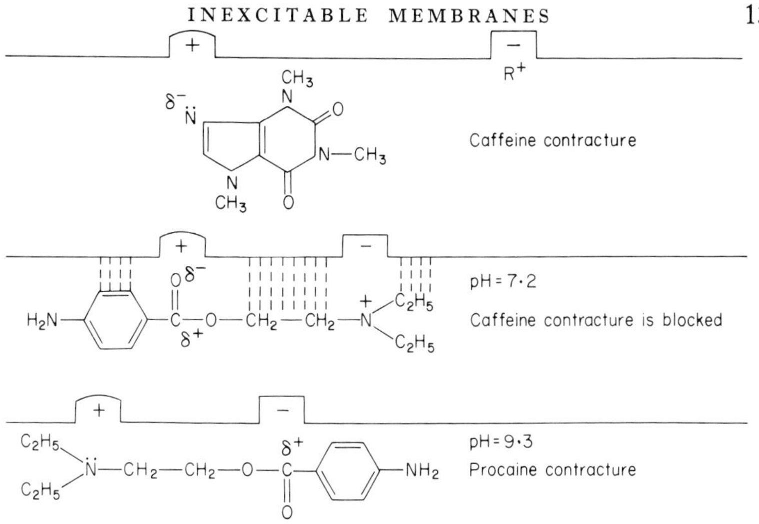


FIG. IV.17. Schematic representation of the interaction of caffeine or procaine with a hypothetical positive site of the muscle sarcoplasmic reticulum. Interaction of the site with the free pair of electrons available on the nitrogen group at pH 9·3 leads to $Ca^{2+}$ release from the terminal cisternae and resultant contracture. The carbonyl oxygen of the charged form of procaine (pH = 7·2) is proposed to compete for the positive site with both caffeine and the uncharged form of procaine and the contracture is blocked. (Bianchi, 1968)

benzocaine, but not lidocaine, it seems reasonable to suggest that the carbonyl oxygen is the group responsible for competing with caffeine for a positive binding site on the SR. Hence, lidocaine may be less effective in competing with caffeine, since either of the methyl groups, in the ortho position to the amide group, might interact sterically with the carbonyl oxygen, or alternatively, due to the closer position of the tertiary amine, a hydrogen bond may be formed between the carbonyl oxygen and the proton of the charged form of lidocaine. Bianchi (1968) has produced a model of the interaction of caffeine and procaine with SR binding sites, which is based on the ideas presented above. This scheme can be seen in Fig. IV.17. Such an hypothesis is yet to be tested by further experiments and the present state of knowledge concerning the alleged membrane binding sites is poor. However, the implication of this model is that $Ca^{2+}$ displacement from its binding site results in release of $Ca^{2+}$ from the SR and that the permeability of the SR membrane to $Ca^{2+}$ does not control exclusively the $Ca^{2+}$ efflux.

General anaesthetics also alter the $Ca^{2+}$ efflux from SR vesicular suspensions (Fiehn and Hasselbach, 1969), and clinical concentrations of halothane can induce contracture in muscle which has been treated previously by caffeine (Strobel and Bianchi, 1971). This latter finding is of

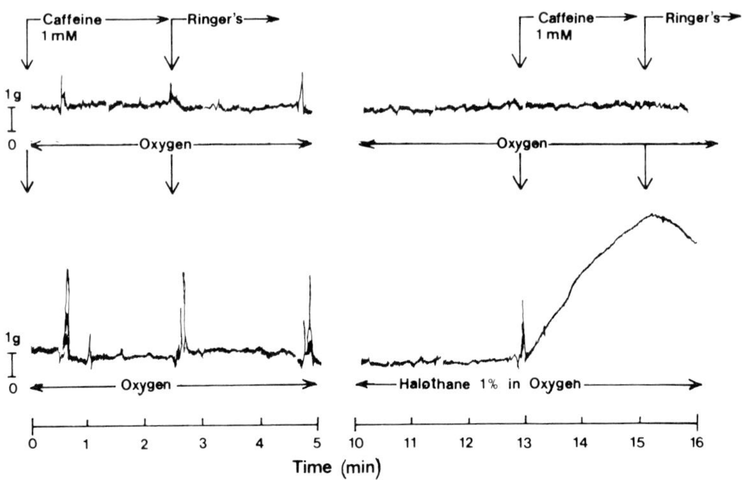


FIG. IV.18. The contracture of frog sartorius muscle produced by a combination of halothane and caffeine. Paired muscles were equilibrated under 1 g resting tension and exposed to 1 mM caffeine without a change in resting tension. One of the pair was then exposed to halothane (1 per cent) and responded to 1 mM caffeine with a contracture. Unsteadiness of the baseline and the "tension spikes" are artefacts associated with the flow of oxygen and solution changes. (Strobel and Bianchi, 1971)

considerable interest since the SR in caffeine-treated muscle has a reduced $Ca^{2+}$ sequestering capacity and hence may behave like the *in vivo* muscle prone to developing rigidity during general anaesthesia. Similarly caffeine, at a concentration which would not normally produce contraction, may induce contraction in a muscle pretreated with halothane (Fig. IV.18). It is clear that in this experiment 1 mM caffeine by itself did not produce any change in resting tension, but in the presence of 1 per cent halothane an 8 g contracture was evoked. This suggests that halothane sensitises the SR $Ca^{2+}$ releasing mechanism to caffeine.

Some of the effects on SR function described in the literature occur at doses of anaesthetic much higher than would be used clinically. Thus, Fiehn and Hasselbach (1969) noted that during treatment with diethyl ether the $Ca^{2+}$ efflux from the SR was increased, this being accompanied at high concentrations by actual removal of lipid from the membrane!

Local anaesthetics displace $Ca^{2+}$ (see also p. 61) from the SR by competing with $Ca^{2+}$ for hypothetical binding sites on the SR membrane (Fig. IV.17). How do the general anaesthetics increase the $Ca^{2+}$ efflux from the SR? A hypothesis which has evolved to explain the release of $Ca^{2+}$ bound to the surface membrane of the muscle cell suggests a change in the association constant of ion pairs following a change in the dielectric constant of the medium surrounding the binding site (Bianchi, 1969). This notion can be developed to describe one possible means by which all anaesthetics may release $Ca^{2+}$ from the SR membrane or, indeed, any membrane. The concept of anaesthetic-induced changes in membrane dielectric constant may have some bearing on other anaesthetic effects observed in both inexcitable and excitable membranes, and consequently will be met again in later parts of the book.

The hydrophobic interior of the SR membrane may be considered to be a region of low dielectric or polarisability, i.e. membrane macromolecules will not orient markedly in response to an applied electric field. Anaesthetics could alter the dielectric constant of the membrane, which in the hydrophobic region may be assumed to be about 3, in a number of ways. If the agent mixes with the membrane interior and the dielectric constant of the agent is much higher than 3, then we would expect that the membrane–anaesthetic mixture would have a significantly higher value. Alternatively, if the concentration of membrane water increased (the dielectric constant of water is about 80) then the membrane dielectric constant would also increase. The degree of membrane hydration might increase if the anaesthetic fluidised the membrane structure.

The Bjerrum ion association theory (Davies, 1962) predicts that an increase in dielectric constant would decrease the ion association constant for an ion pair, so that a rise in the dielectric constant of the environment surrounding the binding site of $Ca^{2+}$ with phospholipid, for example, may produce a decrease in ion association, and resultant $Ca^{2+}$ displacement.

The appropriate experimental tests must be devised for this theory before too much weight is placed on it. Nevertheless, this treatment of anaesthetic action, although speculative, is based on sound physiochemical tenets. Note that the assumption here is once again that increased efflux follows $Ca^{2+}$ displacement, and no membrane permeability changes are implied in the model.

## 5.2 SR ATPase activity

Anaesthetics reduce the phase transition temperature of lipid bilayers (p.111) and experiments with SR membrane have shown how the behaviour of a "real" membrane is correspondingly altered.

Complexes of dipalmitoyllecithin with $Ca^{2+}$, $Mg^{2+}$ dependent ATPase which have been isolated from SR show breaks in the Arrhenius plot of ATPase activity. These breaks are due to the interaction of the lipid molecules with the ATPase (protein) molecules in the complex (Hesketh *et al.*, 1976) What is significant is that benzyl alcohol (50 mM) shifts the break in the Arrhenius plot of ATPase activity to a lower temperature and enhances the ATPase activity up to 10-fold at temperatures below the transition temperature. This effect was fully reversible and was attributed to the action of the alcohol on the lipid–protein interaction. From such studies we can anticipate that in "real" membranes anaesthetics will affect the physical state of the lipid phase and that the activity of penetrant membrane proteins will be correspondingly altered. In other words the membrane protein can also "detect" the presence of the anaesthetic molecule by its influence on the lipid surrounding the protein.

## 6 Conclusions

The anesthetics thus exert a variety of interesting effects on the structure and properties of inexcitable membranes. Studies with lipid bilayers have demonstrated that anaesthetics can fluidise (disorder) the membrane although the composition of the bilayer does modulate this fluidising action. For example, pentobarbitone and the steroid anaesthetic Alphaxalone can actually order bilayers low in cholesterol content. All anaesthetics reduce the phase-transition temperature of lipid bilayers. The permeability characteristics of bilayers are also modified by anaesthetics. The $K^+$ permeability and valinomycin-induced increase in $K^+$ permeability of liposomes is enhanced by a general anaesthetic such as diethyl ether or the alcohol *n*-butanol. The $Na^+$ permeability of liposomes is, however, decreased by local anaesthetics. The latter effect is accompanied by a parallel reduction in the liposome zeta potential.

At low concentrations most anaesthetics protect the erythrocyte membrane against hypotonic haemolysis, but at higher concentrations these anaesthetics produce lysis. Such protection may be explained in terms of a membrane expansion (maximum 7 per cent) accompanying the fluidising action of anaesthetics. Both local and general anaesthetics depress active $Na^+$ transport across the erythrocyte membrane, although at lower concentrations agents such as ether and chloroform can potentiate $Na^+$ transport. The carrier-mediated transport of choline is depressed by benzyl alcohol, which fluidises the erythrocyte membrane, and anaesthetics produce variable effects on the erythrocyte membrane $Na^+$ permeability, all of which can be countered by raising the aqueous $Ca^{2+}$ concentration. The depressant effect of tetracaine on the exchange diffusion of $Cl^-$ across the erythrocyte membrane is not $Ca^{2+}$-dependent, however.

In epithelial membranes it has been proposed that the charged form of local anaesthetics stimulates the $Na^+$ pump and the uncharged form depresses the $Na^+$ pump. General anaesthetics (e.g. cyclopropane, nitrous oxide) stimulate the epithelial $Na^+$ pump at low concentrations and depress at high concentrations. Generally, these agents stimulate the $Na^+$ pump of toad bladder at concentrations close to, or less than, the anaesthetic dose for the toad.

In the muscle sarcoplasmic reticulum the local anaesthetics cause $Ca^{2+}$ release and contracture if present in the uncharged form, but in the charged form block caffeine-induced contracture. The general anaesthetic halothane also promotes $Ca^{2+}$ release from the sarcoplasmic reticulum.

The net effect of an anaesthetic on any inexcitable membrane is clearly a function of the concentration and type of agent, but equally important it is dependent on the organisation of the membrane system under investigation.

## References


Amaranath, L. and Andersen, N. B. (1974). The effects of anaesthetics on permeability to water of the inactivated toad bladder. *Anesthesiol.* **40**, 168–174.

Andersen, N. B. (1968). The effect of local anaesthetic and pH on sodium and potassium flux in human red cells. *J. Pharm. Exp. Therap.* **163**, 393–406.

Andersen, N. B. and Shim, C. Y. (1971). Sodium transport and anaesthetic requirements in the toad. *Anesthesiol.* **34**, 338–343.

Balasubramanian, D. and Wetlaufer, D. B. (1966). Reversible alteration of the structure of globular proteins by anaesthetic agents. *Proc. Nat. Acad. Sci.* **55**, 762–765.

Bangham, A. D., Standish, M. M. and Miller, N. (1965a). Cation permeability of phospholipid model membranes: effect of narcotics. *Nature*, **208**, 1295–1297.

Bangham, A. D., Standish, M. M. and Watkins, J. C. (1965b). Diffusion of univalent ions across the lamellae of swollen phospholipids. *J. Molec. Biol.* **13**, 238–252.

Bennett, P. B., Papadhjopoulos, D. and Bangham, A. D. (1967). The effect of raised pressure of inert gases on phospholipid membranes. *Life Sci.* **6**, 2527–2533.

Bianchi, C. P. (1968). Pharmacological actions on excitation-contraction coupling in striated muscle. *Fed. Proc.* **27**, 126–131.

Bianchi, C. P. (1969). Pharmacology of excitation-contraction coupling in muscle. Introduction: statement of the problem. *Fed. Proc.* **28**, 1624–1628.

Blaustein, M. P. (1967). Phospholipids as ion exchangers: implications for a possible role in biological membrane excitability and anaesthesia. *Biochim. Biophys. Acta*, **135**, 653–668.

Blaustein, M. P. and Goldman, D. E. (1966). Action of anionic and cationic nerve-blocking agents: experiment and interpretation. *Science*, **153**, 429–432.

Boggs, J. M., Young, T. and Hsia, J. C. (1976). Site and mechanism of anaesthetic action. 1. Effect of anaesthetics and pressure on fluidity of spin-labelled lipid vesicles. *Molec. Pharmacol.* **12**, 127–135.

Brewster, E., Collins, S., Funnell, G. R. and Smith, E. B. (1976). The effects of high pressure on the haemolysis of red blood cells. *Undersea Biomed. Res.* **3**, 151–155.

Burt, D. H. and Green, J. W. (1971). The sodium permeability of butanol-treated erythrocytes. The role of calcium. *Biochim. Biophys. Acta*, **225**, 46–55.

CABANTCHIK, Z. I. and ROTHSTEIN, A. (1974). Membrane proteins related to anion permeability of human red blood cells. I. Localisation of disulfonic stilbene binding sites in proteins involved in permeation. *J. Membrane Biol.* **15**, 207–226.

CERBÓN, J. (1972). NMR evidence for the hydrophobic interaction of local anaesthetics. Possible relation to their potency. *Biochim. Biophys. Acta*, **290**, 51–57.

CHAPMAN, D. and WALLACE, D. F. H. (1968). *In* "Biological Membranes: Physical Fact and Function" (D. Chapman, ed.), Chap. 4. Academic Press, London and New York.

CLAYTON, P. and MARTIN, K. (1971). The effect of anaesthetic alcohols on carrier mediated transport in human erythrocytes. *J. Physiol.* **218**, 50P–51P.

CLEMENTS, J. A. and WILSON, K. M. (1962). The affinity of narcotic agents for interfacial films. *Proc. Nat. Acad. Sci.* **48**, 1008–1114.

DAVIES, C. W. (1962). "Ion Association". Butterworths, London.

DEUTICKE, B. and GRUBER, W. (1970). Anion permeability of mammalian red blood cells: Possible relation to membrane phospholipid patterns. *Biochim. Biophys. Acta*, **211**, 369–372.

FEINSTEIN, M. B. (1963). Inhibition of caffeine rigor and radiocalcium movements by local anaesthetics in frog sartorius muscle. *J. Gen. Physiol.* **47**, 151–172.

FEINSTEIN, M. B. (1964). Reaction of local anaesthetics with phospholipids. A possible chemical basis for anaesthesia. *J. Gen. Physiol.* **48**, 357–374.

FENN, W. O. and BOSCHEN, V. P. (1971). Haemolysis under high hydrostatic pressure. *Proc. Soc. Exp. Biol. Med.* **137**, 847–851.

FIEHN, W. and HASSELBACH, W. (1969). The effect of diethyl ether upon the reticulum of the vesicles of sarcoplasmic reticulum. *Eur. J. Biochem.* **9**, 574–578.

FINN, A. L. (1976). Changing concepts of transepithelial sodium transport. *Physiol. Rev.* **56**, 453–464.

GREENE, N. M. (1965). Inhalation anaesthetics and permeability of human erythrocytes to monosaccharides. *Anaesthesiol.* **26**, 731–742.

GUNN, R. B. and COOPER, J. A. (1975). Effect of local anaesthetics on chloride transport in erythrocytes. *J. Membrane Biol.* **25**, 311–326.

GUTKNECHT, J. and TOSTESON, D. C. (1970). Ionic permeability of thin lipid membranes. Effects of *n*-alkyl alcohols, polyvalent cations, and a secondary amine. *J. Gen. Physiol.* **55**, 359–374.

HALE, J., KEEGAN, R., SMITH, E. B. and SNAPE, T. J. (1972). The effect of general anaesthetics on active cation transport in human erythrocytes. *Biochim. Biophys. Acta*, **288**, 107–113.

HALSEY, M. J., SMITH, E. B. and WOOD, T. E. (1970). Effects of general anaesthetics on $Na^+$ transport in human red cells. *Nature*, **225**, 1151–1152.

HAUSER, H. and DAWSON, R. M. C. (1968). The displacement of calcium ions from phospholipid monolayers by pharmacologically active and other organic bases. *Biochem. J.* **109**, 909–916.

HILL, M. W. (1974). The effect of anaesthetic-like molecules on the phase transition in smectic mesophases of dipalmitoyllecithin. 1. The normal alcohol up to C = 9 and three inhalation anaesthetics. *Biochim. Biophys. Acta*, **356**, 117–124.

HESKETH, T. R. SMITH, G. A., HOUSLAY, M. D., MCGILL, K. A., BIRDSALL, N. J. M. METCALFE, J. C. and WARREN, G. B. (1976). Annular lipids determine the ATPase activity of a calcium transport protein complexed with dipalmitoyllecithin. *Biochemistry*, **15**, 4145-4151.

Hubbell, W. L. and McConnell, H. M. (1968). Spin-label studies of the excitable membranes of nerve and muscle. *Proc. Nat. Acad. Sci.* **61**, 12–16.

Hui, S. W. (1976). Direct measurement of membrane motion and fluidity by electron microscopy. *Nature*, **262**, 303–305.

Jacobs, M. H. and Parpart, A. K. (1937). The influence of certain alcohols on the permeability of the erythrocyte. *Biol. Bull.* **73**, 380–381.

Jain, M. K. (1972). "The Bimolecular Lipid Membrane". Van Nostrand Reinhold, New York.

Johnson, S. M. and Bangham, A. D. (1969). The action of anaesthetics on phospholipid membranes. *Biochim. Biophys. Acta*, **193**, 92–104.

Johnson, S. M., Bangham, A. D., Hill, M. W. and Korn, E. D. (1971). Single bilayer liposomes. *Biochim. Biophys. Acta*, **233**, 820–826.

Johnson, S. M., Miller, K. W. and Bangham, A. D. (1973). The opposing effects of pressure and general anaesthetics on the cation permeability of liposomes of varying lipid composition. *Biochim. Biophys. Acta*, **307**, 42–57.

Kwant, W. O. and Seeman, P. (1969). The membrane concentration of a local anaesthetic (chlorpromazine). *Biochim. Biophys. Acta*, **183**, 530–543.

Lee, A. G. (1976). Interaction between anaesthetics and lipid mixtures amines. *Biochim. Biophys. Acta*, **448**, 34–44.

Levine, Y. K. and Wilkins, M. H. F. (1971). Structure of oriented lipid bilayers. *Nature New Biol.* **230**, 69–72.

Lucy, J. A. (1968). Ultrastructure of membranes: micellar organisation. *Brit. Med. Bull.* **24**, 127–129.

Macdonald, A. G. (1978). A dilatometric investigation of the effects of general anaesthetics, alcohols and hydrostatic pressure on the phase transition in smectic mesophases of dipalmitoyl phosphatidylcholine. *Biochim. Biophys. Acta*, **507**, 26-37.

McLaughlin, S. (1975). Local anaesthetics and the electrical properties of phospholipid bilayer membranes. *In* "Molecular Mechanisms of Anaesthesia: Progress in Anaesthesiology" (B. R. Fink, ed.), vol. 1, pp. 193–220. Raven, New York.

Metcalfe, J. C. (1970). Interaction of local anaesthetics and calcium with erythrocyte membranes. *In* "Calcium and Cellular Function" (A. W. Cuthbert, ed.), pp. 219–240. Macmillan, London.

Metcalfe, J. C., Seeman, P. and Burgen, A. S. V. (1968). The proton relaxation of benzyl alcohol in erythrocyte membranes. *Molec. Pharmacol.* **4**, 87–95.

Mikikits, W., Mortara, A. and Spector, R. G. (1970). Effects of drugs on red cell fragility. *Nature*, **225**, 1150–1151.

Miller, K. W. and Pang, K. Y. (1976). General anaesthetics can selectively perturb lipid bilayer membranes. *Nature*, **263**, 253–255.

Mueller, P., Rudin, D. O., Tien, H. T. and Wescott, W. C. (1962). Reconstitution of cell membrane structure *in vitro* and its transformation into an excitable system. *Nature*, **194**, 979–980.

Mullins, L. J. (1954). Some physical mechanisms in narcosis. *Chem. Rev.* **54**, 289–323.

Neal, M. J., Butler, K. W., Polnaszek, C. F. and Smith, I. C. P. (1976). The influence of anaesthetics and cholesterol on the degree of molecular organisation and mobility of ox brain white matter. Lipids in multibilayer membranes:

a spin probe study using spectral simulation by the stochastic method. *Molec. Pharmacol.* **12**, 144–155.

Papahadjopoulos, D. (1970). Phospholipid model membranes. III. Antagonistic effects of $Ca^{2+}$ and local anaesthetics on the permeability of phosphotidylserine vesicles. *Biochim. Biophys. Acta*, **211**, 467–477.

Passow, H. (1969). Passive ion-permeability of the erythrocyte membrane. *Prog. Biophys. Molec. Biol.* **19**, 423–467.

Paterson, S. J., Butler, K. W., Huang, P., Labelle, J., Smith, I. C. P. and Schneider, H. (1972). The effects of alcohols on lipid bilayers: A spin label study. *Biochim. Biophys. Acta*, **266**, 597–602.

Ponder, E. (1948). "Haemolysis and Related Phenomena". Grune and Stratton, New York.

Power, G. G. and Stegall, H. (1970). Solubility of gases in human red blood cell ghosts. *J. Appl. Physiol.* **29**, 145–149.

Rideal, E. and Taylor, F. H. (1957). On haemolysis by anionic detergents. *Proc. Roy. Soc. B*, **146**, 225–241.

Roth, S. and Seeman, P. (1971). All lipid-soluble anaesthetics protect red cells. *Nature New Biol.* **231**, 284–285.

Roth, S. and Seeman, P. (1972). The membrane concentrations of neutral and positive anaesthetics (alcohols, chlorpromazine, morphine) fit the Meyer-Overton rule of anaesthesia; negative narcotics do not. *Biochim. Biophys. Acta*, **255**, 207–219.

Schneider, H. (1968). The intramembrane location of alcohol anaesthetics. *Biochim. Biophys. Acta*, **163**, 451–458.

Schreier-Muccillo, S., Marsh, D., Dugas, H., Schneider, H. and Smith, I. C. P. (1973). A spin probe study of the influence of cholesterol on motion and orientation of phospholipids in oriented multibilayers and vesicles. *Chem. Phys. Lipids*, **10**, 11–27.

Seeman, P. (1966a). Membrane stabilization by drugs: tranquilisers, steroids and anaesthetics. *Int. Rev. Neurobiol.* **9**, 145–221.

Seeman, P. (1966b). Erythrocyte membrane stabilisation by local anaesthetics and tranquilisers. *Biochem. Pharmacol.* **15**, 1753–1766.

Seeman, P. (1969). Temperature dependence of erythrocyte membrane expansion by alcohol anaesthetics. Possible support for the partition theory of anaesthesia. *Biochim. Biophys. Acta*, **183**, 520–529.

Seeman, P. (1972). The membrane actions of anaesthetics and tranquilizers. *Pharmacol. Rev.* **24**, 583–655.

Seeman, P. (1975). The membrane expansion theory of anaesthesia. *In* "Molecular Mechanisms of Anaesthesia: Progress in Anaesthesiology" (B. R. Fink, ed.), vol. 1, pp. 243–251. Raven Press, New York.

Seeman, P. and Kwant, W. O. (1969). Membrane expansion of the erythrocyte by both the neutral and ionized forms of chlorpromazine. *Biochim. Biophys. Acta*, **183**, 512–519.

Seeman, P. and Roth, S. (1972). General anaesthetics expand cell membranes at surgical concentrations. *Biochim. Biophys. Acta*, **255**, 171–177.

Seeman, P., Chau, M., Goldberg, M., Sauks, T. and Sax, L. (1971a). The binding of $Ca^{2+}$ to the cell membrane increased by volatile anaesthetics (alcohols, acetone, ether) which induce sensitization of nerve or muscle. *Biochim. Biophys. Acta*, **225**, 185–193.

SEEMAN, P., KWANT, W. O., GOLDBERG, M. and CHAU-WONG, M. (1971b). The effects of ethanol and chlorpromazine on the passive membrane permeability to $Na^+$. *Biochim. Biophys. Acta*, **241**, 349–355.

SEEMAN, P., KWANT, W. O. and SAUKS, T. (1969a). Membrane expansion of erythrocyte ghosts by tranquilisers and anaesthetics. *Biochim. Biophys. Acta;* **183**, 499–511.

SEEMAN, P., KWANT, W. O., SAUKS, T. and ARGENT, W. (1969b). Membrane expansion of intact erythrocytes by anaesthetics. *Biochim. Biophys. Acta*, **183**, 490–498.

SEEMAN, P., ROTH, S. and SCHNEIDER, H. (1971c). The membrane concentrations of alcohol anaesthetics. *Biochim. Biophys. Acta*, **225**, 171–184.

SEEMAN, P., SHA'AFI, R. I., GALEY, W. R. and SOLOMON, A. K. (1970). The effect of anaesthetics (chlorpromazine, ethanol) on erythrocyte permeability to water. *Biochim. Biophys. Acta*, **211**, 365–368.

SINGER, M. (1975). Effects of local anaesthetics on phospholipid bilayer membranes. *In* "Molecular Mechanisms of Anaesthesia: Progress in Anaesthesiology" (B. R. Fink, ed.), vol. 1, pp. 223–234. Raven, New York.

SINGER, S. J. and NICOLSON, G. L. (1972). The fluid mosaic model of the structure of cell membranes. *Science*, **175**, 720–731.

SKOU, J. C. (1958). Relation between the ability of various compounds to block nervous conduction and their penetration into a monomolecular layer of nerve-tissue lipoids. *Biochim. Biophys. Acta*, **30**, 625–629.

SKOU, J. C. (1961). The effects of drugs on cell membranes with special reference to local anaesthetics. *J. Pharm. Pharmacol.* **13**, 204–217.

SKOU, J. C. and ZERAHN, K. (1959). Investigations on the effect of some local anaesthetics and other amines on the active transport of sodium through the isolated short-circuited frog skin. *Biochim. Biophys. Acta*, **35**, 324–333.

SMITH, N. T. and SMITH, P. (1972). Circulating effects of modern inhalation anaesthetic agents. *In* "Modern Inhalation Anaesthetics" (Handb. Exp. Pharmacol.), vol. 30, pp. 149–241. Springer-Verlag, Berlin.

STROBEL, G. E. and BIANCHI, C. P. (1971). An *in vitro* model of anaesthetic hypertonic hyperpyrexia, halothane-caffeine-induced muscle contractures. *Anesthesiol.* **35**, 465–473.

THORPE, W. R. and SEEMAN, P. (1971). The site of action of caffeine and procaine in skeletal muscle. *J. Pharm. Exp. Therap.* **179**, 324–330.

TRUDELL, J. R., HUBBELL, W. L. and COHEN, E. N. (1973a). The effect of two inhalation anaesthetics on the order of spin-labelled phospholipid vesicles. *Biochim. Biophys. Acta*, **291**, 321–327.

TRUDELL, J. R., HUBBELL, W. L., COHEN, E. N. and KENDIG, J. J. (1973b). Pressure reversal of anaesthesia: The extent of small molecule exclusion from spin labelled phospholipid model membranes. *Anesthesiol.* **38**, 207–211.

TRUDELL, J. R., PAYAN, D. G., CHIN, J. H. and COHEN, E. N. (1975). The antagonistic effect of an inhalation anaesthetic and high pressure on the phase diagram of mixed dipalmitoyl-dimyristoyl-phosphatidylcholine bilayers. *Proc. Nat. Acad. Sci.* **72**, 210–213.

CHAPTER V

# Excitable membranes



## 1 Introduction

The depressant effect of anaesthetics on nerve impulse conduction was recognised as early as the late nineteenth century, when Szpilman and Luchsinger (1881) demonstrated that ether, chloroform and ethanol blocked conduction in frog nerves. The action of ethanol was confirmed by Lucas (1913), and Winterstein (1919) (see Alper and Flacke, 1969, p. 275) summarised the then known actions of anaesthetic agents on excitable tissues, accounting for their depressant effect in terms of a reduction in permeability, while ascribing the irreversible effects to an increase in permeability.

Subsequent work suggested that both ether (Lorente, de Nó, 1947,

Chapters I and VII) and ethyl alcohol (Gallego, 1948) produced conduction block in frog nerve due to their depolarising actions. However, a more precise description of the mode of action of these agents at a membrane level had to await a better understanding of the events underlying the nerve-excitation process. The development of the glass microelectrode for intracellular recording (Graham and Gerard, 1946), the birth of the voltage-clamp technique (Marmont, 1949), and the resultant elegant description by Hodgkin and Huxley of the ionic permeability changes during the action potential (Hodgkin and Huxley, 1952) provided the experimental tools and the appropriate membrane concepts to reinvestigate afresh the nature of anaesthetic action.

The effects of anaesthetic agents on the electrical characteristics of many excitable membranes have thus received considerable attention over the last 20 years or so. In this chapter consideration will be given primarily to results obtained from well-defined preparations, such as the lobster and squid giant axons, single amphibian myelinated fibres, mammalian Purkinje and amphibian skeletal muscle fibres; but in many instances results derived from experiments conducted on vertebrate whole-nerve bundles will be cited where appropriate.

Recent investigations of the action of anaesthetics on the electrical characteristics of axonal or muscle preparations have been undertaken, often in the vain hope that the results may lead to an understanding of the nature of the anaesthetic state. As a consequence, some authors subscribe to the perhaps rather unorthodox view that local anaesthetic *block* and the general anaesthetic *state* share a common mechanism (Frank, 1972; Seeman, 1972a). That is to say, both local and general anaesthetics decrease excitability by inhibiting the regenerative increase in permeability underlying action potential depolarisation (Thesleff, 1956).

It is evident also that many excitable membrane studies using anaesthetics have been motivated by a desire to characterise the nature of the interaction of the anaesthetic agent with the membrane. This approach could be potentially fruitful, in that it may furnish, in addition, molecular information about the basic structure of excitable membranes, and perhaps the chemical nature of the ionic channels. Certainly one possible way of tackling the problem of channel structure is to examine the nature of a range of drugs which may block, or at least modify, the channel's ionic permeability. This approach has been exploited in channel studies previously using toxins, such as tetrodotoxin (TTX) (Hille, 1968a), or the quaternary ammonium ions (Armstrong, 1975). The possibilities which this line of attack offer may be best illustrated by considering simple examples.

Let us assume that the dependence of the ionic permeability of a membrane channel on the presence of charged anaesthetic molecules is linked to the ability of these agents to bind to a site within the channel. The relative potencies of a series of agents, or analogues of agents, will yield, therefore, data concerning the nature of the binding site, which could involve a group located at the mouth of a channel, or perhaps (as in the case of TTX) may involve a group within the channel (Hille, 1975a). The pH dependence of the blocking potency yields further useful information about the binding sites, and the relative susceptibilities of the $Na^+$ and $K^+$ permeabilities will suggest differences in charged groups in or close to the channel mouth. More generally, the depressant effects of anaesthetics on permeability can indicate the dependence of the ion channel gating characteristics on its hydrophobic surround, which would be altered by the presence of lipid-soluble anaesthetic agents.

Molecular models of the interaction of anaesthetics with excitable membrane channels have been proposed recently (e.g. Lee, 1976) and it seems not unlikely that in the future anaesthetics will prove increasingly useful probes in determining the organisation of membrane macromolecules.

## 1.1 INDEX OF POTENCY

According to the Meyer–Overton rule of anaesthetic action, the nerve-conduction block produced by an agent should occur when the membrane concentration of the agent reaches a certain critical value (see Seeman, 1972b; Seeman, 1975). In order to test the applicability of this rule, it is necessary to measure both the concentration of a particular agent in the membrane phase, and its potency in producing a membrane-based effect in the same membrane system. Unfortunately, this exercise seems to have been performed only with the erythrocyte membrane.

It is possible to isolate erythrocyte membranes, measure radioactively the uptake of labelled anaesthetic agents, and to compare the degree of adsorption, and hence membrane concentration, with the ability of the agent to protect the erythrocyte from hypotonic haemolysis (p. 114). The erythrocyte membrane has been used extensively as a model system for anaesthetic action, and results obtained from these studies may have bearing on excitable membrane work, although conceptually it is perhaps difficult to view conduction block and protection from hypotonic haemolysis as effects mediated by the same membrane process.

Nevertheless, the membrane/buffer partition coefficients of anaesthetics in nerve, muscle and erythrocyte membranes are similar (Roth and Seeman, 1972), and it is known that the anaesthetic concentration producing an antihaemolytic effect is correlated with the concentration producing conduction block (Seeman, 1972b). It is thus plausible that broadly similar mechanisms are involved in these actions.

Seeman has calculated for a range of anaesthetic molecules the membrane concentrations producing a 50 per cent reduction in haemolysis ($AH_{50}$) of the erythrocyte. These concentrations vary over about a fortyfold range, showing that the Meyer–Overton rule does not hold. Such calculations do yield, however, interesting figures for the density of packing of anaesthetic molecules in the membrane, at either local or general anaesthetic level. These values are presented in Table V.1. The estimates of 25 000–250 000 molecules $\mu m^{-2}$ are obviously many orders of magnitude larger than the current figures for the Na channel density of nerve membranes which range from between 6 $\mu m^{-2}$ for the olfactory nerve fibre to 27 $\mu m^{-2}$ for rabbit nerve (Ritchie, 1975), although the Na channel density of the squid axon membrane may be about 500 $\mu m^{-2}$ (Keynes and Rojas, 1974). If the number of anaesthetic molecules per $\mu m^{-2}$ in an excitable membrane during

TABLE V.1

The concentration and calculated volumes of anaesthetics in human erythrocyte membranes at clinical doses; the associated changes in erythrocyte membrane area and volume are also presented. The calculated number of anaesthetic molecules per Na channel of the giant squid axon membrane are also shown. N.B. the tenfold difference in concentrations for general and local anaesthetics. (Adapted from Seeman, 1975)

| Concentration of anaesthetic in the membrane phase | General anaesthesia | Local anaesthesia |
|---|---|---|
| (a) Expressed as molality (kg of dry membrane) | 4 mmol $kg^{-1}$ | 40 mmol $kg^{-1}$ |
| (b) In terms of number of anaesthetic molecules per $\mu m^2$ of membrane area | 25 000 sites $\mu m^{-2}$ | 250 000 sites $\mu m^{-2}$ |
| (c) In terms of number of anaesthetic molecules per Na channel in the squid axon membrane | 50 per channel | 500 per channel |
| Volume of anaesthetic in the membrane phase (derived from membrane/buffer partition coefficient and calculated as in text) | 0·02% V/V (0·2 ml (kg membrane)$^{-1}$) | 0·2% V/V (2 ml (kg membrane)$^{-1}$) |
| Apparent membrane area increase (obtained from haemolysis experiment) | 0·4% | 2–4% |
| Membrane volume increase (derived from density measurements) | 0·5–0·6% | 4–5% |

conduction block is similar to Seeman's estimate for the erythrocyte membrane, then hundreds (see Table V.1) of anaesthetic molecules are present when the $Na^+$ channel is blocked, so presumably the $Na^+$ channel does not possess a high affinity for the anaesthetic.

If, as Seeman (1972b) showed, the "volume of occupation" of the anaesthetic in the membrane is calculated, then for the same anaesthetic this volume varies only about eightfold. It should be noted that the "volume of occupation" was calculated from figures for the "solid state" atomic radii and only provide an estimate of the partial molar volume of the anaesthetic within a membrane. Nevertheless, the eightfold variation in estimated volume at $AH_{50}$ is smaller than the variation seen for membrane concentration. Thus, as suggested by Mullins (1954), this parameter may be a more suitable index of narcotic potency. It is however significant that the equinarcotic volumes of occupation of negatively charged anaesthetic molecules are considerably less than that of neutral or positive anaesthetic agents (Roth and Seeman, 1972).

## 1.2 MEMBRANE EXPANSION

It has been suggested that penetration of membranes by anaesthetic molecules causes membrane expansion (Seeman, 1972b). This is discussed more fully in the previous chapter. Such membrane expansion in erythrocytes is at present considered to be the basis for the antihaemolytic effect of these agents.

When anaesthetics become "buried" in a bulk phase such as rubber or olive oil, these phases swell by an amount equivalent to the molar volume of the anaesthetic entering the phase. However, in experiments investigating the antihaemolytic effect of anaesthetics, it is suggested that local anaesthetic concentrations increase the erythrocyte membrane area by 2–4 per cent (Seeman *et al.*, 1969), whereas general anaesthetics at surgical concentrations produce about a 0·4 per cent increase (Seeman and Roth, 1972). This is a considerably greater increase than would be expected from the alleged occupying volume of the anaesthetic agent. The values for the change in the erythrocyte membrane geometry in the presence of a drug concentration suffice to produce local or general anaesthesia, in addition to the occupying volume of the anaesthetic, are shown in Table V.1. It is clear that the increase in membrane volume is approximately twenty times or more the occupying volume of the agent producing the swelling. This excessively large expansion may be accounted for if it is proposed that the "dissolving" of the anaesthetic molecule in a hydrophobic portion of the membrane results in a disordering of the membrane.

As a corollary, it is not unreasonable to suggest that a similar expansion may occur in excitable membranes, and any model of anaesthetic action should not ignore this. Certainly, fluidisation of the membrane components of erythrocyte and synaptosome membranes occurs at about the same anaesthetic concentration (Metcalfe and Burgen, 1968).

### 1.3 NATURE OF THE ANAESTHETIC–MEMBRANE INTERACTION

Anaesthetic-induced effects are generally reversible. It is, of course, true that prolonged administration of some agents (e.g. ethanol) classified as anaesthetics may produce symptoms of withdrawal, or lead to addiction in the human. These phenomena could be due to structural changes in membranes, or in membrane receptors, of select nerve cells in the central nervous system, or may be due to subtle changes in the neurochemistry of central nerve cells. In the case of the haloalkylamines, irreversible electrical changes in excitable membranes are brought about which are associated with actual alkylation of the membrane structure (Ehrenpreis and Rosen, 1974). Although potentially of immense interest, it is not the intention of this section to describe such structural changes. Accordingly, when considering the action of anaesthetics on excitable membranes should we ask whether reversibility implies a particular kind of interaction? An anaesthetic could associate in a non-specific way with membrane components. Such an interaction might be a weak one and this may be the case, for example, for the inert gases (p. 54). Alternatively, anaesthetics such as the tertiary amines (e.g. local anaesthetics) could *bind* specifically to membrane components. Finally, certain substances can *bond* to membrane components, i.e. a chemical reaction takes place between the membrane and the anaesthetic molecule (e.g. the haloalkylamines).

The latter *bonding* reaction is not readily reversible and as stated above will not be dealt with here. Can we, however, distinguish between non-specific *association* and specific *binding* from our knowledge of the reversibility of anaesthetic action? If we reflect on the effects of saxitoxin on nerve membranes then this question can be settled quickly. The action of saxitoxin on myelinated fibres is readily reversible (within 30 sec) yet there seems no doubt as to the selective binding characteristics of this agent (Hille, 1968a).

The blocking action of an anaesthetic on the nerve action potential is correlated with the membrane/buffer partition coefficient (Staiman and Seeman, 1974). In Fig. V.1 are presented data for a range of agents. The concentration of the anaesthetic producing either a 50 per cent reduction in the compound action potential height of the rat phrenic nerve or that concentration producing a total block of conduction in the frog sciatic nerve is compared with the partition coefficient for each agent. The results

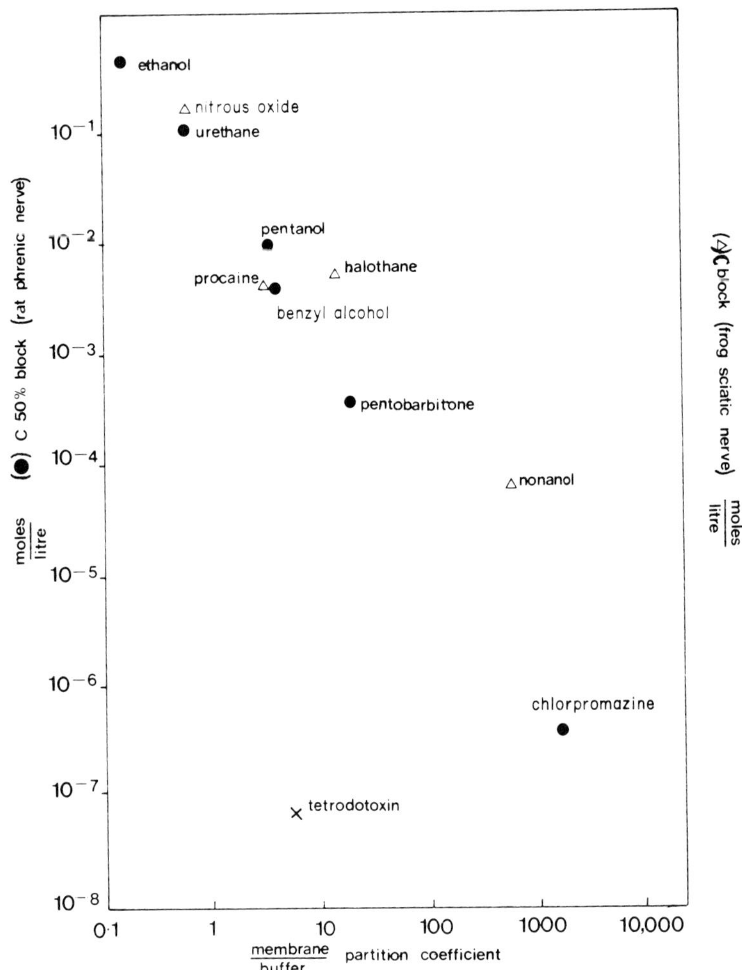


FIG. V.1. The relationship between the concentrations of anaesthetic producing a 50 per cent nerve block (C 50 per cent block) in the rat phrenic nerve (●), or block (C block) in the frog sciatic nerve (△) and the membrane/buffer partition coefficients of the anaesthetics. The concentration of tetrodotoxin producing 50 per cent nerve block in the rat phrenic nerve is also shown. It appears that the toxin does not conform to the Meyer–Overton rule. (Adapted from Seeman, 1972b; Staiman and Seeman, 1974)

of this study suggest that the efficiency with which these agents produce block is related to their ability to "*partition*" in a hydrophobic phase. It is significant that tetrodotoxin, a non-hydrophobic substance, was more potent in blocking the action potential, and its blocking concentration did not fall into a range that could be related to its partition coefficient. This suggests immediately that the blockage produced by anaesthetics and tetrodotoxin has a different basis.

That the anaesthetic interaction within the membrane involves association with a hydrophobic region has been confirmed by the finding that the free energy change accompanying the adsorption of agents, such as the alcohols (Schneider, 1968) and the tranquilliser chlorpromazine (Kwant and Seeman, 1969), is similar in magnitude to the free energy change involved in moving these agents from an aqueous to a non-polar environment. For example, Schneider (1968) has pointed out that the free energy change accompanying transfer of methylene ($CH_2$) groups from an aqueous to a non-polar environment is in the range 750–887 cal $mol^{-1}$, which compares favourably with the free energy of binding values associated with the depression by anaesthetic agents of a variety of membrane-related functions, such as block of axonal conduction.

Needless to say, such free energy changes provide no clue to the identity of the membrane macromolecules with which the agent is associated. Thus the anaesthetic molecule could attach to the hydrophobic region of the membrane protein structure, or combine (associate) with the hydrocarbon tails of the lipid moiety of the membrane.

## 2 Documented effects

### 2.1 THE TRANSMEMBRANE POTENTIAL

In 1940, Guttmann reported that the alkaline earth cations reduced the depolarising action of $K^+$ or veratrine. These cations may therefore be considered to *stabilise* the membrane, in the electrical sense. Moreover, the concept of membrane stabilisation has now been extended, so that diverse agents including anaesthetics and many other drugs are now considered to fall into the category of membrane stabilisers (Shanes, 1958; Seeman, 1966a). Undoubtedly there are marked differences in the mode of action of stabilising agents, and it is worth stating that although the divalent actions are generally considered to be both structural and electrical stabilisers, the anaesthetic agents may be *electrical stabilisers*, but should perhaps be termed *structural labilisers*, in that they can fluidise or disorder membranes. Also, it should be emphasised that the electrical stabilisation produced by anaesthetics is concentration-dependent, and high concentrations will produce electrical breakdown in the excitable membrane.

Some of the early studies on the demarcation potentials of frog nerve revealed that excitability could be blocked by anaesthetics with little or no change (2–3 mV) in the steady state potential (Bishop, 1932). If the concentrations tested are sufficiently high, depolarisation is produced by ether, chloroform, the alcohols and the inert gases (Carpenter, 1954; Mullins and Gaffey, 1954). A significant observation is that lower members of the aliphatic alcohols depolarise frog nerve fibres, whereas higher members produce hyperpolarisation (Berney and Posternak, 1956). Unfortunately, it is not possible to derive any information concerning the *membrane* concentrations of the alcohols producing these effects in this study, so that it is consequently difficult to say anything about the basis of such differential actions. It is surprising, therefore, that this finding has not been corroborated and analysed systematically with intracellular measurements. It is likely that even higher concentrations of anaesthetics will produce irreversible structural rearrangements in the membrane, with resultant loss of both the steady state potential and the input resistance of the cell under investigation. This concentration may correspond to the value at which the lytic action of these agents on the erythrocyte is elicited (see Seeman, 1972b, Fig. 3).

Intracellular recordings of the resting membrane potential of frog muscle fibres have shown that ethanol at a bath concentration of 0·2 M or higher produces significant depolarisation after about 15–20 min., although it is not clear how reversible this effect is (Knutsson, 1961). Typical data are shown in Fig. V.2. This depolarisation was accompanied by a reduction in input resistance of the muscle membrane to about 50 per cent of the control value. In Fig. V.3 are shown current-voltage records obtained from skeletal muscle fibres before (A) and during the administration of 1·0 M ethanol (B)–(E). Ethanol (1·0 M) clearly reduces the input resistance ($2V/I$) of this fibre, although it should be noted that in this example the input resistance was measured over a different voltage range in the control and test solutions, since the membrane potential fell from a control value of 88 mV to 66 mV in the ethanol solution. Since the depolarisation occurred in the presence but not in the absence of $Na^+$ ions in the bathing medium, it was ascribed to an increase in $Na^+$ permeability (Knutsson and Katz, 1967). These same authors also reported that ethanol did not affect the resting $K^+$ permeability of skeletal muscle fibres. Such a selective action of ethanol would not be expected if the agent simply caused a "loosening up" of membrane macromolecules. It is curious that in the $Na^+$-free Ringer solution, where choline was the replacement cation, depolarisation was not

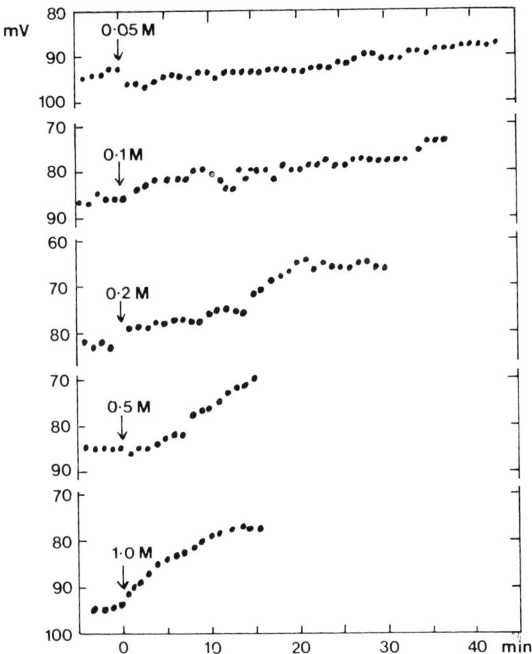


FIG. V.2. Continuous plots of resting membrane potential values of frog sartorius muscle fibres during exposure to different bath concentrations of ethanol (added at arrows). Different muscles were used for each drug concentration. (Knutsson, 1961)

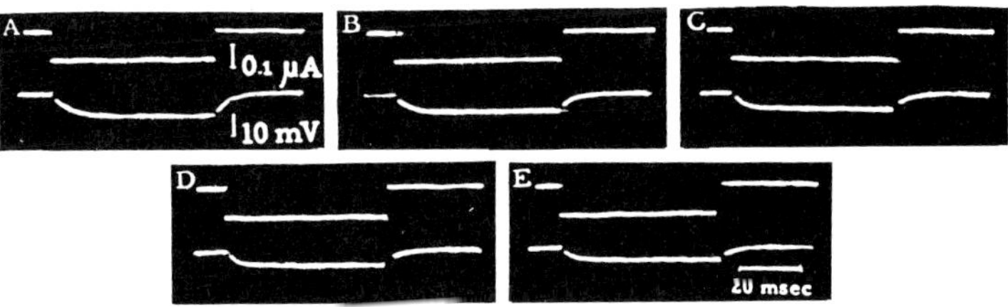


FIG. V.3. The effect of ethanol on the input resistance (2V/$I$) of frog sartorius muscle fibres. In all records the top trace represents current and the bottom trace voltage. Inward (hyperpolarising) current is in a downward direction. Vertical calibrations represent 0·1 μA and 10 mV. (A) immediately before; (B)–(E) 5, 10, 15 and 20 min, respectively, after the administration of ethanol at a bath concentration of 1·0 M. (Knutsson, 1961)

induced by ethanol, since normally the muscle membrane is known to be as permeable to choline as to $Na^+$ (Renkin, 1961).

Reference to the data of Inoue and Frank (1967), however, indicates that depolarisation or a reduction in input resistance are not necessarily the primary effects produced by a low concentration of ethanol. As is shown in Table V.2, 0·33–0·57 M ethanol produces a marked increase in both threshold depolarisation and threshold current, at a time when depolarisation is not seen. It is therefore not depolarisation block which underlies the reduction in excitability as suggested by Gallego (1948).

The alcohols can also produce membrane depolarisation in lobster axon (Houck, 1969) and in squid axon (Armstrong and Binstock, 1964). In the case of lobster axon, the depolarising aqueous concentration increases as the chain length increases. The short-chain alcohols ($C_2$–$C_5$) always produce depolarisation of squid axon, whereas octyl alcohol can produce hyperpolarisation. Without knowledge of the corresponding membrane concentrations of the alcohols it is fruitless to speculate on the molecular mechanisms underlying these effects, although notably a long-chain alcohol such as octyl alcohol is less effective in altering steady state conductance despite its higher membrane/buffer partition coefficient. For the short-chain alcohols, the depolarisation is 7–8 mV at concentrations which reduce the maximum $Na^+$ conductance activated during the action potential by 60–70 per cent, whereas octyl alcohol hyperpolarises by about 2 mV at concentrations which decrease the maximum $Na^+$ conductance by the same amount (Armstrong and Binstock, 1964). In squid axon the steady state potential is determined by the ratio of $g$K:$g$ leak (Armstrong and Binstock, 1964), and the depolarisation is accompanied by a reduction in this ratio, this being reflected in a reduction in resting membrane conductance. This study indicated that the major contributing factor is a marked reduction in the $g$K value (tenfold decrease) measured near the resting potential. A small effect of the alcohols on $g$ leak cannot be entirely discounted, however. One final observation worth mentioning is that in the lobster axon the resting membrane potential changes occur after the depression in action potential height (Houck, 1969), whereas work on squid axon has shown the reverse to be true (Armstrong and Binstock, 1964).

We can conclude that the depolarisation produced by the alcohols thus seems to be mediated via different permeability mechanisms in skeletal muscle and squid axon. The alcohols depolarise frog skeletal muscle fibres by increasing the resting $Na^+$ conductance but depolarise the squid axon by virtue of their depressant effect on the steady state $K^+$ conductance.

Lorente, de Nó (1947) suggested that the block in excitability produced by ether was due to its depolarising action. Intracellular work has since

TABLE V.2

Effects of ethyl alcohol on some membrane electrical properties of frog sartorius muscle fibres (Inoue and Frank, 1967)

| Ethanol concentration % (mol) | Observations[a] (No.) | Resting potential (mV)[b] | Threshold depolarization (mV)[b] | Threshold current ($\times 10^{-7}$ A)[b] | $\tau_m$ (msec)[b] | Effective resistance (KΩ)[b] | Overshoot potential (mV)[b] |
|---|---|---|---|---|---|---|---|
| (A) (5 muscles) | | | | | | | |
| 0% | 42 | 94 ± 1·2 | 45 ± 1·0 | 2·1 ± 0·1 | 14 ± 1·1 | 254 ± 15 | 27 ± 1·6 |
| 0·5% (0·08) | 26 | 93 ± 1·9 | 50 ± 0·6 | 2·2 ± 0·1 | 11 ± 0·7 | 230 ± 11 | 22 ± 1·8 |
| 2·0% (0·33) | 15 | 92 ± 2·5 | 56 ± 2·5 | 2·8 ± 0·2 | 9 ± 1·5 | 226 ± 14 | 10 ± 3·4 |
| (B) (6 muscles) | | | | | | | |
| 0% | 16 | 92·5 ± 2·1 | 38·8 ± 1·6 | 1·4 ± 0·1 | | | 29·5 ± 1·1 |
| 3·5% (0·57) | 12 | 88·0 ± 1·6 | 67·0 ± 3·0 | 2·3 ± 0·1 | | | 2·6 ± 3·6 |
| | 17 | 87·0 ± 1·2 | | | | | |

[a] All observations were obtained after the muscles had been exposed to the specified solution for at least 30 min.

[b] Mean ± standard error of mean.

TABLE V.3

Effect of ether on resting potentials, on thresholds and on time constants ($\tau_m$) of frog sartorius muscle fibres (Inoue and Frank, 1965)

| Series | Ether concn. (%) | Resting potential (mV) | Threshold depolarisation (mV) | Threshold current ($\times 10^{-7}$ A) | $\tau_m$ (msec) | Fibres ($n$) | Muscles ($n$) |
|---|---|---|---|---|---|---|---|
| (A) | 0 | 87 ± 1·1 | 38 ± 0·9 | 1·9 ± 0·14 | 14·1 ± 1·3 | 48 | 7 |
| | 0·2 | 83 ± 2·1 | 39 ± 0·2 | 1·6 ± 0·56 | 15·1 ± 1·2 | 27 | 6 |
| | 0·5 | 82 ± 1·6 | 39 ± 0·6 | 1·5 ± 0·10 | 17·6 ± 1·3 | 21 | 7 |
| | 2·0 | 81 ± 1·2 | 49[a] | 2·5[a] | 18·9 ± 1·7 | 22 | 6 |
| (B) | 0 | 91·5 ± 2·3 | 42·1 ± 0·7 | 1·9 ± 1·3 | 13·4 ± 1·9 | 14 | 3 |
| | 1·0 | 86·6 ± 1·9 | 54·5 ± 3·3 | 2·1 ± 0·25 | 15·5 ± 1·9 | 12 | 3 |
| | 3·0 | 82·3 ± 1·0 | — | — | 16·5 ± 1·5 | 15 | 3 |

[a] Three fibres in two muscles.

shown that in skeletal muscle, at least, application of 2–3 per cent ether produced a maximum depolarisation of about 6–9 mV (Inoue and Frank, 1965). These data are presented in Table V.3. In this study ether increased the input resistance of the muscle fibre membrane, so that in some fibres the threshold current decreased slightly although the threshold depolarisation remained constant (see Table V.3). The action of ether on the current-voltage relationship is shown in Fig. V.4. This increase in input resistance (in the hyperpolarising voltage range) may be responsible for the slight increase in time constant observed with ether application (Table V.3).

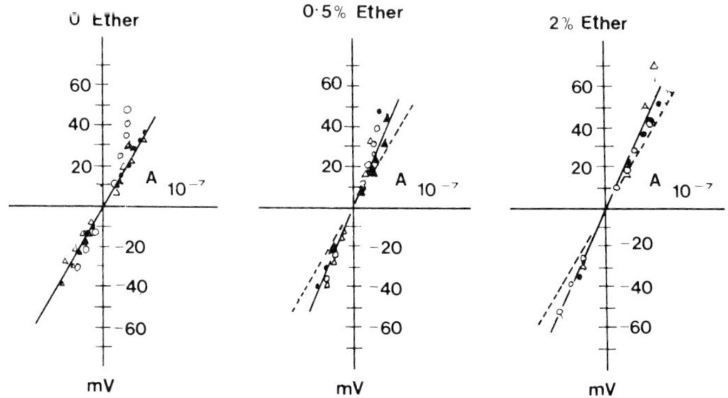


FIG. V.4. The action of ether on the current-voltage relation of frog sartorius muscle fibres. The different symbols on each curve represent data from individual muscle fibres from the same muscle. The solid linear regression lines were drawn to fit the data in 0 ether, 0·5 per cent ether and 2 per cent ether. The broken lines represent the control regression line (0 ether) superimposed to show the effects of ether treatment. The current-voltage relationship in the hyperpolarising direction is shown in the bottom left quadrant. (Inoue and Frank, 1965)

Thesleff (1956) also reported that a variety of anaesthetic agents produced little or no depolarisation of skeletal muscle fibres. Only sodium pentobarbitone (3 × $10^{-4}$ M) produced significant depolarisation, which was accompanied by about a 20 per cent increase in input resistance. The basis of the depolarisation was not investigated, however. Chloralose and paraldehyde also increased the input resistance by about 20 per cent, whereas urethane and tribromethanol decreased the input resistance by about 15 per cent. Once again no investigation of the ionic basis of the resistance change was undertaken.

In summing up, the transmembrane potential and input resistance of excitable cells are comparatively resistant (c.f. data on lipid bilayers) to anaesthetic agents. Indeed, in one study by Gage *et al.* (1975) it was observed that the input resistance values (and presumably transmembrane potentials) of amphibian skeletal muscle fibres exposed to 0·5 M ethanol were still within the normal range. It is unfortunate, nevertheless, that more attention has not been paid to the basis of the small effects that have been observed. It would be interesting to know whether the action of an agent on the transmembrane potential or steady state conductance could be correlated with, for example, its hydrophobicity. The short-chain alcohols which depolarise primarily have a lower membrane/buffer partition coefficient than the longer-chain alcohols which appear to depolarise less effectively. Also, can the depolarisation be opposed by varying the $Ca^{2+}$ concentration? If a $Na^{+}$ permeability increase is suspected, it would be of interest to know whether the channel opened were sensitive to tetrodotoxin or not. In addition, because of the known interaction between ethanol and temperature on the conductances activated during the action potential (see below) it would seem profitable to characterise the temperature dependence of observed effects, and certainly temperature should be considered more systematically in future work.

## 2.2 MEMBRANE CAPACITANCE

Whereas ether increases the time constant of the skeletal muscle membrane (Table V.3), ethanol reduces this parameter (Table V.2). These effects suggest that in skeletal muscle fibres these agents are altering the input capacitance, the input resistance, or both of these factors. Little systematic work has been done on the action of anaesthetic agents on the membrane capacitance of excitable cells, yet small changes might occur if the membrane geometry, in particular its thickness, or the membrane dielectric were altered. The change in membrane dielectric constant produced by the alcohols is clearly too small to produce any detectable change in the capacitance of either the muscle membrane (Gage *et al.*, 1975) or the squid axon membrane (Armstrong and Binstock, 1964). Presumably very high membrane concentrations of agents with very high dielectric constants could cause changes in the membrane dielectric constant and hence capacitance. Depending on how the agent affects the membrane thickness, we would expect the membrane capacitance should be increased. Measurements of the electrical capacitance of lipid bilayers suggest however that the *n*-alkanes and benzyl alcohol actually reduce the hydrocarbon region capacitance by increasing the bilayer thickness (Haydon *et al.*, 1977; Ashcroft *et al.*, 1977).

## 2.3 THE ACTION POTENTIAL

The depressant effect of anaesthetic agents on excitability is well documented. No attempt will be made to summarise all of the available literature, and attention will only be given to the principal effects produced.

Action potential block is brought about by the inert gases (Carpenter, 1954), the local anaesthetics and derivatives (Weidmann, 1955; Shanes *et al.*, 1959; Narahashi *et al.*, 1969b; Kiss and Vadasz, 1973; Strichartz, 1973), the general anaesthetics (Thesleff, 1956; Inoue and Frank, 1965), the barbiturates (Blaustein, 1968a) and the alcohols (Armstrong and Binstock, 1964; Inoue and Frank, 1967; Houck, 1969).

The concentration of the agent required to produce conduction block in the preparation tested will depend on several factors. Thus the cable characteristics of the cell, in particular its space constant which will be related to its size, will determine its susceptibility to block. In the case of nerve fibres, it might be anticipated, therefore, that smaller diameter fibres would be blocked preferentially, due to their lower safety factor for action potential propagation. Indeed, this has been shown to be the case for a variety of anaesthetic molecules (Nathan and Sears, 1961; Staiman and Seeman, 1974). Further, should we anticipate that the blocking concentration will be dependent on the density of $Na^+$ channels on the fibre membrane? We should note, therefore, that present estimates of the $Na^+$ channel density vary from about 500 $\mu m^{-2}$ for the squid giant axon (Keynes and Rojas, 1974) to only 6 $\mu m^{-2}$ for the 0·2 $\mu$m diameter axons of the garfish olfactory nerve (Ritchie, 1975).

The change in membrane resistance accompanying the action potential of the frog muscle fibre is also reduced by general anaesthetics such as urethane, ether or chloroform (Yamaguchi and Okumura, 1963). This finding indicated that these agents reduce the changes in ionic permeability underlying the action potential. The change in action potential shape produced by anaesthetics also implied that these compounds were affecting both $Na^+$ and $K^+$ permeability (Thesleff, 1956; Shanes *et al.*, 1959). Figure V.5 shows the progressive effect on the muscle action potential of increasing the concentration of sodium pentobarbitone in the Ringers solution bathing sartorius muscle. It is evident that this agent produces a reduction in the rate of rise and the overshoot of the action potential, and also slows down the rate of repolarisation, so that the action potential is considerably prolonged.

It is possible to restore the rate of rise of the action potential depressed by ethanol if the $Na^+$ concentration of the bathing medium is increased (Inoue and Frank, 1967). The results of such an experiment are shown in Fig. V.6. Thus, if the electrochemical gradient for the $Na^+$ ion is

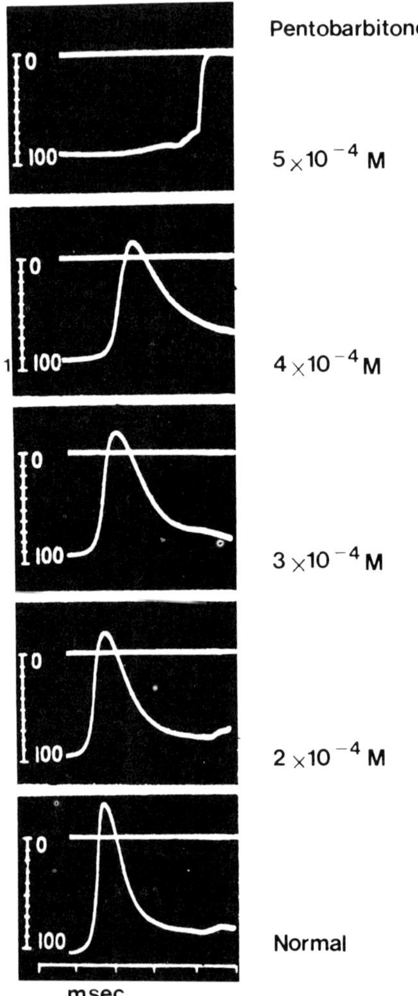


FIG. V.5. The effect of increasing concentrations of sodium pentobarbitone on the action potential of frog sartorius muscle fibres. The ordinate represents the voltage in mV. (Thesleff, 1956)

increased, the block of the $Na^+$ permeability induced by ethanol may be restored partially.

Shanes *et al.* (1959) reported that in squid axon, cocaine and procaine reduced the rate of repolarisation more than depolarisation and increased the size and duration of the after-potential. They suggested that one possible basis for the effects on the after-potential might be that the $Na^+$ permeability mechanism was switched off more rapidly and that this inactivation was then maintained for a longer period. In this respect, it is significant that Yamaguchi and Okumura (1963) have observed that the refractory period of the muscle fibre membrane is increased up to sevenfold in the presence of an anaesthetic such as chloroform.

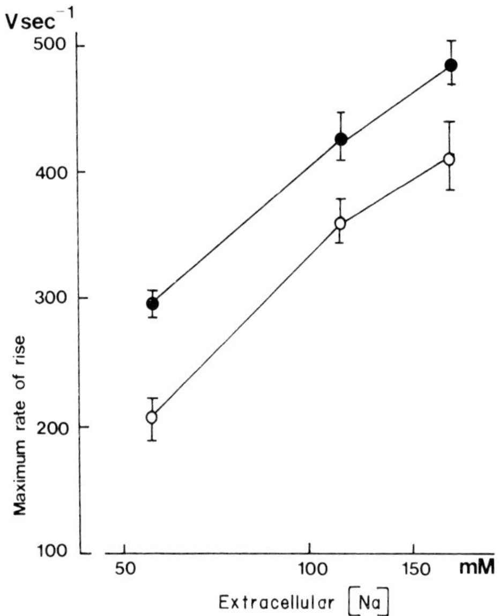


FIG. V.6. The relationship between the maximum rate of rise of the action potential of frog sartorius muscle fibres and the extracellular sodium concentration in the absence (●) or presence (○) of alcohol (1 per cent). Points represent the means, and the bar height represents 1 × SE. The osmolarity of the $Na^+$-deficient Ringer was adjusted with sucrose. (Inoue and Frank, 1967)

As can be seen from Table V.2, the block produced by the alcohols is preceded by a rise in both threshold current and threshold depolarisation.

Similar data exist for the general anaesthetic agents (Thesleff, 1956), but in the case of ether the threshold current is often reduced although the threshold depolarisation remains the same (Inoue and Frank, 1965). This can be accounted for by the known effect of ether on membrane resistance (see Fig. V.4).

Additional findings with the alcohols are worth consideration. Some of the alcohols ($C_2$–$C_4$) can produce repetitive firing in lobster axons (Houck, 1969), and this effect often accompanies delayed repolarisation of the membrane. This effect has not been systematically investigated, but the delayed repolarisation can be mimicked by varying the leakage conductance in computer-simulated action potentials (Dalton and Fitzhugh, 1960).

Another finding that appears to have been overlooked by many authors is the opposing actions of low temperature and ethanol. Thus, Spyropoulos (1957) demonstrated that the reduction in action potential height produced by ethanol at 22°C could be counteracted by lowering the temperature to 4°C. This effect is shown in Fig. V.7. The action potential height is reduced

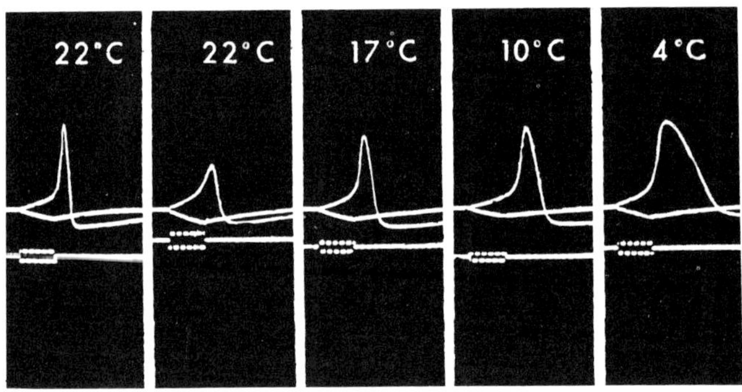


Fig. V.7. The effect of a reduction in temperature and 3 per cent ethanol on the squid axon action potential. Time marking 5 KHz. (Spyropoulos, 1957)

markedly in the presence of 3 per cent ethanol at 22°C. However, reducing the temperature to 17°C partially opposes this action and at 4°C the height (but not the shape) is restored. Anaesthetics may fluidise the lipid environment of the ionic channel (see previous chapter), whereas decreasing temperature would reduce fluidity, thus producing the observed interaction. It is evident that the effect produced by ethanol, and perhaps many anaesthetic agents, depends on temperature, and it is unfortunate that no thorough study of the temperature–anaesthetic interaction has been performed.

## 2.4 VOLTAGE-CLAMP STUDIES

Voltage-clamp experiments with select excitable membrane preparations have demonstrated how anaesthetics block excitability. The results described here are derived almost entirely from experiments on single axons of lobster or squid, or from studies on single myelinated fibres of amphibia. This work has been carried out at different temperatures so that although qualitative comparison between the data is feasible, quantitative differences in results could be ascribed to lack of consideration of the temperature at which the experiments were conducted.

One of the early studies on squid axon by Shanes *et al.* (1959) demonstrated that 0·1 per cent procaine and cocaine reduced the peak inward ($Na^+$) current ($I_{Na}$), and the maximum steady state outward ($K^+$) current ($I_K$) and also reduced the maximum rate of rise of these currents. These experiments were carried out at 25°C, at which temperature these agents affected the $Na^+$ and $K^+$ conductances equally. Subsequent work has substantiated the depressant effects of local anaesthetics on both the $Na^+$ and the $K^+$ current (Taylor, 1959; Hille, 1966; Narahashi *et al.*, 1967; Århem and Frankenhaeuser, 1974) and has shown that the alcohols (Armstrong and Binstock, 1964) and also the barbiturates (Blaustein, 1968a) act in a similar fashion. The typical effect of external application of procaine is shown in Fig. V.8. The voltage-clamp currents were obtained from a squid axon preparation, the action of procaine being examined 10–20 min after application. The most striking effect is the marked reduction in the amplitude of both the early transient peak current and the steady state current over the entire voltage range studied (D). Recovery was not complete after about 20 min (A).

It is necessary to be cautious when comparing results from different workers, not only because the experiments may have been conducted at different temperatures, but also it is probable that the depression of conductance is voltage-dependent. Hence it is imperative that we consider the voltage range over which the depression occurs. Nevertheless, it has been stated that in squid axon a low concentration of procaine (1 mM) produces a similar (50 per cent) reduction in both the early transient and late steady state conductance if the drug is applied internally (Narahashi *et al.*, 1967), whereas often external application of a higher concentration of procaine (e.g. 3·7 mM) can produce a large reduction in the early transient conductance compared with the steady state conductance (Taylor, 1959).

Generally speaking, the maximum potassium conductance ($\bar{g}K$) seems to be less susceptible to anaesthetic action than the peak sodium conductance ($\bar{g}Na$). In fact, Hille (1966) asserts that xylocaine possesses a similar specificity to that of tetrodotoxin (TTX). Certainly there are clear

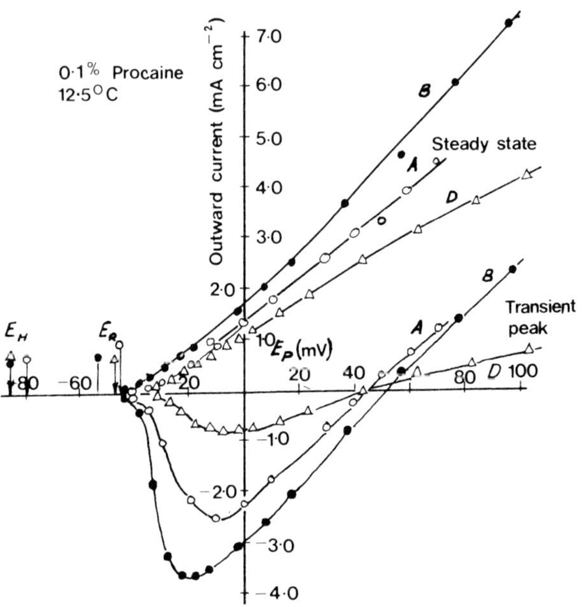


FIG. V.8. Squid axon membrane current in a typical voltage-clamp experiment. The lower set of curves shows the current at the time of the transient peak, i.e. $I_{Na}$; the upper set of curves shows the late steady state current, i.e. $I_K$. Voltage-clamp step was from $E_H$ to $E_P$. Resting membrane potential values indicated by $E_R$. (B) before, (D) during and (A) after the external application of 0.1 per cent procaine at pH 7·9. Temperature was 12·5°C. (Taylor, 1959)

differences in the concentrations of a local anaesthetic required to block the Na compared with the $K^+$ conductance of amphibian myelinated fibres (Århem and Frankenhaeuser, 1974). This is shown in Fig. V.9.

As is evident, the dose-response curve is S-shaped, and could be described by an equation of the form

$$\frac{P_0}{P_x} = \frac{k}{k + [X]} \tag{1}$$

where $[X]$ is the concentration of anaesthetic; $k$ is the concentration of anaesthetic at which the permeability is half normal, $P_0$ is the control permeability, and $P_x$ is the permeability at concentration of anaesthetic $X$. The $k$ values for the $Na^+$ system are evidently between about 1 and 10 per cent of the $k$ values of the $K^+$ system for the agents tested.

## 2.5 THE KINETICS OF THE $Na^+$ CONDUCTANCE SYSTEM ($g$Na)

Most experimental work has shown that a variety of agents including the

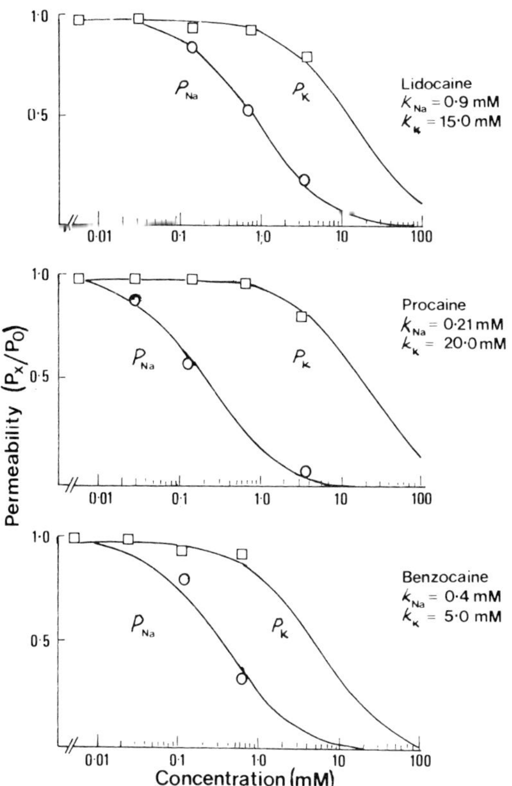


FIG. V.9. The effect of three local anaesthetics (externally applied) on the maximum peak $P_{Na}$ and steady state $P_K$ of an amphibian myelinated fibre. The smooth curves are solutions of equation 1 in the text for the $K^+$ values indicated. Temperature was 15°C. (Århem and Frankenhaeuser, 1974)

alcohols (Armstrong and Binstock, 1964), external procaine (Taylor, 1959), internal procaine (Narahashi *et al.*, 1967), and the barbiturates (Blaustein, 1968a) produce a depolarising shift in the relationship between $Na^+$ conductance and membrane potential. The shift in the normalised $Na^+$ conductance curve may be only of the order of 3–6 mV, however (e.g. Taylor, 1959).

During the course of blockage of the $Na^+$ channel by TTX no change in the time to peak of the early transient ($Na^+$) current was observed (see Narahashi *et al.*, 1967). However, in the squid axon, procaine increased the time to peak of the early transient ($Na^+$) current if applied externally (Taylor, 1959), or internally (Narahashi *et al.*, 1967). Also, the time to peak transient current is increased by internal or external application of barbiturates (Narahashi *et al.*, 1969a), or external application of the alcohols (Armstrong and Binstock, 1964, see their Fig. 4). Significantly, in squid axon external or internal dibucaine did not alter the time to peak transient current (Narahashi *et al.*, 1969a), and in amphibian myelinated fibres external application of procaine, lidocaine, benzocaine or xylocaine did not alter the kinetics of the $Na^+$ activation (Hille, 1966; Århem and Frankenhaeuser, 1974). Depending on the system, then, and the anaesthetic agent, the mechanism underlying the block produced by the compounds may be quite different from that of TTX. However, the full significance of the differential effects of these agents on the kinetics of channel opening remain to be investigated.

The published effects of anaesthetics on the $Na^+$ inactivation processes are at first sight confusing. Weidmann (1955) investigated the action of cocaine on the rate of rise of the action potential recorded in single cardiac Purkinje fibres. This agent was found to shift the relationship between the maximum rate of rise of the action potential and the clamp potential in a hyperpolarising direction. This is shown in Fig. V.10. The S-shaped relationship between the rate of rise and potential reflects the progressive inactivation of the $Na^+$ permeability mechanism with depolarisation. The shift in the relationship with cocaine was attributed to the drug's effect on the $Na^+$ inactivation process i.e. the $Na^+$ inactivation curve was shifted in the hyperpolarising direction.

Later work with nerve fibres however ruled out an effect of anaesthetics on the steady state $Na^+$ inactivation. No change was reported for the local anaesthetics (Taylor, 1959; Hille, 1966; Arhem and Frankenhaeuser, 1974), the alcohols (Armstrong and Binstock, 1964), or the barbiturates (Blaustein, 1968a). In Weidmann's study the preparation was subjected to the action of cocaine for the comparatively long time of an hour, and more recently it has been demonstrated that "repetitive use" of the $Na^+$ channel promotes a local anaesthetic induced change of the $Na^+$ inactivation process (Courtney, 1975; Hille, 1977), and Khodorov *et al.*, (1976) have shown that the depression of $Na^+$ currents by repetitive stimulation in the presence of procaine and trimecaine is due to a slow inactivation process. There is now strong evidence that many of the local anaesthetics or analogues can alter the steady state $Na^+$ inactivation process tested under particular conditions (see Hille, 1977) but it is not clear whether

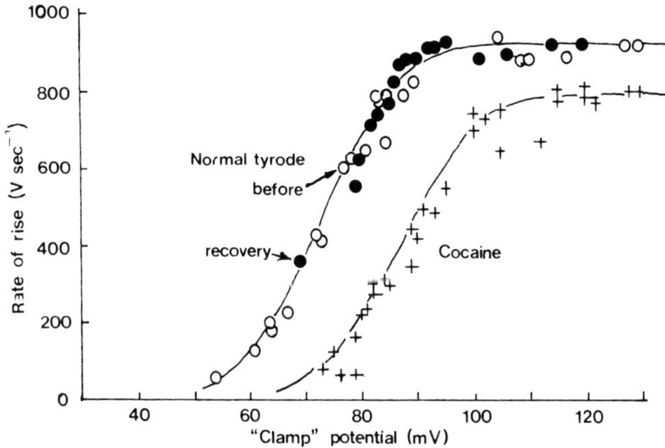


FIG. V.10. The effect of cocaine (aqueous concentration 0·05 mM) on the relationship between membrane "clamp" potential and the maximum rate of rise of the action potential of a mammalian Purkinje fibre. The smooth curves are drawn according to the equation

$$(h) \text{ steady state} = \frac{1}{1 + \exp (V_h - V)/7}$$

where $h$ is the inactivation variable, $V$ is in mV and $V_h$ is the value of $V$ at which $h = \frac{1}{2}$ in the steady state. (Weidmann, 1955)

"repetitive use" of the $Na^+$ channel can promote changes in $Na^+$ inactivation in the presence of other anaesthetics.

## 2.6 THE KINETICS OF THE $K^+$ CONDUCTANCE SYSTEM ($g$K)

As shown by Fig. V.8, procaine also slows slightly the onset of the steady state current. Taylor (1959) reported that the reduction in steady state $g$K produced by procaine was voltage-independent. Thus the steady state $g$K associated either with a step depolarisation to 0 or +60 mV was depressed to the same extent by 0·1 per cent procaine. The effects of procaine on the kinetics of the $K^+$ conductance were however voltage-dependent, the time constant of the $K^+$ current associated with a clamp pulse to 0 mV being increased by 70 per cent whereas the time constant of the $K^+$ current following a clamp pulse to +60 mV was increased by only 40 per cent.

A significant feature of the action of the local anaesthetic dibucaine was the appearance of a hump on the steady state $K^+$ current (Narahashi *et al.*,

1969a). This effect is shown in Fig. V.11 External (a) or internal (b) application of dibucaine produces a blockage of both the inward and outward currents, but in addition the late current develops a characteristic hump from which outward current then declines to a steady state value. This inactivation of the $K^+$ current has also been produced in the squid axon by the internal application of the tertiary form of a tropine compound (Narahashi *et al.*, 1969a), and lobster axon by the external application of both the tertiary and quaternary tropines (Blaustein, 1968b). Such voltage-dependent inactivation resembles the effect produced by certain quaternary ammonium ions (Armstrong, 1975).

The alcohols produce variable effects on the steady state *g*K, which depends on their concentration and chain length (Armstrong and Binstock, 1964). Low concentrations of octyl alcohol which block the $Na^+$ current have little effect on *g*K, whereas higher blocking concentrations (for the $Na^+$ current) block *g*K as do the other alcohols ($C_2$–$C_5$), although the block only occurs at small depolarisations. Blockage of *g*K over the entire voltage range requires higher concentrations of the alcohols, which then produce some 10 mV or more depolarisation.

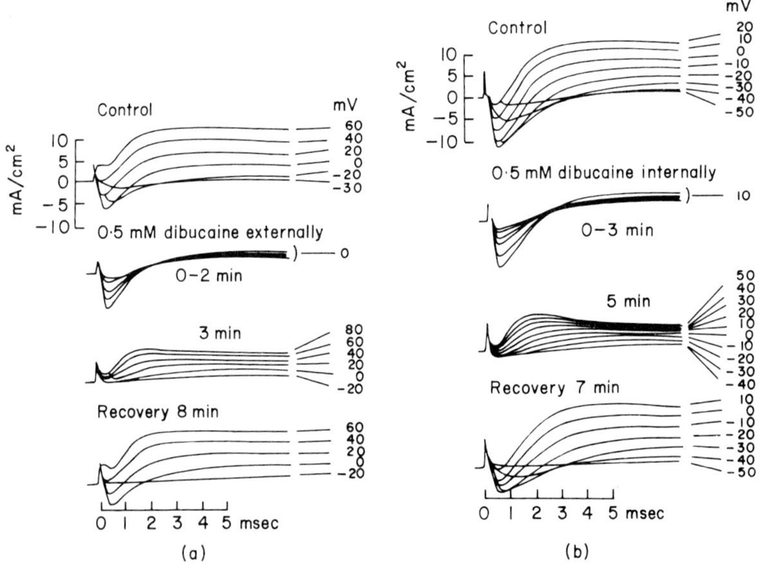


FIG. V.11. Squid axon membrane currents associated with step depolarisations before and during the application of 0·5 mM dibucaine applied externally (a) or internally (b), and after washing with normal media. The second series of records in each experiment shows changes in current associated with a clamp pulse to 0 or −10 mV during the course of the blockage. (Narahashi *et al.*, 1969a)

### 2.7 LEAKAGE CHANNELS

Intuitively it might be anticipated that the anaesthetics would cause an increase in the leakage channel conductance, due to their general fluidising or disordering effect on membrane structure (Hubell and McConnell, 1968; Metcalfe, 1970). Available evidence, although slim, suggests that this is not the case. Neither the alcohols (Armstrong and Binstock, 1964) nor xylocaine (Hille, 1966) produce detectable changes in the leakage channel conductance. The ionic selectivity of this channel in the squid axon is not well characterised, although it appears to favour $K^+$ (Adelman and Taylor, 1961). Armstrong and Binstock (1964) reported that the alcohols do increase the $gK:g$ leak ratio, but attribute this primarily to an effect on $gK$. However, as they point out, an effect on $g$ leak cannot be entirely discounted. A systematic study of the action of anaesthetics on the leakage conductance may unmask, therefore, a small effect.

### 2.8 EFFECTS ON GATING (ASYMMETRY) CURRENTS

In 1952 Hodgkin and Huxley predicted the existence of a small current associated with the opening and closing of ionic channels. One of the most exciting developments over the last several years has been the experimental realisation of these currents, which are now commonly called gating or asymmetry currents (Meves, 1975).

Gating current occurs as the result of the movement of the mobile charged gating particles which respond to changes in the electric field in the membrane, and control (or gate) the $Na^+$ ion movement, i.e. open and close the channel. These are measured in squid axon by subjecting the membrane to alternate equally matched depolarising and hyperpolarising pulses. The resultant currents are added so that the equal capacitative and leakage currents cancel, and the remaining asymmetry current is thought to represent, in part, the $Na^+$ gating currents. If the clamp pulses are repeated (perhaps thirty-two times) and the current transients subjected to signal-averaging techniques, then the asymmetry current takes the form of an initial outward transient on depolarisation and a transient inward current on repolarisation. In order to detect this current, the ionic current must be removed by perfusing the squid axon inside and out with impermeant ions and blocking agents. Despite these precautions it is of course difficult to separate "real" gating current from current associated with the rearrangement of other non-gating membrane macromolecules. This is not surprising in view of the sparse density of ionic channels. One means of linking the asymmetry current to $Na^+$ ion movement is to examine the effects on this current of agents which are known to block the $Na^+$ permeability mechanism. To this end, Keynes and Rojas (1974) employed procaine.

The effect of 1·0 per cent procaine on asymmetry current is shown in Fig. V.12.

The principal effect was to reduce the total charge displacement to about 40 per cent of the control value, and to reduce the time constant of switching on and off of the charge displacement. It might be expected that procaine would have a dual effect on membrane processes. Thus it could speed up the orientation of dipoles in an electric field by virtue of a fluidising effect on the environment surrounding the dipoles or, alternatively, by increasing the dielectric constant of the environment it might slow down any reaction sequence involving molecules with dipole moments (see Chapter VI). The effect of procaine on the time constant of the charge displacement is certainly consistent with the fluidising action of this agent, and the reduction in charge displacement may underlie its action on the Na permeability mechanism. It is interesting that procaine increases the rate at which the gating process controlling the $Na^+$ permeability switches on, but of course, as discussed previously, slows down the time to the peak of the early transient ($Na^+$) current. The latter affect could reflect the

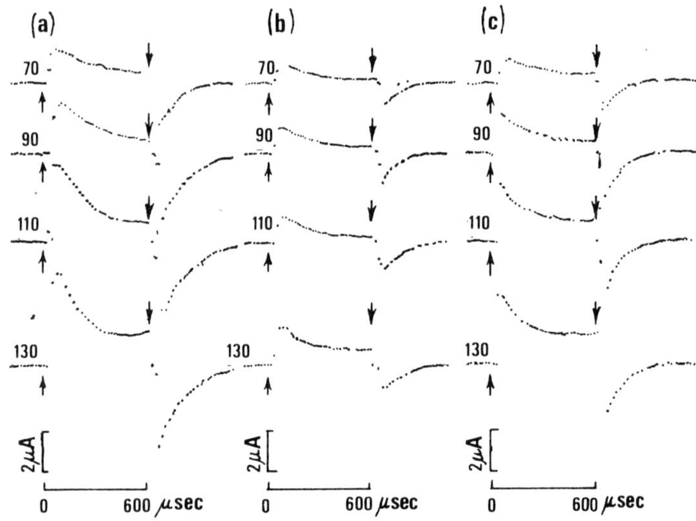


FIG. V.12. The effect of externally applied procaine (1 per cent) on charge displacement in a squid axon. The external solution was $Na^+$- and $K^+$-free artificial seawater containing 300 nM tetrodotoxin at pH 8·0. (a) control; (b) 10 min after the application of 1 per cent procaine; (c) 10 min after removal of procaine. The holding potential was −90 mV throughout and the size of the voltage-clamp pulses in mV is shown against each record. Vertical calibration bars 2 μA, horizontal bars 600 μsec. Temperature 5·3–5·5°C. (Keynes and Rojas, 1974)

depressant action of procaine on the magnitude of the gate current associated with a potential step.

Future investigations into the action of anaesthetic agents on asymmetry currents should prove fruitful. Even if the recorded current is not pure gate current, an analysis of the action of a range of agents (perhaps with different dielectric constants) might provide novel data about membrane macromolecules and their movement in an electric field.

## 3 Active form of the local anaesthetics and the barbiturates and possible site of action

It is appropriate at this stage to consider which is the active form of particular agents such as the local anaesthetics or the barbiturates, i.e. is it the charged or uncharged form of the molecule? Also, it is pertinent to ask whether these agents act on the inside or outside surfaces of the membrane. Further attention, however, will be given to the possible sites of action in the next section, when models of anaesthetic action will be discussed.

The arguments relating to the active form of the local anaesthetic agents have been discussed by Ritchie and Greengard (1966) and Narahashi (1971) and will only be dealt with briefly here. Local anaesthetic agents are a combination of secondary or tertiary amines and lipophilic hydrocarbon groups. Their general structure has been discussed in Chapter IV but the structure of some agents is shown again in Fig. V.13. At physiological pH, these agents exist in both the cationic and uncharged form. The ratio of the two forms obviously depends on the pH of the surrounding medium and also the pK of the agent.

Investigations undertaken in an attempt to define which form of the local anaesthetic agents is the active one have therefore examined the effects of manipulating both the intracellular and external pH values and measured the blocking potency of various agents over a prescribed pH range. Early work suggested that the blocking potency of local anaesthetics was greater at alkaline pH (e.g. Skou, 1954), implying that the uncharged form was the active one. It is now generally accepted, however, that the uncharged molecule gains access to its site of action more readily, since it crosses any barriers to diffusion, such as the external nerve sheath, whereas the cationic form penetrates less easily (Ritchie and Greengard, 1966). Under conditions where the penetration of the agent into its site of action can be ignored, the cationic form of certain agents may be more potent.

This has been shown convincingly in the simple experiment of Ritchie and Greengard (1961), using the long-acting agent dibucaine. The blocking effects of this anaesthetic remain after the preparation has been returned to an anaesthetic-free medium. Ritchie and Greengard (1961 pretreated desheathed nerve fibres with this agent, removed the anaesthetic from the bathing medium, and then proceeded to test the degree of block at different pH values. This experiment showed that at pH 7·2 the block remained, but at pH 9·6 the block was removed. The dramatic effect of varying pH is shown in Fig. V.14.

The simplest interpretation of these data is that varying the pH of the bathing medium changed the ratio of the charged to the uncharged form

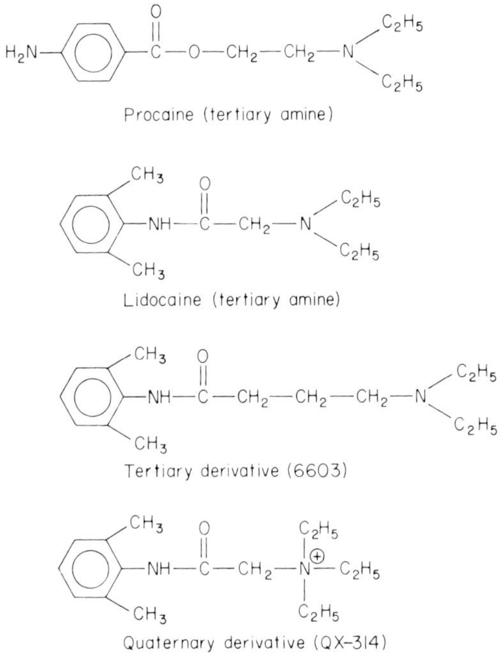

Procaine (tertiary amine)

Lidocaine (tertiary amine)

Tertiary derivative (6603)

Quaternary derivative (QX-314)

FIG. V.13. The structure of procaine, lidocaine and a tertiary and quaternary derivative of lidocaine. Note that the local anaesthetics procaine, lidocaine and the tertiary derivative of lidocaine are shown in their uncharged form. The quaternary derivative, however, only exists in the charged form.

of agent at the blocking site, and so altered the blocking potency of drug.

Some later work supported the notion that the cationic form of certain local anaesthetics or derivatives is the more potent (Dettbarn, 1962; Narahashi *et al.*, 1969b). The situation is not quite so simple as this, however. As Ritchie and Ritchie (1968) point out, benzocaine can be at least as potent as procaine, and this agent cannot exist in the cationic form. Also, procaine is more effective in producing block of desheathed vagus fibres at alkaline pH (Ritchie and Ritchie, 1968), and in the frog nodal membrane Århem and Frankenhaeuser (1974) reported that the blocking potency of both lidocaine and procaine was greater at alkaline pH. In the latter study it was concluded that the potency could not be accurately predicted if it was assumed that either the charged or the uncharged form alone was the active one.

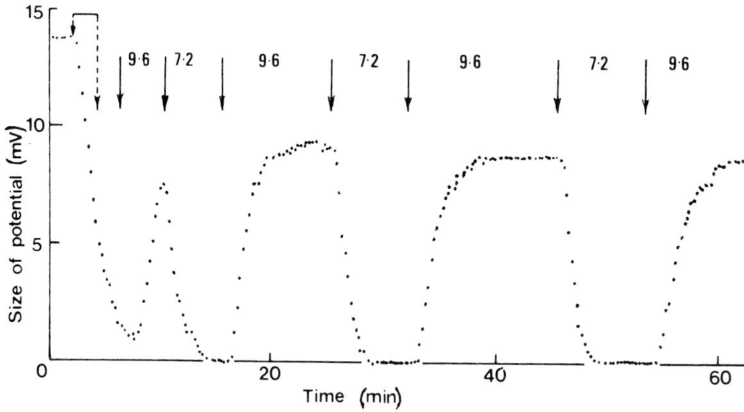


FIG. V.14. The effect of pH on the action potential of a rabbit cervical vagus nerve pretreated with dibucaine. The ordinate shows the height of the C elevation of the monophasic compound action potential. The perfusing solution contained 1 mM dibucaine between the broken arrows. From time 0 to the first solid arrow the pH of the perfusing solution was 7·2. Thereafter the pH of the perfusing solution was alternated, at the arrows, between 9·6 and 7·2. Temperature 28°C. (Ritchie and Greengard, 1961)

It seems unwise, therefore, to accept that the cationic head is *essential* for local anaesthetic activity. This activity probably depends on the aromatic hydrophobic part of the molecule, and in some cases, but not all, is augmented by the presence of a cationic head (Fig. V.13).

There is still disagreement about the active form of the barbiturates. Blaustein (1968a) reported that sodium pentobarbitone blocked the lobster axon action potential more effectively at pH 8·5 than at pH 6·7, and therefore concluded that the anionic form of the drug was more potent. In contrast, Krupp *et al.* (1969) observed that both sodium pentobarbitone and sodium phenobarbitone were more potent in blocking the action potential of desheathed frog nerve at an external pH of 6·8 than at a pH of 8·8. Similarly, Narahashi *et al.* (1971) reported that in squid axon sodium pentobarbitone was more effective at low pH. These studies therefore concluded that the unionised form of barbiturates was more potent. It is difficult to dispute the data and corroborative calculations of Narahashi *et al.* (1971). Knowing the pH of the phases in which the anaesthetic is dissolved, these authors calculated for several pH values the concentration of the anionic and uncharged form of the agent during nerve block. This is readily done using the Henderson–Hasselbach equation. Such calculations show that the concentration of the anionic form of the barbiturates at a nerve-blocking dose increases with increasing pH, as the concentration of the uncharged form decreases. Thus, if the active form were the uncharged species then the aqueous concentration of barbiturate required to block the nerve should increase as the pH rises or, alternatively, the blocking potency of the barbiturate should decrease as pH is raised. This is essentially what is seen experimentally with the perfused axon, suggesting that the uncharged form is the more potent.

Experiments with internally and externally perfused axons have suggested that although anaesthetics can exert effects when applied externally, these agents are at least as effective internally and probably act on the inside surface of the excitable membrane. This has been shown for sodium pentobarbitone (Narahashi *et al.*, 1971) and for the local anaesthetics and many derivatives (Narahashi *et al.*, 1967; Narahashi *et al.*, 1969b; Strichartz, 1973). Indeed, the permanently charged quaternary derivatives of local anaesthetics have little effect when applied externally, presumably because they cannot penetrate to the inside surface of the membrane. The blocking potency of the cationic form of a local anaesthetic applied externally is decreased by raising the internal pH, yet the potency of the anaesthetic is independent of external pH changes. This is compatible with the view that these agents block from the inside of the membrane (Narahashi *et al.*, 1969b).

Although this point will be dealt with again below, it appears that the site of action of local anaesthetics could lie deep within the $Na^+$ channel itself (Strichartz, 1973; Hille, 1977). It would be of real interest to know whether internal application of many other agents was as effective as external application.

## 4 Models of action

### 4.1 ROLE OF MEMBRANE-BOUND $Ca^{2+}$

The possible role of calcium in cellular narcosis is mentioned in Chapter

II. There are obvious similarities between the effects produced in nerve axons by a $Ca^{2+}$-rich bathing medium and the actions of certain anaesthetic agents. Thus local anaesthetics and a high $Ca^{2+}$ medium produce similar changes in the threshold behaviour of squid nerve fibres, which can be ascribed to a reduced sensitivity to depolarisation of the $Na^+$ conductance mechanism (Shanes *et al.*, 1959). As stated previously, many anaesthetics produce a small shift in the depolarising direction of the curve relating $Na^+$ conductance to potential, and this effect is observed classically with $Ca^{2+}$-rich media (Frankenhaeuser and Hodgkin, 1957) or polyvalent cations (Blaustein and Goldman, 1968). In addition, the time to peak transient current is increased in the presence of polyvalent cations and $Ca^{2+}$ (Blaustein and Goldman, 1968). Once again similar effects have been reported for many anaesthetic agents (see above).

On the other hand, local anaesthetics and $Ca^{2+}$-rich medium may produce opposite electrical effects. For example, a rise in external $Ca^{2+}$ concentration increases the rate of rise of the action potential (Weidmann, 1955; Shanes *et al.*, 1959).

These findings coupled with the fact that cationic anaesthetics and $Ca^{2+}$ compete for binding sites on phospholipids (Feinstein, 1964; Blaustein and Goldman, 1966a) have led to the suggestion that membrane-bound $Ca^{2+}$ may be important in anaesthetic action. An interaction between $Ca^{2+}$ and an anaesthetic was suggested by the work of Simon and Szelöczey (1928), who reported an augmented release of $Ca^{2+}$ from rabbit sciatic nerve treated with cocaine. More recent work has shown that in frog spinal ganglion cells the reduction in the amplitude of the action potential produced by procaine and xylocaine was opposed by increasing the external $Ca^{2+}$ (Aceves and Machne, 1963). This finding has been extended by Seeman *et al.* (1974) who demonstrated that, in rat phrenic nerve, the blockade produced by cationic, anionic or uncharged anaesthetic agents is also opposed by raising the external $Ca^{2+}$ concentration. This finding is significant since it means that the membrane-bound level of $Ca^{2+}$ probably does not dictate the degree of block produced, for although the cationic agents decrease the amount of $Ca^{2+}$ binding to phospholipids, the anionic agents increase the amount, and the uncharged have presumably little effect (Blaustein and Goldman, 1966a; Blaustein, 1967).

Blaustein and Goldman (1966b) suggested that the actions of $Ca^{2+}$ and procaine were competitive at an excitable membrane level, since the actual decrease in peak $g$Na produced by procaine was dependent on the external $Ca^{2+}$ concentration. This is shown in Fig. V.15. The curve relating sodium conductance to potential is shifted to the right by 4 mM procaine (●), but the reduction in peak $g$Na is less in the presence of 50 mM $Ca^{2+}$(●) compared with 25 mM $Ca^{2+}$ (□). Such data have been obtained in lobster

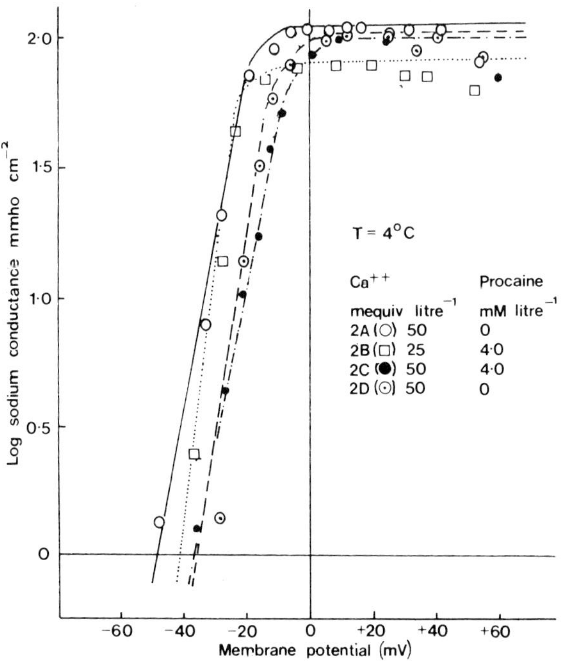


FIG. V.15. The relationship between the logarithm of the peak sodium conductance (obtained from voltage-clamp pulses) and the membrane potential for a lobster axon. The curves were constructed in seawater of normal calcium concentration (50 mM), before (○) and after (⊙) exposure to procaine. The axon was exposed to seawater containing 4·0 mM procaine and either 25 mM (□) or 50 mM (●) calcium. (Blaustein and Goldman, 1966b)

axon, but in myelinated fibres raising the external $Ca^{2+}$ level does not oppose the reductions in peak $P_{Na}$ produced by procaine (Århem and Frankenhaeuser, 1974). The notion of a direct competition between $Ca^{2+}$ and anaesthetic becomes even less appealing in view of the finding that a $Ca^{2+}$-rich medium can partially reverse the nerve blockade produced by a variety of agents which enhance $Ca^{2+}$ binding or do not affect it (Seeman *et al.*, 1974). Since these authors demonstrated that the nerve block produced by TTX is slightly offset by raising the external $Ca^{2+}$ concentration, it is more likely that the divalent cation is affecting the $Na^+$ conductance system more generally, e.g. by removing the $Na^+$ inactivation via its effect on the transmembrane electric field profile.

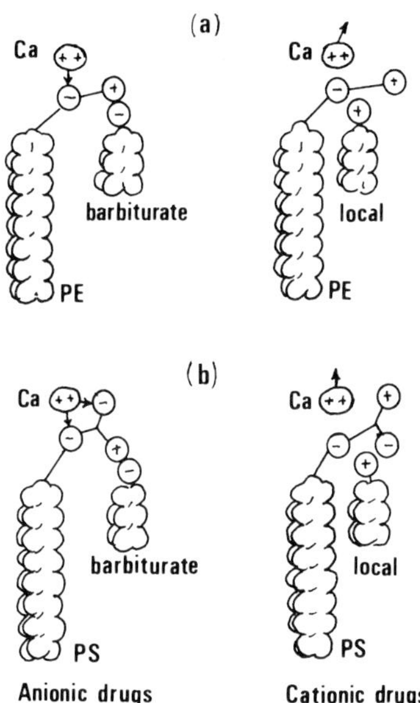


FIG. V.16. Molecular models of interactions between the divalent cation $Ca^{2+}$, anionic (barbiturate) and cationic (local) drugs and phospholipid polar heads. In (a) is shown the interaction with phosphatidyl ethanolamine (PE), in (b) the interaction with phosphatidyl serine (PS). (Blaustein and Goldman, 1966a)

Blaustein and Goldman (1966a) have proposed a simple but useful model of the interaction of charged anaesthetics with the phospholipid components of the membrane. This is shown in Fig. V.16. In this model, cationic agents such as procaine titrate out the negative charges on the phospholipid heads, thus making less binding sites available for $Ca^{2+}$, which is therefore displaced. The anionic anaesthetics such as the barbiturates could by virtue of their negative charge neutralise the cationic nitrogen of the phospholipid, thus somehow increasing the negative sites available for $Ca^{2+}$ binding. In both cases the hydrophobic portion of the anaesthetic agent is associated with the apolar hydrocarbon chain of the phospholipid. The suggested effect of this interaction is to reduce the freedom, or mobility, of the polar heads, which may otherwise move under the influence of a changing electric field. If, as suggested by Goldman (1964), movement of these heads is one of the molecular rearrangements associated with the activation of the $Na^+$ and $K^+$ conductance mechanisms, then it might be anticipated that the restricted mobility imposed by the presence of the anaesthetic agent would therefore slow down the kinetics of the conductance changes, e.g. increase the time to peak transient current as observed experimentally. Binding of polyvalent cations to such polar groups could also reduce their ability to orient and therefore alter the kinetics correspondingly.

The reduction in peak *g*Na produced by anaesthetics may be related to the fact that the $Na^+$ channel is held in the open configuration by a rigid lipid environment. Disordering or fluidisation of the lipid surround by the penetration of the anaesthetic might therefore reduce the size of some $Na^+$ channels, and consequently reduce peak conductance.

From the model it is possible to make other predictions. Frankenhaeuser and Hodgkin (1957) proposed that $Ca^{2+}$ removal from a membrane binding site may be one of the first steps in membrane depolarisation. One such binding site may be the polar head of a phospholipid molecule. Anionic anaesthetics increase the amount of $Ca^{2+}$ bound to phospholipid bilayers (Blaustein, 1967), and as a consequence the initial stage of depolarisation in nerve would perhaps require more $Ca^{2+}$ removal from the membrane. Nerve block by anionic anaesthetics might, therefore, be accomplished more readily, if indeed $Ca^{2+}$ removal from the membrane is a necessary part of the excitation process. If the partial molar volume of the anaesthetic in the membrane is deemed a suitable index of potency see Table 1.2), then in accordance with this prediction this value is less for the anionic agents (Roth and Seeman, 1972).

The model also predicts that uncharged agents should not necessarily alter the kinetics of channel opening. This would be interesting to test systematically. The alcohols probably do have a small effect on the time to peak transient current (Armstrong and Binstock, 1964, Fig. 4). In this sense the charged and uncharged parts of certain agents, e.g. procaine, probably both influence potency, the uncharged portion of the agent therefore affecting the peak conductance alone (by a fluidising effect), the charged part influencing the kinetics. As stated previously, the anaesthetic potency of procaine lies in the aromatic hydrophobic portion of the molecule, but is augmented if the agent is charged.

The proposed model of interaction fails to explain why dibucaine does not affect the $Na^+$ conductance kinetics in squid axon (Narahashi *et al.*, 1969a) and the lack of effect of local anaesthetics on the $Na^+$ conductance kinetics in myelinated nerve (Hille, 1966; Århem and Frankenhaeuser, 1974). In addition, one final point of interest is whether the model can describe in molecular terms the basis of the shifts in the relationship between $Na^+$ conductance and voltage with anaesthetic agents. Raising external $Ca^{2+}$ is thought to reduce the external negative surface charge density, producing an effective increase in the transmembrane electric field, which is then "sensed" by the voltage-sensitive $Na^+$ gates. The resultant effect is a shift in the Na conductance curve in the depolarising direction. Do cationic anaesthetics also reduce the negative surface charge density despite displacing $Ca^{2+}$? What happens when an anionic agent increases $Ca^{2+}$ binding—is the depolarising shift greater? Further, it may be asked whether uncharged anaesthetics shift the $Na^+$ conductance curve. It would undoubtedly be of immense interest to compare the actions of the general anaesthetics or inert gases with the effects described here using the charged anaesthetics. It is unfortunate that more experiments of this nature have not been performed. There is clearly much scope for additional work in this field and many challenging questions remain unanswered.

## 4.2 PRESSURE–ANAESTHETIC INTERACTIONS

It is implicit in the membrane expansion or critical volume hypothesis of anaesthetic action (Mullins, 1954; Seeman, 1972b) that raising the hydrostatic pressure could counteract the effect produced by an anaesthetic. That is the alleged volume expansion of biological (excitable) membranes induced by the anaesthetic would be opposed by high hydrostatic pressure.

Pressure alteration of an anaesthetic effect could, however, be ascribed to a variety of factors. Thus, pressure reversal of an anaesthetic effect could be simply due to removal of the anaesthetic from the membrane phase. In this case, pressure would be altering the membrane partitioning of the agent. Anaesthetic and pressure effects might be produced because these experimental variables alter the physical characteristics of the membrane, e.g. by changing the membrane fluidity. If anaesthetics fluidised nerve membranes and pressure reduced fluidity we might expect that their actions would be opposing on a particular nerve membrane function. On the other hand, the target for anaesthetic and pressure action might be a membrane protein, in which case opposite conformational changes in channel proteins might underlie the actions of anaesthetics and pressure if these two treatments produced opposing effects.

We must first look at the available experimental findings on pressure–anaesthetic interactions in nerve membranes and then decide which of the above possibilities is consistent with such data.

Hydrostatic pressure opposes the reduction in the amplitude of the squid axon action potential which is produced by ethanol (Spyropoulos, 1957). Pressure also counteracts the depressant action of several volatile anaesthetics (chloroform, ether and halothane) on the frog sciatic nerve action potential (Roth, 1975) and reverses the reduction by halothane of the amplitude of the action potential of the rat sympathetic preganglionic nerve (Kendig and Cohen, 1975). These studies have been performed over the temperature range room up to 37°C.

In contrast, Roth (1975) reported no effect of pressure on the conduction block of frog sciatic nerve produced by non-volatile agents such as procaine or the alcohols. What do these studies tell us, if anything, about the nature of the anaesthetic action?

At present there is no reason to believe that pressure alters the membrane/buffer partition coefficient of the anaesthetic. Electron spin resonance (ESR) work with the spin-labelled molecule TEMPO suggests that although high pressure displaces this agent from phospholipid vesicles, this effect is not large enough to account for the pressure reversal of the anaesthetic effect (Trudell *et al.*, 1973). Experiments with certain anaesthetics are consistent with the data using TEMPO. For example, high-pressure dilatometry studies with dipalmitoyl phosphatidylcholine liposomes treated with methoxyfluorane or butanol show clearly that pressures up to 300 atm do not alter the partition coefficient of these substances (Macdonald, 1978). Strictly speaking, we should bear in mind, however, that phospholipid vesicles do not act in the same way as nerve membranes (see Hsia and Boggs, 1973). Also equally important is the fact that TEMPO does produce nerve block but not necessarily by the same mechanism as a conventional anaesthetic. The latter agents could produce block by associating to a different membrane site and may therefore be displaced by pressure. Latest work has shown that the action of a spin-labelled local anaesthetic on membrane current does differ from that of the parent compound without the spin label (Yeh *et al.*, 1975).

Do anaesthetics and pressure produce their characteristic effects by altering membrane fluidity? Anaesthetic action could be related simply to a change in fluidity of nerve membranes, which would be opposed by high pressure. In model systems such as phospholipid vesicles it is possible that pressure and anaesthetic effects are simply opposed, but if nerve-conduction block by anaesthetics is brought about purely by a change in the freedom of movement of hydrocarbon tails of membrane lipid, it is difficult to explain the lack of effect of pressure on the anaesthetic block produced by the local anaesthetic procaine (Roth, 1975).

Perhaps it is most realistic to invoke membrane protein as a target for both anaesthetics and pressure. Membrane proteins undoubtedly play a key role in excitable membrane transport processes. For example, channel gating depends on lipoprotein (Armstrong *et al.*, 1973). That membrane protein conformational changes accompany the interaction of the anaesthetic agent with biological membranes may be inferred from the excessively large membrane expansion relative to the volume of occupation of the anaesthetic (see Table V.1). Although this expansion has not been measured in excitable membranes, it has been suggested that a membrane protein conformational change underlies the action of anaesthetics. The case for the lipoprotein conformation change theory has been argued recently by Woodbury *et al.* (1975). On this basis, then, anaesthetics would inactivate some channel protein and pressure would reverse the anaesthetic-induced conformational change if it occurred with a volume increase (See Chapter III, 5.1).

However, conformational changes may involve either no volume change or a decrease in volume, in which case pressure might have no effect, as Roth (1975) observed, or would produce the same effect as the anaesthetic. This could explain the potentiating effect of He pressure on the nerve block produced by the spin-label TEMPO (Boggs *et al.*, 1976a). ESR work with nerve membranes suggests that pressure causes both slight displacement of TEMPO from its apolar binding site and also causes TEMPO to bind to another induced polar site (Hsia and Boggs, 1973). Since pressure induction of a polar binding site is not observed with studies on phospholipid vesicles, caution should be exercised when extrapolating data from such systems. It would be consequently of great interest to test whether a conduction block by conventional anaesthetics can also be enhanced by hydrostatic pressure. Such data do not seem to be available as of yet.

It seems likely that such a diverse group of agents would produce nerve block by attacking a variety of sites on the nerve membrane, and therefore the effect of pressure on such a system would be variable. In this respect it should prove profitable to look into the pressure dependence of block induced by local anaesthetics at different pH values. Depending on the ratio of the charged/uncharged form of the agent, the main site of attack might be a hydrophobic (apolar) lipid or protein site or, alternatively, the anionic (polar) head of lipid or the polar portion of lipoprotein. Further, it would be informative to compare the pressure dependence of a block induced by benzocaine (uncharged molecule) with procaine in the cationic form. The former agent might block primarily by producing expansion, whereas procaine block might not necessarily involve such expansion.

### 4.3 THE MEMBRANE RECEPTOR?

That a single membrane model of anaesthetic action which fits all the experimental data exists is doubtful, and such alleged models should be accepted with some reservation. The anaesthetics are such a diverse group of compounds that it is unlikely that a common mechanism can be invoked to explain their many different actions.

Many authors have relied heavily on correlations of potency with some physical property of the agent, such as its hydrophobicity, or, alternatively, the stability of the hydrate (see p. 56). Such correlations were always of limited value in explaining mode of action and have now probably outlived their scientific usefulness. Perhaps, taking the opposite view, we should now focus not on the uniform effect produced by these agents but rather should stress the differences in action, in the hope that it is feasible to attach molecular significance to these data.

There may be marked differences in the mode of action of local and general anaesthetics. In the case of the former, the depressant effects are normally opposed by raising the external $Ca^{2+}$ concentration (Seeman *et al.*, 1974); however, this manipulation potentiates the depressant effects of a general anaesthetic such as halothane (Diamond *et al.*, 1975). Similarly, replacement of $H_2O$ by $D_2O$ increases the depressant action of general anaesthetics (halothane and enfluorane) but not the local agents (Diamond *et al.*, 1975).

It is apparent that TTX produces nerve block despite its low lipid solubility. By all accounts, then, hydrophobicity is a useful, but not an essential, property for a blocking agent to possess. When discussing the blockage of the $K^+$ channels by quaternary ammonium ions (e.g. TEA), no particular heed is paid to the hydrophobicity of the agents, save to describe the possible nature of binding to the membrane (Armstong, 1975). No attempt has been made to compare the physical characteristics of these agents, viz TTX and TEA, in order to discuss mechanism of action, and it is clear that the block is achieved by different molecular mechanisms. Despite the structural diversity of the anaesthetic agents it still seems possible that these compounds could act in a specific, as yet undefined, way. After all, it seems reasonable that a variety of agents could block the $Na^+$ channel with varying degrees of success, depending on their precise site of action.

That the $Na^+$ channel itself is the receptor for anaesthetic action is an appealing proposition. Let us consider this possibility first. Despite the comment that certain anaesthetics can possess a specificity similar to TTX (Hille, 1966), it is evident that fundamental differences exist between their mechanism of action. For example, TTX is inert when applied internally, although as discussed previously many anaesthetics agents are active. Consider also the pH dependence of the block. In the case of external TTX, changes in intracellular pH do not modify the block, whereas the block produced by certain derivatives of lidocaine may be affected by varying the intracellular pH (Narahashi *et al.*, 1969b; Narahashi, 1971).

The case for the $Na^+$ channel acting as the receptor is strongest for the local anaesthetics and derivatives. Experiments with optically active enantiomers demonstrate that although the uptake and partition coefficient of certain agents is the same, their potency differs. For example, Hille *et al.* (1975) reported that the tertiary amine local anaesthetics RAC 109 I and II, which are enantiomers, differed in potency by a factor of about 2. Such experimental data argues for stereospecific binding of these compounds to a membrane receptor. It is significant that other *in vitro* studies also detect this potency difference (Seeman, 1972b, Table 5).

Other studies have demonstrated that the membrane receptor is the same for the local anaesthetics and their derivatives (Strichartz, 1973; Hille *et al.*, 1975) and that the binding site is approximately half-way down the $Na^+$ channel electrical gradient (Strichartz, 1973). Thus the block is promoted by conditioning depolarising clamp pulses which open the channels, and the block is reversed, i.e. the channel cleared, by small depolarising pulses. The effect of conditionary depolarising pulses on the inhibition of $Na^+$ currents by internal QX–314 is shown in Fig. V.17. In

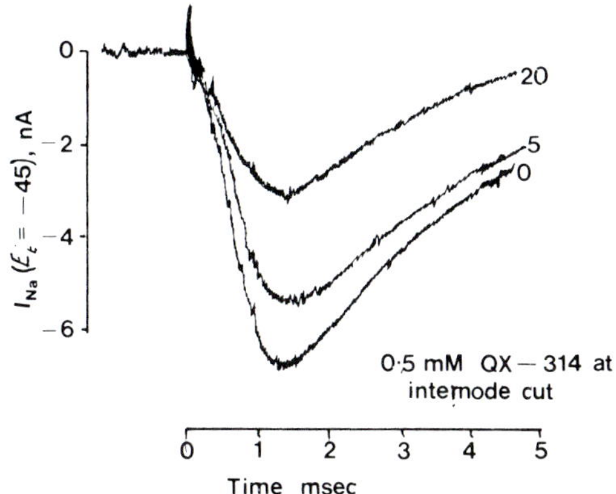


FIG. V.17. Records of the time-course of the sodium current of a single amphibian myelinated fibre measured in the presence of a quaternary derivative of lidocaine, QX–314. The drug was present inside the nerve and the currents were measured during a depolarising "test" pulse to −45 mV. The largest peak current occurred after the membrane potential was held at −75 mV (0). Peak currents during the test pulse were diminished when their measurement was preceded by five depolarisations (5) to +75 mV, and further decreased by preceding the test pulse by twenty of these depolarising pulses (20). The depolarising pulses were 5 msec long. (Strichartz, 1973)

this experiment a single myelinated fibre was mounted in a four-compartment chamber and the recordings taken at a nodal region in one of the central compartments. The nerve was bathed in Ringers' containing QX–314 (0·5 mM) and an internodal region of the nerve cut in one of the outer compartments of the chamber. The drug then gained access to the internal side of the membrane at the recording nodal site by diffusing along the inside of the nerve. The finding that the frequency of channel opening modulates the subsequent block is consistent with the suggestion that the binding site for the anaesthetic is within the $Na^+$ channel. In addition, Strichartz (1973) also concluded that the binding reaction is voltage-dependent. Thus, since QX–314 is a cation, depolarisation will tend to drive the ion (from the intracellular fluid) into the channel and increase the binding of the drug to the membrane receptor. Hence the drug–membrane interaction may be represented schematically as in Fig. V.18.

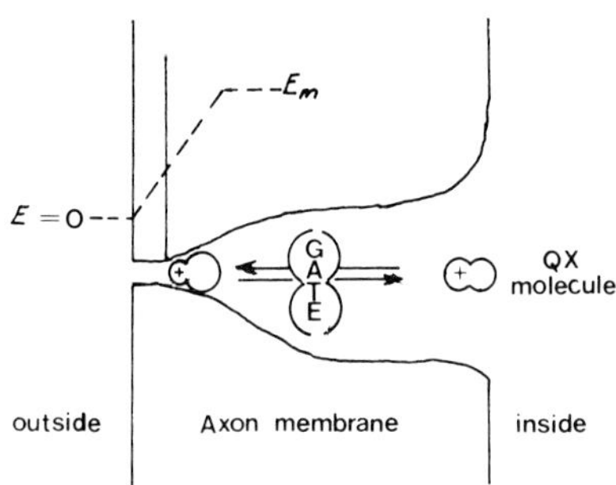


Fig. V.18. Schematic diagram of a quaternary lidocaine molecule (QX) blocking a sodium channel. The channel is blocked when the QX molecule binds within it. The channel "gate" must be open for the binding reaction both to occur and to reverse. The binding reaction may be influenced by the electrical potential profile through the sodium channel, the profile of which is represented by the dashed line. (Strichartz, 1975)

If we accept this model, then presumably imposed hyperpolarising pulses could displace the drug from the membrane, depending on its position with respect to the channel gate. Such a mechanism may underlie the alleged removal of $Na^+$ inactivation reported by Weidmann (1955) for cocaine. If the receptor for the drug is beyond the channel gate, it is perhaps easier to see how varying the external $Ca^{2+}$ concentration could oppose the action of cationic local anaesthetics. The proposal of competitive binding between $Ca^{2+}$ and an anaesthetic molecule at a hypothetical site also becomes feasable although it may be necessary to assume that the drug can be repelled electrostatically by the presence of a $Ca^{2+}$ ion in the outer mouth of the channel. The interaction of $K^+$ ions with the nonyltriethyl ammonium ion has been explained previously in this way (Armstrong, 1975).

Although$^+$ undoubtedly anaesthetics interact with the $Na^+$ channel, generally they do not possess a specificity analogous to TTX. In the case of procaine, the permeability constant $\bar{g}$Na is affected more than $\bar{g}$K if the agent is applied externally (Taylor, 1959) but both are affected equally by internal application (Narahashi *et al.*, 1967).Can we learn anything about channel structure from the knowledge that in myelinated fibres $\bar{g}$K is much less sensitive to local anaesthetic action than $\bar{g}$Na (Århem and Frankenhaeuser, 1974)?

In these fibres, increasing the external $Ca^{2+}$ concentration produces a variable effect on the curve relating $K^+$ permeability to membrane potential (Hille, 1968b; Brismar and Frankenhaeuser, 1972). In one study an *e*-fold increase in $Ca^{2+}$ concentration produced an 8·7 mV shift in the $Na^+$ activation curve, but no shift in the $K^+$ activation curve (Hille, 1968b). This implies that the surface charge density close to the mouth of the $K^+$ channel is less than that of the $Na^+$ channel. Such surface charge may be neutralised by the cationic head of the local anaesthetic and derivatives, hence shifting the $\bar{g}$Na or $\bar{g}$K curves in a depolarising direction.

This surface charge may also be necessary for binding of the anaesthetic, in order that it may exert its maximum effect in reducing either the peak $Na^+$ or $K^+$ conductance. A low surface-charge density in the vicinity of the $K^+$ channel may therefore render it less vulnerable to attack by the charged local anaesthetics.

As noted previously (Fig. V.11) dibucaine produces inactivation of $K^+$ currents, an effect analagous to that of the pentyltriethylammonium ion (Armstrong, 1975). The ability of an agent to produce this effect may be linked to its rate of dissociation from the blocking site and might, therefore, also be dependent on the ability to "lock" to surface charges. In the case of the quaternary ammonium ions, increasing the chain length increases the magnitude of inactivation, which probably reflects improved binding. Hence, inactivation is absent with the methyltriethylammonium ion (loose binding) but is marked with nonyltriethylammonium ion (almost irreversible binding). It would be expected that strongly hydrophobic general anaesthetic agents like halothane or trichloroethylene would produce marked inactivation also.

Most anaesthetics, therefore, probably affect both the $Na^+$ and $K^+$ channel permeability, so that the notion of the $Na^+$ channel as a specific receptor is a limited one. From a functional viewpoint these agents do not discriminate between channels very efficiently; do they discriminate at a molecular level? Is membrane lipid the receptor? One admittedly naive possibility is that volatile, and perhaps all uncharged, anaesthetics migrate into the membrane lipid, causing general fluidisation and resultant expansion of the membrane and that their actions are therefore pressure-sensitive. There are serious difficulties with this hypothesis.

First, clinical concentrations of a range of anaesthetic agents do not produce fluidisation of bilayer vesicles (Boggs *et al.*, 1976b). Secondly, disordering, or fluidisation of lipid bilayers, also depends on the composition of these artificial systems. Generally, bilayers low in cholesterol (5 per cent) have a low degree of order and are *ordered*, not fluidised, by anaesthetics (Neal *et al.*, 1976). On the other hand, bilayers with higher cholesterol contents are often fluidised by anaesthetics (Miller and Pang, 1976). Consequently, we simply do not know what happens to general fluidity in a nerve membrane, and in particular cannot predict what happens to fluidity in the lipid environment close to channels. Lee (1976) has argued that nerve-blocking concentrations of anaesthetics would not change membrane bulk fluidity and suggests that anaesthetics cause a phase transition (see Chapter IV) in the lipid surrounding the $Na^+$ channel.

If it is assumed that the channel is held in the open position by a rigid lipid surround, then it may indeed "relax" back to its equilibrium state when the lipid is changed from the gel to the fluid, liquid-crystalline state. Hence, channel conductance would decrease. Note that the hypothesis applies only to the $Na^+$ channels opened during the action potential since it is obviously difficult to reconcile the alleged increase in resting $Na^+$ permeability produced by ethanol with this model (Knutsson and Katz, 1967). Whether fluidisation or a phase transition of membrane lipid is central to anaesthetic action on excitable membranes must be resolved by future experiments.

The final difficulty with the membrane-expansion theory of anaesthetic action resides in the finding that the nerve-blocking action of TEMPO is enhanced by pressure as stated previously.

The most we can do at present is to look at the density of anaesthetic molecules and guess what membrane macromolecules the anaesthetic molecules might be associated with. Seeman's figures suggest that the number of anaesthetic molecules in the membrane may vary from 1–10 every 4000 $Å^2$ (see Table V.1). If a membrane phospholipid molecule occupies a surface area of about 60 $Å^2$ (Levine and Wilkins, 1971), then this would mean that there might be as few as one anaesthetic molecule per 65–70 lipid molecules. The cross-sectional area of a high molecular weight protein may be about 5000 $Å^2$ (Bretscher, 1973). Correspondingly, anaesthetic molecules must be close to membrane protein throughout the membrane. These calculations of course assume homogeneous distribution of phospholipid, protein and anaesthetic molecules, which is clearly an oversimplification, yet they do provide some conceptual feel for the possible molecular interactions.

In summary it would seem fair to conclude that generally the anaesthetic–excitable membrane interaction does not involve the classical drug-receptor or "key in lock" type chemistry. The weight of available evidence suggests a more broad-based interaction with membrane macromolecules. Whether protein conformational changes, or lipid transformations are important, can only be resolved by further investigation. Whatever their primary site of action is, however, anaesthetics should prove particularly useful as membrane probes. The precise molecular mechanisms by which they produce effects on excitable membranes are at present obscure. But we can anticipate that in the process of unravelling this particular problem we can hope to learn much more about the molecular basis of excitability itself.

## 5 Conclusions

The actions of anaesthetics on excitable membranes are now well established. These agents may be termed electrical stabilisers since they can block changes in membrane permeability with little or no change in the steady state electrical characteristics of the cell. At high concentrations they can, however, produce alterations in the steady state electrical parameters of excitable membranes.

Thus ethanol, ether and sodium pentobarbitone produce depolarisation, and ethanol produces a reduction in input resistance of the frog skeletal muscle membrane. Ethanol also depolarises the squid axon and lobster axon membrane. This effect is accompanied by an increase in input resistance in the case of the squid axon. It has been suggested that the ethanol-induced depolarisation of the frog skeletal muscle membrane is mediated via a rise in $Na^+$ permeability, whereas the ethanol-induced depolarisation of the squid axon membrane is produced by a reduction in $K^+$ permeability.

At lower concentrations the anaesthetics produce a block in excitability. Such block is generally preceded by an increase in threshold depolarisation and threshold current. The early work showed that the rate of depolarisation and repolarisation of the action potential and the action potential height were depressed by the anaesthetics. The decrease in the rate of rise of the action potential of the frog muscle membrane produced by ethanol may be offset if the electrochemical gradient for $Na^+$ is increased. The depression of the action potential amplitude of the squid axon membrane induced by ethanol may be countered if the temperature is reduced to 4°C.

Voltage-clamp studies with squid and lobster axons have demonstrated that the local anaesthetics, the alcohols and the barbiturates reduced the peak inward ($Na^+$) currents and the maximum steady state outward ($K^+$)

currents. These agents also increased the time to peak $Na^+$ current in the squid axon membrane, and procaine can also increase the time constant of the $K^+$ current of squid axon. An agent such as the local anaesthetic dibucaine has been shown to produce inactivation of the $K^+$ current in the squid axon. In amphibian myelinated fibres the kinetics of the $Na^+$ activation process are unaffected by local anaesthetics, and also the permeability constant $\bar{g}K$ is less sensitive to the local anaesthetics than the permeability constant $\bar{g}Na$. This suggests that the surface charge density in the vicinity of the $K^+$ channel is less than at the mouth of the $Na^+$ channel.

It has been suggested that agents such as the barbiturates or the local anaesthetics act on the inside surface of the membrane. In the case of the quaternary derivative of lidocaine, QX–314, there is evidence that the compound binds within the $Na^+$ channel itself. The uncharged form of the barbiturates is the more potent but in the case of the local anaesthetics a positive charge is thought to increase the potency.

The action of an anaesthetic agent on the kinetics of the ion conductance systems of excitable membranes may be a result of an interaction of a charged portion of the molecule with the polar head group of a phospholipid molecule. The action of an anaesthetic agent on the peak magnitude of an ion conductance may reflect its fluidising effect on the lipid environment of the membrane channel.

At present our limited knowledge of the molecular mechanisms accompanying anaesthetic action is due in part to our ignorance of the molecular processes governing excitability. Perhaps judicious use of the anaesthetic molecule as a membrane probe will help define such processes.

## References


Aceves, J. and Machne, X. (1963). The action of calcium and local anaesthetics on nerve cells, and their interaction during excitation. *J. Pharm. Exp. Therap.* **140**, 138–148.

Adelman, W. J. and Taylor, R. E. (1961). Leakage current rectification in the squid giant axon. *Nature*, **190**, 883–885.

Alper, M. H. and Flacke, W. (1969). The peripheral effects of anaesthetics. *Ann. Rev. Pharmac.* **9**, 273–296.

Århem, P. and Frankenhaeuser, B. (1974). Local anaesthetics: effects on permeability properties of nodal membrane in myelinated nerve fibres from Xenopus. Potential clamp experiments. *Acta Physiol. Scand.* **91**, 11–21.

Armstrong, C. M. (1975). Potassium pores of nerve and muscle membranes. *In* "Membranes" (G. Eisenman, ed.), vol. 3, pp. 325–358. Dekker, New York.

Armstrong, C. M. and Binstock, L. (1964). The effect of several alcohols on the properties of the squid axon. *J. Gen. Physiol.* **48**, 265–277.

Armstrong, C. M., Bezanilla, F. and Rojas, E. (1973). Destruction of sodium conductance inactivation in squid axons perfused with pronase. *J. Gen. Physiol.* **62**, 375–391.

Ashcroft, R. G., Coster, H. G. L. and Smith, J. R. (1977). Local anaesthetic benzyl alcohol increases membrane thickness. *Nature*, 269, 819–820.

Berney, J. and Posternak, J. (1956). Modifications de l'excitabilité nerveuse par des narcotiques. *Helv. Physiol Pharmacol. Acta*, **14**, C5.

Bishop, G. H. (1932). Action of nerve depressants on potential. *J. Cell Comp. Physiol.* **1**, 177–194.

Blaustein, M. P. (1967). Phospholipids as ion exchangers: implications for a possible role in biological membrane excitability and anaesthesia. *Biochim. Biophys. Acta*, **135**, 653–668.

Blaustein, M. P. (1968a). Barbiturates block sodium and potassium conductance increases in voltage-clamped lobster axons. *J. Gen. Physiol.* **51**, 293–307.

Blaustein, M. P. (1968b). Action of certain tropine esters on voltage-clamped lobster axon. *J. Gen. Physiol.* **51**, 309–319.

Blaustein, M. P. and Goldman, D. E. (1966a). Action of anionic and cationic nerve-blocking agents: experiment and interpretation. *Science*, **153**, 429–432.

Blaustein, M. P. and Goldman, D. E. (1966b). Competitive action of calcium and procaine on lobster axon. *J. Gen. Physiol.* **49**, 1043–1063.

Blaustein, M. P. and Goldman, D. E. (1968). The action of certain polyvalent cations on the voltage-clamped lobster axon. *J. Gen. Physiol.* **51**, 279–291.

Boggs, J. M., Roth, S. H., Yoong, T., Wong, E. and Hsia, J. C. (1976a). Site and mechanism of anaesthetic action. II. Pressure effect on the nerve conduction–blocking activity of a spin label anaesthetic. *Molec. Pharmacol.* **12**, 136–143.

Boggs, J. M., Yoong, T. and Hsia, J. C. (1976b). Site and mechanism of anaesthetic action. 1. Effect of anaesthetics and pressure on fluidity of spin-labelled lipid vesicles. *Molec. Pharmacol.* **12**, 127–135.

Bretscher, M. S. (1973). Membrane structure: some general principles. *Science*, **181**, 622–629.

Brismar, T. and Frankenhaeuser, B. (1972). The effect of calcium on the potassium permeability in the myelinated nerve fibre of Xenopus Laevis. *Acta Physiol. Scand.* **85**, 237–241.

Carpenter, F. G. (1954). Anaesthetic action of inert and unreactive gases on intact animals and isolated tissues. *Amer. J. Physiol.* **178**, 505–509.

Courtney, K. R. (1975). Mechanism of frequency-dependent inhibition of sodium currents in frog myelinated nerve by the lidocaine derivative GEA 968. *J. Pharm. Exp. Therap.* **195**, 225–236.

Dalton, J. C. and Fitzhugh, R. (1960). Applicability of Hodgkin–Huxley model to experimental data from the giant axon of lobster. *Science*, **131**, 1533–1534.

Dettbarn, W. F. (1962). The active form of local anaesthetics. *Biochim. Biophys. Acta*, **57**, 73–76.

Diamond, B. I., Havdala, H. S. and Sabelli, H. C. (1975). Differential membrane effects of general and local anaesthetics. *Anesthesiol.* **43**, 651–660.

Ehrenpreis, S. and Rosen, G. M. (1974). Neuromuscular blocking action of an alkylating local anaesthetic: site of action and effects of temperature and calcium ions. *Nature*, **250**, 576–578.

Feinstein, M. B. (1964). Reaction of local anaesthetics with phospholipids. *J. Gen. Physiol.* **48**, 357–374.

Frank, G. B. (1972). *In* "Cellular Biology and Toxicity of Anaesthetics" (B. R. Fink, ed.), pp. 29–38. Williams and Wilkins, Baltimore.

Frankenhaeuser, B. and Hodgkin, A. L. (1957). The action of calcium on the electrical properties of squid axons. *J. Physiol.* **137**, 218–244.

GAGE, P. W., MCBURNEY, R. N. and SCHNEIDER, G. T. (1975). Effects of some aliphatic alcohols on the conductance change caused by a quantum of acetylcholine at the toad end plate. *J. Physiol.* **244**, 409–429.

GALLEGO, A. (1948). On the effect of ethyl alcohol upon frog nerve. *J. Cell. Comp Physiol.* **31**, 97–106.

GOLDMAN, D. E. (1964). A molecular structural basis for the excitation properties of axons. *Biophys. J.* **4**, 167–188.

GRAHAM, J. and GERARD, R. W. (1946). Membrane potentials and excitation of impaled muscle fibres. *J. Cell Comp. Physiol.* **28**, 99–117.

GUTTMANN, R. (1940). Stabilisation of spider crab nerve membranes by alkaline earths as manifested in resting membrane potential measurements. *J. Gen. Physiol.* **23**, 343–364.

HAYDON, D. A., HENDRY, B. M., LEVINSON, S. R. and REQUENA, J. (1977). The molecular mechanisms of anaesthesia. *Nature*, **268**, 356–358.

HILLE, B. (1966). Common mode of action of three agents that decrease the transient change in sodium permeability in nerves. *Nature*, **210**, 1220–1222.

HILLE, B. (1968a). Pharmacological modifications of the sodium channels of frog nerve. *J. Gen. Physiol.* **51**, 199–219.

HILLE, B. (1968b). Charges and potentials at the nerve surface. *J. Gen. Physiol.* **51**, 221–236.

HILLE, B. (1975a). An essential ionised acid group in sodium channels. *Fed. Proc.* **34**, 1318–1321.

HILLE, B. (1975b). Ionic selectivity of Na and K channels of nerve membranes. *In* "Membranes" (G. Eisenman, ed.), vol. 3, pp. 255–323. Dekker, New York.

HILLE, B. (1977). Local anaesthetics: hydrophilic and hydrophobic pathways for the drug receptor reaction. *J. Gen. Physiol.* **69**, 497–515.

HILLE, B., COURTNEY, K. and DUM, R. (1975). Rate and site of action of local anaesthetics in myelinated nerve fibres. *In* "Molecular Mechanisms of Anaesthesia" (B. R. Fink, ed.), vol. 1, pp. 13–20. Raven, New York.

HODGKIN, A. L. and HUXLEY, A. F. (1952). A quantitative description of membrane current and its application to conduction and excitation in nerve. *J. Physiol.* **117**, 500–544.

HOUCK, D. J. (1969). Effect of alcohols on potentials of lobster axons. *Amer. J. Physiol.* **216**, 364–367.

HSIA, J. C. and BOGGS, J. M. (1973). Pressure effect on the membrane action of a nerve-blocking spin label. *Proc. Nat. Acad. Sci.* **70**, 3179–3183.

HUBELL, W. L. and MCCONNELL, H. M. (1968). Spin-label studies of the excitable membranes of nerve and muscle. *Proc. Nat. Acad. Sci.* **61**, 12–16.

INOUE, F. and FRANK, G. B. (1965). Action of ether on frog skeletal muscle. *Can. J. Physiol. Pharmacol.* **43**, 751–761.

INOUE, F. and FRANK, G. B. (1967). Effects of ethyl alcohol on excitability and on neuromuscular transmission in frog skeletal muscle. *Brit. J. Pharm. Chem.* **30**, 186–193.

KENDIG, J. and COHEN, E. N. (1975). Neural sites of pressure–anaesthesia interactions. *In* "Molecular Mechanism of Anaesthesia" (B. R. Fink, ed.), vol. 1, pp. 421–427. Raven, New York.

KEYNES, R. D. and Rojas, E. (1974). Kinetics and steady-state properties of the charged system controlling sodium conductance in the squid giant axon. *J. Physiol.* **239**, 393–434.

KHODOROV, B., SHISHKOVA, L., PEGANOV, E. and REVENKO, S. (1976). Inhibition of sodium currents in frog Ranvier node treated with local anaesthetics. Role of slow sodium inactivation. *Biochim. Biophys. Acta*, **433**, 409–435.

Kiss, I. and Vadasz, I. (1973). The effects of local anaesthetics on the activity generation and chemical sensitivity of giant neurones. *Ann. Inst. (Tihany) Hung. Acad. Sci.* **40**, 73–84.

Knutsson, E. (1961). Effects of ethanol on the membrane potential and membrane resistance of frog muscle fibres. *Acta Physiol. Scand.* **52**, 242–253.

Knutsson, E. and Katz, S. (1967). The effect of ethanol on the membrane permeability to sodium and potassium ions in frog muscle fibres. *Acta Pharmacol. Toxicol.* **25**, 54–64.

Krupp, P., Bianchi, C. P. and Suarez-Kurtz, G. (1969). On the local anaesthetic effect of barbiturates. *J. Pharm. Pharmacol.* **21**, 763–768.

Kwant, W. O. and Seeman, P. (1969). The membrane concentration of a local anaesthetic (chlorpromazine). *Biochim. Biophys. Acta*, **183**, 530–543.

Lee, A. G. (1976). Model for action of local anaesthetics. *Nature*, **262**, 545–548.

Levine, Y. K. and Wilkins, M. H. F. (1971). Structure of oriented lipid bilayers. *Nature New Biol.* **230**, 69–76.

Lorente, de Nó, R. (1947). A study of nerve physiology. "The Studies from the Rockefeller Institute for Medical Research", vol. 131.

Lucas, K. (1913). The effect of alcohol on the excitation, conduction and recovery processes in nerve. *J. Physiol.* **46**, 470–505.

Macdonald, A. G. (1978). A dilatometric investigation of the effect of general anaesthetics, alcohols and hydrostatic pressure on the phase transition in smectic mesophases of dipalmitoyl phosphatidylcholine. *Biochim. Biophys. Acta*, **507**, 26–37.

Marmont, G. (1949). Studies on the axon membrane. I. A new method. *J. Cell Comp. Physiol.* **34**, 351–382.

Metcalfe, J. C. (1970). Interaction of local anaesthetics and calcium with erythrocyte membranes. *In* "Calcium and Cellular Function" (A. W. Cuthbert, ed.), pp. 219–240. Macmillan, London.

Metcalfe, J. C. and Burgen, A. S. V. (1968). Relaxation of anaesthetics in the presence of cyto-membranes. *Nature*, **220**, 587–588.

Meves, H. (1975). Asymmetry currents in intracellularly perfused squid giant axons. *Phil. Trans. Roy. Soc. (Lond.) B*, **270**, 493–500.

Miller, K. W. and Pang, K. Y. (1976). General anaesthetics can selectively perturb lipid bilayer membranes. *Nature*, **263**, 253–255.

Mullins, L. J. (1954). Some physical mechanisms in narcosis. *Chem. Rev.* **54**, 289–323.

Mullins, L. J. and Gaffey, C. T. (1954). The depolarisation of nerve by organic compounds. *Proc. Soc. Exptl. Biol. Med.* **85**, 144–149.

Narahashi, T. (1971). Neurophysiological basis for drug action: ionic mechanism, site of action and active form in nerve fibres. *In* "Biophysics and Physiology of Excitable Membranes" (W. J. Adelman, ed.). pp. 423–462. Van Nostrand Reinhold, New York.

Narahashi, T., Anderson, N. C. and Moore, J. W. (1967). Comparison of tetrodotoxin and procaine in internally perfused squid giant axons. *J. Gen. Physiol.* **50**, 1413–1428.

Narahashi, T., Frazier, D. T., Deguchi, T., Cleaves, C. A. and Ernau, M. C. (1971). The active form of pentobarbital in squid giant axons. *J. Pharm. Exp. Therup.* **177**, 25–33.

NARAHASHI, J., MOORE, J. W. and POSTON, R. N. (1969a). Anaesthetic blocking of nerve membrane conductances by internal and external applications. *J. Neurobiol.* **1**, 3–22.

NARAHASHI, T., YAMADA, M. and FRAZIER, D. J. (1969b). Cationic forms of local anaesthetics block action potentials from inside the nerve membrane. *Nature*, **223**, 748–749.

NATHAN, P. W. and SEARS, T. A. (1961). Some factors concerned in differential nerve block by local anaesthetics. *J. Physiol.* **157**, 565–580.

NEAL, M. J., BUTLER, K. W., POLNASZEK, C. F. and SMITH, I. C. P. (1976). The influence of anaesthetics and cholesterol on the degree of molecular organisation and mobility of ox brain, white matter. *Molec. Pharmacol.* **12**, 144–155.

RENKIN, E. M. (1961). Permeability of frog skeletal muscle cells to choline. *J. Gen. Physiol.* **44**, 1159–1165.

RITCHIE, J. M. (1975). Binding of tetrodotoxin and saxitoxin to sodium channels. *Phil. Trans. Roy. Soc. (Lond.) B*, **270**, 319–336.

RITCHIE, J. M. and GREENGARD, P. (1961). On the active structure of local anaesthetics. *J. Pharm. Exp. Therap.* **133**, 241–245.

RITCHIE, J. M. and GREENGARD, P. (1966). On the mode of action of local anaesthetics. *Ann. Rev. Pharmacol.* **6**, 405–430.

RITCHIE, J. M. and RITCHIE, B. R. (1968). Local anaesthetics: effect of pH on activity. *Science*, **162**, 1394–1395.

ROTH, S. H. (1975). Anaesthetics and pressure: antagonism and enhancement. *In* "Molecular Mechanisms of Anaesthesia" (B. R. Fink, ed.), vol. 1, pp. 405–420. Raven Press, New York.

ROTH, S. H and SEEMAN, P. (1972). The membrane concentrations of neutral and positive anaesthetics (alcohols, chlorpromazine, morphine) fit the Meyer–Overton rule of anaesthesia; negative narcotics do not. *Biochim. Biophys. Acta*, **255**, 207–219.

SCHNEIDER, H. (1968). The intramembrane location of alcohol anaesthetics. *Biochim. Biophys. Acta*, **163**, 451–458.

SEEMAN, P. (1966a). Membrane stabilisation by drugs: tranquilisers, steroids and anaesthetics. *Int. Rev. Neurobiol.* **9**, 145–221.

SEEMAN, P. (1966b). Erythrocyte membrane stabilisation by local anaesthetics and tranquilisers. *Biochem. Pharmacol.* **15**, 1753–1766.

SEEMAN, P. (1972a). The effects of anaesthetics and tranquilisers on cell membranes. *In* "Cellular Biology and Toxicity of Anaesthetics" (B. R. Fink, ed.), pp. 149–159. Williams and Wilkins, Baltimore.

SEEMAN, P. (1972b). The membrane actions of anaesthetics and tranquilisers. *Pharmacol. Rev.* **24**, 583–655.

SEEMAN, P. (1975). The membrane expansion theory of anaesthesia. *In* "Molecular Mechanisms of Anaesthesia" (B. R. Fink, ed.), vol. 1, pp. 243–251. Raven Press, New York.

SEEMAN, P. and ROTH, S. (1972). General anaesthetics expand cell membranes at surgical concentrations. *Biochim. Biophys. Acta*, **255**, 171–177.

SEEMAN, P., CHEN, S. S., CHAU-WONG, M. and STAIMAN, A. (1974). Calcium reversal of nerve blockade by alcohols, anaesthetics, tranquilisers, and barbiturates. *Can. J. Physiol. Pharmacol.* **52**, 526–534.

SEEMAN, P., KWANT, W. O. and SAUKS, T. (1969). Membrane expansion of erythrocyte ghosts by tranquilisers and anaesthetics. *Biochim. Biophys. Acta*, **183**, 499–511.

SHANES, A. M. (1958). Electrochemical aspects of physiological and pharmacological action in excitable cells. *Pharmacol. Rev.* **10**, 59–164.

SHANES, A. M., FREYGANG, W. H., GRUNDFEST, H. and AMATNIEK, E. (1959). Anaesthetic and calcium action in the voltage-clamped squid giant axon. *J. Gen. Physiol.* **42**, 793–802.

SIMON, A. and SZELÖCZEY, J. (1928). Untersuchungen über Kalium-und Calciumgehalt der peripheren Nervenfasern. *Biochem. Zeitschrift*, **193**/**194**, 393–399.

SKOU, J. C. (1954). Local anaesthetics. I. The blocking potencies of some local anaesthetics and of butyl alcohol determined on peripheral nerves. *Acta Pharmacol. Toxicol.* **10**, 280–291.

SPYROPOULOS, C. S. (1957). The effects of hydrostatic pressure upon the normal and narcotised nerve fibre. *J. Gen. Physiol.* **40**, 849–857.

STAIMAN, A. and SEEMAN, P. (1974). The impulse blocking concentrations of anaesthetics, alcohols, anticonvulsants, barbiturates, and narcotics on phrenic and sciatic nerves. *Can. J. Physiol. Pharmacol.* **52**, 535–550.

STRICHARTZ, G. (1973). The inhibition of sodium currents in myelinated nerve by quaternary derivatives of lidocaine. *J. Gen. Physiol.* **62**, 37–57.

STRICHARTZ, G. (1975). Inhibition of ionic currents in myelinated nerves by quaternary derivative of lidocaine. *In* "Molecular Mechanisms of Anaesthesia" (B. R. Fink, ed.), vol. 1, pp. 1–11. Raven Press, New York.

SZPILMAN, J. and LUCHSINGER, B. (1881). Zur Beziehung von Leitungs- und Erregungsvermögen der Nervenfaser. *Pflügers Archiv.* **24**, 347–357.

TAYLOR, R. E. (1959). Effect of procaine on electrical properties of squid axon membrane. *Amer. J. Physiol.* **196**, 1071–1078.

THESLEFF, S. (1956). The effect of anaesthetic agents on skeletal muscle membrane. *Acta Physiol. Scand.* **37**, 335–349.

TRUDELL, J. R., HUBBELL, W. L., COHEN, E. N. and KENDIG, J. J. (1973). Pressure reversal of anaesthesia: the extent of small molecule exclusion from spin labelled phospholipid model membranes. *Anesthesiol.* **38**, 207–211.

WEIDMANN, S. (1955). Effects of calcium ions and local anaesthetics on electrical properties of Purkinje fibres. *J. Physiol.* **129**, 568–582.

WINTERSTEIN, H. (1919). "Die Narkose". Springer, Berlin.

WOODBURY, J. W., D'ARRIGO, J. S. and EYRING, H. (1975). Physiological mechanism of general anaesthesia: synaptic blockade. *In* "Molecular Mechanisms of Anaesthesia" (B. R. Fink, ed.), vol. 1, pp. 53–87. Raven Press, New York.

YAMAGUCHI, T. and OKUMURA, H. (1963). Effects of anaesthetics on the electrical properties of the cell membrane of the frog muscle fibre. *Annot. Zool. Jap.* **36**, 109–117.

YEH, J. Z., Takeno, K., ROSEN, G. M. and NARAHASHI, T. (1975). Ionic mechanism of action of a spin-labelled local anaesthetic on squid axon membranes. *J. Memb. Biol.* **25**, 237-247.

## CHAPTER VI

# Synapses



## 1 Introduction

Since Sherrington (1906) concluded that anaesthetics depress spinal reflexes by interfering with the process of synaptic transmission, it has often been proposed that these agents produce their characteristic central nervous system effects primarily by depressing synaptic events. In view of the known susceptibility of synaptic transmission to other drugs and changes in environmental conditions, such as alterations in temperature or ionic composition, this is perhaps a justifiable proposition. It is, however, not the purpose of this chapter to debate whether the state of anaesthesia is brought about primarily by a depression of certain target synapses, but rather to outline the principal effects of anaesthetics on the synaptic process.

Indeed, many anaesthetics have been shown to exert profound effects at a synaptic level. It is evident that not all agents act in a uniform way at the same synapse; nor should it be anticipated that the same anaesthetic agent produces an identical effect at different types of synapse. The geometry of the synaptic contact alone will determine how vulnerable the various components of the process are to anaesthetic action. Thus, factors such as the size and arrangement of the presynaptic endings, and the morphology and hence cable properties of the postsynaptic cell, govern the safety factor involved in transmission at these sites. Further, the concentration of the agent tested modifies the result obtained. Low concentrations may suggest a presynaptic action whereas higher concentrations may unmask a secondary postsynaptic effect.

In many cases we will be dealing with the action of anaesthetics on a single synapse. However, as pointed out by Burgen (1971), the more synapses there are in a "target area" the more susceptible the system becomes to an anaesthetic. For example, if the probability of the anaesthetic blocking a single synapse in a neuronal chain is $P$, and if the number of synapses in the chain is $n$, then the overall probability of a block in the chain is $1 - (1 - P)^n$, if it is assumed that each synapse is equally sensitive. This fact may help explain the differential susceptibility of different neural systems to anaesthetics.

As discussed by Wall (1967), there are many possible stages of the synaptic process which could conceivably be affected by anaesthetics, and it is the aim of this chapter to focus attention on these aspects of the process which are clearly sensitive to attack by anaesthetic molecules.

To facilitate discussion, the actions of anaesthetics on presynaptic functions will be considered initially, then the postsynaptic effects of these compounds will be described. It should become apparent to the reader that the postsynaptic changes may be ascribed in some cases to a presynaptic effect of the anaesthetic, and that, since all the processes to be dealt with are interlocked, often one aspect of synaptic function is altered because of an action of the agent on a supporting process.

In addition to discussing the effects of anaesthetics on the process of excitatory synaptic transmission, their effects on both pre- and postsynaptic inhibitory processes will be described. Finally, brief reference will be made to the interaction of anaesthetics with high pressure at a synaptic level.

Before taking up discussion of the action of these agents on synaptic transmission, however, it is worth asking whether anaesthetics do act selectively at synapses or not. It is all too often stated that anaesthetics depress synaptic transmission with little or no effect on the process of axonal conduction. In support of this many authors cite the work of Larrabee and Posternak (1952). These workers used either the perfused cat stellate ganglion or the excised superior cervical ganglion of the rat, and employed a variety of anaesthetic agents in an attempt to define the basis of any selectivity. The preparations offered the advantage that agents could be applied to a comparatively simple synaptic system, but also, since other axons enter the ganglion and traverse it without synaptic interruption, the effects of agents on "direct" axonal conduction and synaptic transmission could be examined simultaneously. The anatomy of the pathways involved and the positions of the stimulating and recording electrodes used in some of these experiments are shown in Fig. VI.1.

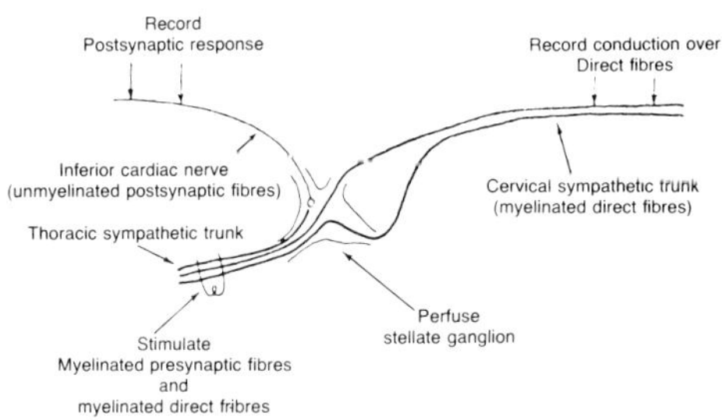


FIG. VI.1. Diagram of the cat stellate sympathetic ganglion, showing the positions of the stimulating and recording electrodes used in the study. (Larrabee and Posternak, 1952).

The principal finding revealed by this study was that the degree of selectivity (defined by the ratio of the concentration of the agent required to halve the "direct" action potential to the concentration required to halve the synaptically evoked action potential) was highly dependent on the agent under consideration. Hence, sodium pentobarbitone was most selective, in that the synaptically activated action potential was depressed 50 per cent by one-tenth of the concentration required to reduce the amplitude of the "direct" action potential by a corresponding amount. This is shown in Figs VI.2(a) and VI.2(b). It is clear from Fig. VI.2(a) that sodium pentobarbitone at a concentration of 0·25 mM and 0·5 mM produced a marked reduction in the postsynaptic response with no effect on the "direct" action potential. Figure VI.2(b) shows that during recovery from a block produced by 5 mM pentobarbitone the direct fibre action potential returned to normal at a concentration of 2 mM (○), compared with the delayed recovery of the postsynaptic response at 0·2 mM (●).

Other agents, such as chloretone, chloroform, ether, cocaine, amyl alcohol and hexanol displayed similar but lesser selectivity, but urethane and the short-chain alcohols showed no selective action. In the case of urethane and some alcohols, it is interesting to note that in certain bathing concentrations these agents were found to produce exactly the opposite

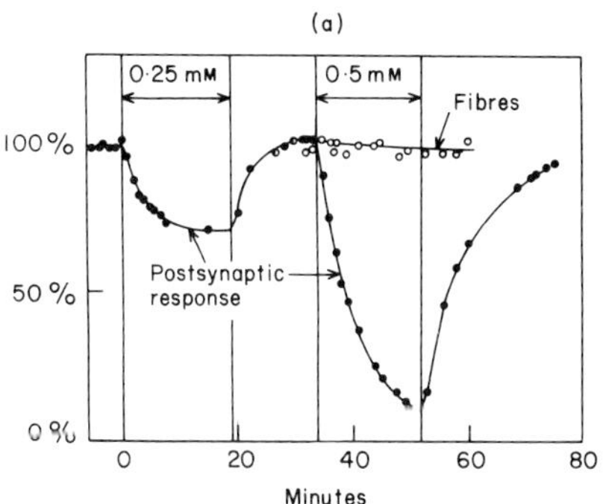


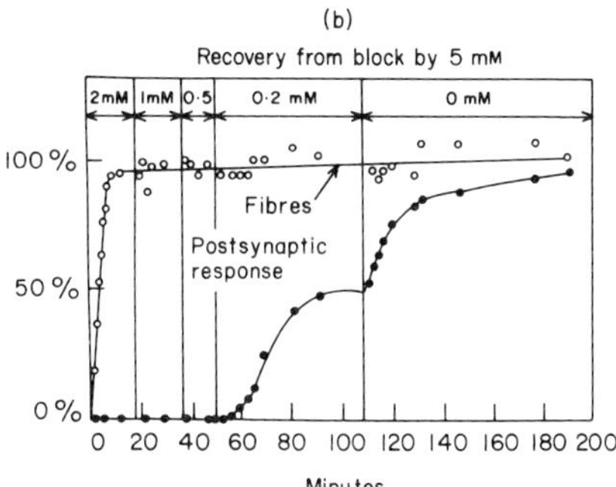


FIG. VI.2. Selective action of sodium pentobarbitone on synapses in the perfused stellate ganglion of the cat. Stimulation was applied to the preganglionic nerve (see Fig. VI.1). Ordinates represent the heights of action potentials which are synaptically induced ("postsynaptic response") or directly conducted ("fibres"). (a) shows the application and removal of two concentrations of pentobarbitone; (b) shows stepwise reductions in concentration after the application of 5 mM pentobarbitone which blocked conduction over the direct fibres as well as the postsynaptic response. (Larrabee and Posternak, 1952)

results, i.e. a depression of axonal conduction at a lower concentration than that required to block the synaptically evoked response.

Larrabee and Posternak (1952) concluded that all the anaesthetics with a high selectivity for synapses were of high molecular weight, so that in the case of the alcohols the degree of selective action increased as the number of C atoms in the chain increased. Rather surprisingly, it has proved possible to correlate the degree of selectivity of these agents with their octanol–$H_2O$ partition coefficient. This exercise has been performed recently by Barker (1975b) using the original data of Larrabee and Posternak (1952). The result of this analysis is shown in Fig. VI.3, in which the log of the ratios of the concentrations to block direct and synaptically evolved action potentials is plotted on the ordinate against the log of the partition coefficient on the abscissa. The conclusion reached is that the

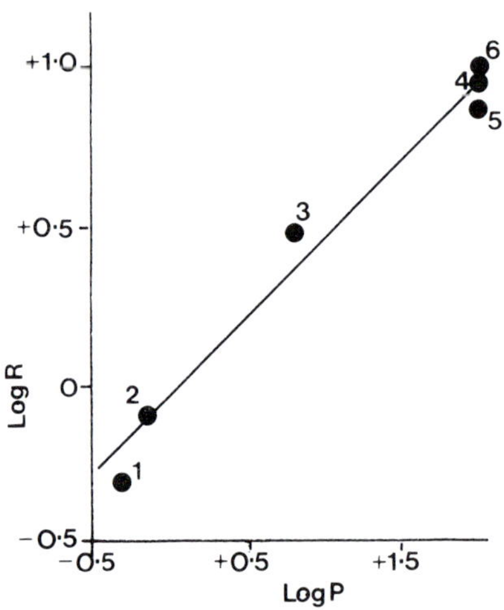


FIG. VI.3 Plot of the ratio of the concentration (Log R) depressing (by equal percentage) pre- and postsynaptically derived compound action potentials in the cat stellate ganglion as a function of the hydrophobicity (Log P) of the agents. 1 = ethanol, 2 = urethane, 3 = ether, 4 = chloretone, 5 = chloroform, 6 = pentobarbitone. (Barker, 1975b)

ratio of concentrations is significantly correlated with hydrophobicity. This has obvious immediate implications. It means, first, that one should bear in mind the hydrophobic nature of an agent when considering the kind of effect it might produce. The analysis predicts, also, that highly hydrophobic agents such as halothane and trichloroethylene should be more selective for synapses than, for example, nitrous oxide. One study on the dog stellate ganglion suggests that halothane is certainly more effective than cyclopropane or $N_2O$ in blocking ganglionic transmission, which is at least consistent with this notion (Price and Price, 1967). Further, it could be interesting to extend the analysis to the processes underlying synaptic transmission. Thus, is it possible to correlate hydrophobicity with depression of release of transmitter or depression of receptor sensitivity? The results documented below may begin to answer this question.

## 2 Presynaptic actions

### 2.1 RELEASE OF TRANSMITTER

Many studies have concluded that anaesthetics depress excitatory synaptic transmission by reducing the amount of transmitter released from the presynaptic terminals. For example, in the cat spinal cord, Løyning *et al.* (1964) and Weakly (1969) proposed that barbiturates depress transmitter release. Further, Richards and White (1975) suggested that volatile anaesthetics depressed synaptic transmission in the dentate gyrus of the guinea pig hippocampus by reducing the amount of transmitter released from the perforant path fibres.

Alternatively, there is good evidence that ethanol, pentobarbitone, chloral hydrate (a hypnotic), chloroform and chlorpromazine (a tranquilliser) increase spontaneous transmitter release at the mammalian neuromuscular junction (Gage, 1965; Quastel *et al.*, 1972) although the depolarisation-induced release is depressed by most of these agents (Quastel *et al.*, 1972). Furthermore, some barbiturates increase both the spontaneous and evoked release of acetylcholine at the frog neuromuscular junction (Thomson and Turkanis, 1973). Direct measurement of the amount of transmitter released in the presence of anaesthetics is technically difficult, but the results of a few studies are worth noting.

Matthews and Quilliam (1964) performed a comprehensive study of the effect of a range of agents including the local anaesthetic procaine, the anticonvulsant trimethadione and the hypnotic agent chloral hydrate on transmitter release from the perfused cat superior cervical ganglion and the isolated rat phrenic nerve-diaphragm preparation. Some of their results are shown in Table VI.1. The acetylcholine released was assayed biologically by measuring the reduction in blood pressure produced by an injection of a 0·3 ml sample of the test effluent into the rat femoral vein, or by monitoring the contraction of the leech dorsal muscle. It can be seen that at some concentrations all of the agents produced a reduction of acetylcholine output. Two points emerge from this study. In not all of the cases was the effect readily reversible, e.g. in the experiments with chloral hydate. In addition, choline in the perfusing fluid reduced the depressant effect of these agents, suggesting that one possible site of action was on the synthesis of the transmitter.

In comparison with this work, Gergis *et al.* (1975) reported no reduction

TABLE VI.1

Effect of central depressant drugs on release of acetycholine from perfused cat ganglia. Perfusions were with Locke solution, Locke + choline $(10^{-6})^a$ or Locke and plasma $(10\%)^b$. (From Matthews and Quilliam, (1964)

| Drug | Drug concentration (g ml$^{-1}$) | No. of experiments | Acetylcholine output (% depression) |
|---|---|---|---|
| Amylobarbitone | $5 \times 10^{-5}$ | 3 | 39 |
| | $10^{-4}$ | 2 | 52 |
| | $10^{-4}$ | $2^a$ | 33 |
| Chloral hydrate | $5 \times 10^{-4}$ | 1 | 0 |
| | $10^{-3}$ | 3 | 77 |
| | $10^{-3}$ | $2^a$ | 59 |
| Trichloroethanol | $5 \times 10^{-4}$ | 1 | 54 |
| Methylpentynol | $5 \times 10^{-4}$ | 1 | 49 |
| | $10^{-3}$ | 3 | 73 |
| | $10^{-3}$ | $2^a$ | 44 |
| | $2 \times 10^{-3}$ | $1^b$ | 100 |
| | $2 \times 10^{-3}$ | $1^a$ | 76 |
| Methylpentynol carbamate | $10^{-3}$ | 1 | 100 |
| Paraldehyde | $2 \times 10^{-3}$ | 1 | 39 |
| | $4 \times 10^{-3}$ | 4 | 31 |
| | $2 \times 10^{-3}$ | $1^a$ | 10 |
| | $4 \times 10^{-3}$ | $1^{a,b}$ | — |
| Trimethadione | $5 \times 10^{-3}$ | 4 | 61 |
| | $5 \times 10^{-3}$ | $1^a$ | 28 |
| | $3 \times 10^{-3}$ | $1^a$ | 4 |
| Procaine | $5 \times 10^{-5}$ | 1 | 77 |
| | $10^{-4}$ | 1 | 88 |

in acetylcholine release from the frog sciatic nerve–sartorius muscle preparation during exposure to four anaesthetic agents. The acetylcholine was assayed biologically, again in this case using the guinea pig ileum. Halothane (10 per cent) had no effect on the amount of acetylcholine released or the amplitude of the muscle contraction elicited by indirect stimulation, whereas twitch tension was reduced significantly in the presence of diethyl ether, methoxyfluorane and thiopentone. None of these agents, however, affected the acetylcholine release over the concentration range tested, although the concentrations used were probably higher than that associated with surgical anaesthesia. The motor nerve presynaptic terminals were thus quite resistant to anaesthetic action. In one other study, ether and chloroform were found to depress the acetylcholine release elicited by transmural stimulation of the guinea pig ileum, but trichloroethylene enhanced the resting release (Speden, 1965). In this study, the acetylcholine release was again monitored by measuring the contraction of the guinea pig ileum.

The work so far described has been performed on cholinergic synapses. It has also been reported that methoxyfluorane decreased the spontaneous and evoked release of catecholamines from the adrenal medulla (Dreyer *et al.*, 1974). Unfortunately, it was not possible to determine whether the depression of release was mediated via a change in the acetylcholine release from the preganglionic nerve terminals or not, and these authors concluded that the reduction in catecholamine output was due in part to a depressed sensitivity of the chromaffin cells for acetylcholine.

The effect of barbiturates on both spontaneous and electrically induced release of [$^3$H] GABA efflux has been studied recently in rat cerebral cortical slices (Cutler and Dudzinski, 1974; Cutler *et al.*, 1974). Pentobarbitone produced a dose-dependent depression of the spontaneous efflux of [$^3$H] GABA. In contrast, the spontaneous efflux of [$^{14}$C] glutamate was affected to a much smaller degree. In Fig. VI.4 are shown typical data displaying the suppression of the efflux of GABA by pentobarbitone and amylobarbitone. The differential effect on the efflux of GABA and glutamate is quite obvious over the full range of concentrations tested.

An action of the barbiturates on the electrically induced [$^3$H] GABA efflux was only evident at higher concentrations, since $10^{-4}$ M pentobarbitone produced no change and $5 \times 10^{-3}$ M produced a 46 per cent

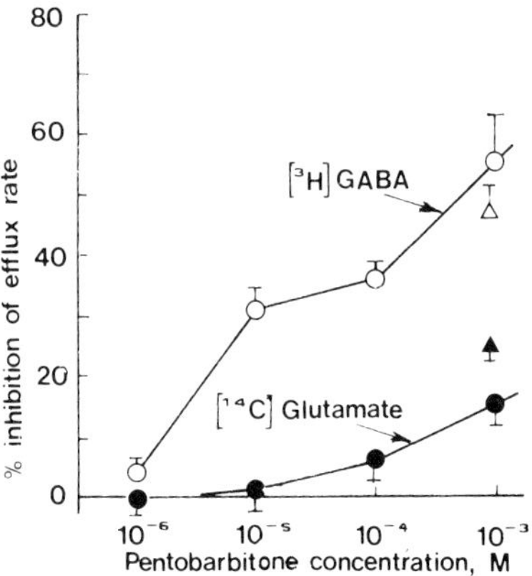


FIG. VI.4. The effect of pentobarbitone (circles) and amylobarbitone (triangles) on the rate of spontaneous efflux of [$^3$H] GABA (open symbols) and [$^{14}$C] glutamate (closed symbols) from rat cortex slices. Following 20 min of superfusion in plain medium, the barbiturate was added to the medium and the superfusion was continued for 20 min. Points are means and vertical bars are 1 × SEM of 5–9 slices. (Cutler *et al.*, 1974)

increase in the efflux (Cutler and Dudzinski, 1974), that is the inverse effect on the evoked efflux. Pentobarbitone and amylobarbitone, however, both inhibited the accelerated efflux of GABA produced by ouabain and raised $[K]_0$ (Cutler *et al.*, 1974). Of significance is the observation that phenobarbitone and hexobarbitone did not influence either the spontaneous or depolarisation-induced $[^3H]$ GABA efflux.

These barbiturates possess a ring structure as a substituted side chain at the $C_5$ position on the barbituric acid ring, whereas pentobarbitone and amylobarbitone possess 4-carbon chains. Since the GABA release mechanism from nerve terminals is thought to involve a carrier-mediated membrane transport system, occupation of the GABA transport site by a barbiturate may require the presence of this 4-carbon chain.

Many electrophysiological studies imply that presynaptic release mechanisms are susceptible to anaesthetics. In some work, terminal excitability is used as an index of the membrane potential of presynaptic terminals, which is well accepted as being a primary determinant of transmitter release. Thus, Krnjević and Morris (1976) have reported that in the cat cuneate nucleus, terminal excitability is lowered by the intravenous injection of sodium pentobarbitone. This was interpreted in terms of a presynaptic hyperpolarisation which would imply enhanced transmitter release and improved transmission. Such an increase in the efficiency of transmission was observed in this study. However, some caution should be exercised when correlating terminal excitability with membrane potential and hence transmitter release. Nicoll (1975) observed that in the frog spinal cord primary afferent terminal excitability increased in the presence of $4 \times 10^{-4}$ M pentobarbitone. This was accompanied by a small depolarisation of these terminals. However, in the presence of $4 \times 10^{-3}$ M pentobarbitone, excitability of the terminals decreased although the depolarisation was still maintained.

Definitive electrophysiological evidence that anaesthetics affect release mechanisms has been provided by intracellular recording at the neuromuscular junction, or in the spinal cord. Gage (1965) demonstrated that at the mammalian neuromuscular junction, ethanol (0·24 per cent v/v) produced an increase in the end-plate potential size and its quantal content. Quastel *et al.* (1972) reported that chlorpromazine, pentobarbitone, chloral hydrate and chloroform produced an increase in miniature end-plate potential frequency at the mammalian neuromuscular junction. In this respect, chlorpromazine was the most effective, producing a sixfold increase relative to control. These same authors, however, report that the same concentration of chlorpromazine (10 $\mu$M) depressed the presynaptic transfer function so that the increase of miniature end-plate potential (m.e.p.p.) frequency in response to depolarising current applied focally to an end-plate was depressed. In the presence of a constant level of depolarisation (20 mM $K^+$ in the bathing medium) the increase in m.e.p.p. frequency produced by changing $Ca^{2+}$ from 0·125 mM to 0·5 mM was reduced from twenty-sixfold in the control situation to twofold in the presence of 20 μM chlorpromazine. This is shown in Fig. VI.5. The peak frequency in 0·5 mM $Ca^{2+}$ is not reduced in the presence of chlorpromazine but the basal m.e.p.p. frequency in 0·125 mM $Ca^{2+}$ is significantly increased, so that the change induced by a rise in $Ca^{2+}$ is markedly reduced.

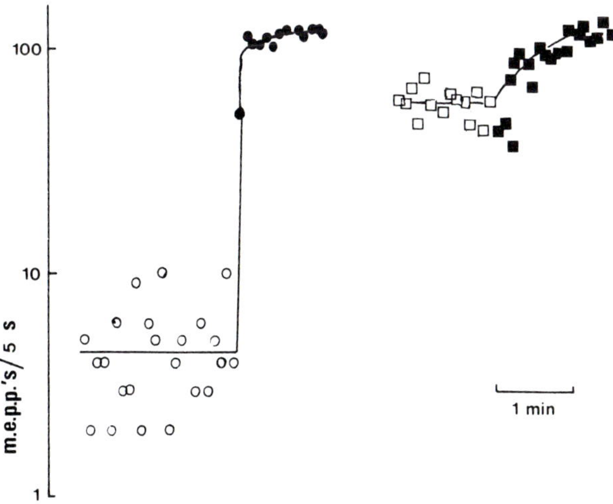


FIG. VI.5. The response of m.e.p.p. frequency in the mouse diaphragm to an increase of $Ca^{2+}$ from 0·125 mM (open symbols) to 0·5 mM (closed symbols) in the presence of 20 mM $K^+$. Circles are control values, squares refer to data obtained in 20 μM chlorpromazine (10 min exposure). (Quastel *et al.*, 1972)

The other agents tested also depressed the m.e.p.p. frequency change induced by raising $Ca^{2+}$. These authors consequently concluded that anaesthetic agents act by reducing the sensitivity of the $Ca^{2+}$-dependent release mechanism to depolarisation, and although the concentrations of anaesthetics used were often high these authors calculate that if the coupling between release and depolarisation were reduced by only 3 per cent, then the evoked release of transmitter by an action potential could be reduced by about 30 per cent.

Intracellular analysis of the quantal content of excitatory postsynaptic potentials (EPSPs) in cat motoneurones have demonstrated that reductions in transmitter release can occur in the presence of barbiturates (Weakly, 1969). In this study, EPSPs from triceps surae motoneurones of the cat were evoked by stimulation of afferent fibres in the muscle nerve and the amplitude of the "unit" EPSP, $v_1$, and the mean quantal content, $m$, were estimated in the presence and absence of thiopentone and pentobarbitone. The drugs were injected intravenously, the dose being 10 mg kg$^{-1}$. The quantal content, $m$, is defined as $m = \bar{V}/\bar{v}_1$, where $\bar{V}$ is the mean amplitude of the "compound" EPSP and $\bar{v}_1$ is the mean amplitude of the unit EPSP. The size of $\bar{v}_1$ is usually dependent primarily on postsynaptic factors whereas the size of $m$ depends on presynaptic changes, since it indicates the average number of "packets" of transmitter making up the "compound" EPSP. Weakly's data show clearly that in the concentrations employed the barbiturates produced a reduction in the mean EPSP amplitude, $\bar{V}$, which was accompanied by a parallel reduction in mean quantal content, $m$. No change in the size of the "unit" EPSP was observed, $\bar{v}_1$. A summary diagram of his results is shown in Fig. VI.6. The relationship between $\bar{V}$, $m$ and $\bar{v}_1$ may be written $\bar{V} = m\bar{v}_1$, so that the

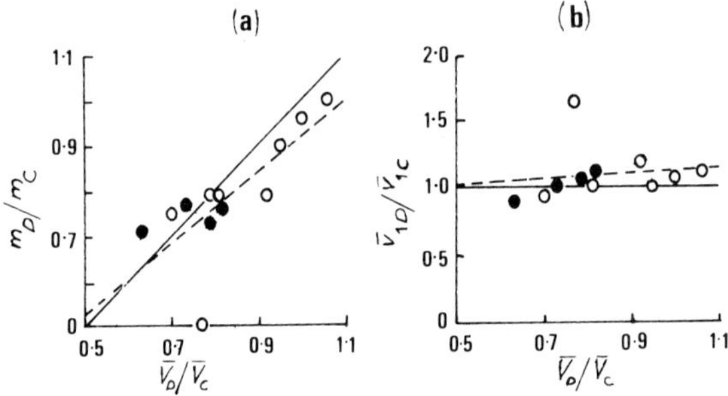


FIG. VI.6. (a) The relationship between relative changes in $m$ and mean amplitude of EPSP ($\bar{V}$) produced by thiopentone (5–20 mg kg$^{-1}$, intravenous, open circles) or pentobarbitone (10–15 mg kg$^{-1}$, intravenous, closed circles). The data were obtained from cat motoneurones. Ordinate, ratio of $m$ after the drug ($m_D$) to $m$ before the drug ($m_C$); $m$ defined as in text. Abscissa, ratio of mean EPSP amplitude after the drug ($\bar{V}_D$) to mean EPSP amplitude before the drug ($\bar{V}_C$). Solid line represents expected relationship between $m_D/m_C$ and $\bar{V}_D/\bar{V}_C$ if depression of the EPSP is due solely to a decrease in $m$. The interrupted line is the calculated regression line for the observed points. (b) Relationship between relative changes in mean unit potential amplitude ($\bar{v}_1$) and $\bar{V}$ produced by thiopentone (open circles) or pentobarbitone (closed circles). Ordinate, ratio of $\bar{v}_1$ after the drug ($\bar{v}_{1D}$) and $\bar{v}_1$ before the drug ($\bar{v}_{1C}$). Abscissa, same as in (a). Continuous line represents expected relationship between $\bar{v}_{1D}/\bar{v}_{1C}$ and $\bar{V}_D/\bar{V}_C$ if changes in $\bar{V}$ produced by the drugs are not accompanied by changes in $\bar{v}_1$. The interrupted line is the calculated regression line for the observed points. (Weakly, 1969)

ratio of the amplitudes of the compound EPSP in the presence and absence of barbiturate may be written

$$\frac{\bar{V}_D}{\bar{V}_C} = \frac{m_D \bar{v}_{1D}}{m_C \bar{v}_{1C}}$$

where the subscripts $C$ and $D$ indicate the absence (control) and presence of the drug, respectively. Referring to Fig. VI.6, it can be seen that in (a) the reduction in the ratio $\bar{V}_D/\bar{V}_C$ can be accounted for by a similar reduction in the ratio $m_D/m_C$ and that the reduction is not accompanied by any change in the ratio $\bar{v}_{1D}/\bar{v}_{1C}$ (b).

The reported effects of two barbiturates on transmitter release at the frog neuromuscular junction provide an interesting companion study to Weakly's results. Thomson and Turkanis (1973) examined the action of a convulsant barbiturate, 5-(2-cyclohexylideneëthyl)-5-ethyl barbituric acid (CHEB), and phenobarbitone on the quantal content, m.e.p.p. amplitude and frequency. Both agents at a concentration of 200 $\mu$M increased mean quantal content markedly and decreased the m.e.p.p. amplitude. With CHEB a significant increase in m.e.p.p. frequency was also observed. Extracellular recording of the nerve terminal action potential showed that this was prolonged in the presence of both barbiturates (Fig. VI.7). The action potential duration increased after a 15 min exposure to the agent and continued to increase for up to 60 min. At this time, in the presence of CHEB, the mean quantal content, mean duration and amplitude of the action potential were 140, 130 and 82 per cent of control values. Thus, this agent increased $m$ and the duration of the action potential correspondingly, but decreased the action potential amplitude. Phenobarbitone often failed to alter the amplitude of the terminal action potential. These results, however, suggest that the barbiturate-induced increase in quantal content is due to a prolongation of the terminal action potential.

The results presented above suggest that barbiturates may both stimulate and depress transmitter release. The discrepancy between the results of Weakly (1969) and Thomson and Turkanis (1973) may be due to the different synaptic preparations used, or the result of different anaesthetic concentrations, or it may reflect genuine differences in the mechanism of action of these barbiturates. It has been shown, for example, that CHEB does not affect presynaptic inhibition like the other barbiturates (Nicoll, 1975).

## 2.2 MICROTUBULE INVOLVEMENT

The fact that transmitter release is often depressed by anaesthetic agents raises the possibility that microtubular disruption may be somehow in-

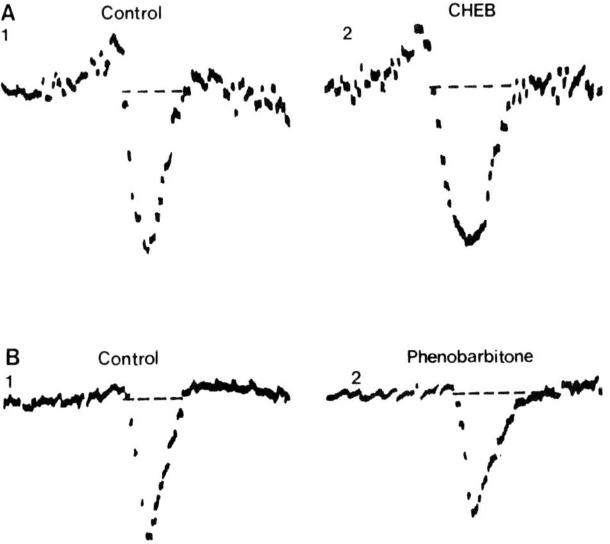


FIG. VI.7. Effects of 5-(2-cyclohexylideneëthyl)-5-ethyl barbituric acid (CHEB) and phenobarbitone on the amphibian nerve-terminal action potential. Concentration of drugs, 200 $\mu$M. Potentials recorded extracellularly with glass microelectrodes. Each trace represents the sum of 300 action potentials. The dashed line indicates the width of the negative phase of the action potential measured at the level of the baseline. A1, controls; A2, 120 min after CHEB; B1, controls; B2, 120 min after phenobarbitone. (Thomson and Turkanis, 1973)

volved in this action. Microtubules have been implicated in the release of catecholamines from the adrenal medulla (Poisner and Bernstein, 1971) and it has been suggested that they participate in the transport of protein (Schmitt, 1968) and noradrenaline storage vesicles (Banks and Till, 1975). The relationship between microtubules and synaptic vesicles, and the possible role of microtubules in the transport and release of vesicles, has been described by Smith (1971).

Allison and Nunn (1968) first proposed that general anaesthetics reversibly disaggregate microtubules, producing synaptic effects which could contribute to the state of anaesthesia. Such disaggregation of microtubules has been observed in protozoan axopods and other cells (see Chapter II and Allison, 1974). Since some anaesthetics cause displacement of $Ca^{2+}$ from membranes (Seeman, 1972) and the assembly of microtubules is known to be inhibitied by even small increases in free $Ca^{2+}$ (Weisenberg, 1972), it is feasible that disruption is brought about via a change in intracellular $Ca^{2+}$.

Present studies suggest that an anaesthetic effect on transmitter release which was mediated via a change in microtubule structure would only occur at relatively high concentrations. Thus, Hinkley and Green (1971) report that concentrations of halothane which were high enough to block conduction in rabbit vagus nerve, increased significantly the number of microtubules per $\mu m^2$ of axonal area. Experiments with saturated concentrations of halothane caused the microtubules to disappear but also disrupted the axonal and Schwann cell membranes and the myelin layers of the larger axons. It is consequently likely that the basis of disaggregation was as much due to the changed ionic environment of the microtubule assembly. If Barker's prediction about hydrophobicity is correct (see above), then halothane would be expected to block transmission at a much lower concentration than that required to produce axonal block. Such lower concentrations would presumably have little effect on microtubules in the nerve axon or its terminals. It can be concluded, therefore, on current evidence, that the depression of release is not mediated via tubular disaggregation. Similarly, *in vitro* studies suggest that 3·25 per cent halothane (considered a lethal dose equal to about twice the blood level in anaesthetised rabbits) produced no reduction in the rapid axoplasmic transport of labelled protein in rabbit vagus nerves (Kennedy *et al.*, 1972), implying that the microtubules were functionally intact. In fact, these same authors showed that axonal microtubules had a normal appearance in the presence of 7·8 per cent halothane, even after a 3 hr exposure. It would be difficult at present, therefore, to ascribe any action of anaesthetics on transmitter mobilisation to an effect on the microtubular transport process.

### 2.3 EFFECTS ON THE MITOCHONDRION

The effects of anaesthetics on the respiration of mitochondria from inexcitable cells are mentioned in Chapter III. Pertinent studies on the effects of anaesthetics on nerve terminal mitochondria do not exist, yet any decrease in mitochondrial activity may have profound effects on synaptic activity. It has been proposed (e.g. Borle, 1973) that in liver and kidney cells the mitochondria regulate the intracellular level of free $Ca^{2+}$. In these cells, the inner membrane area of mitochondria is estimated to be thirty times that of the surface membrane, and since the pump of the two membrane systems has the same transporting capacity for $Ca^{2+}$ it is clear that the intracellular free $Ca^{2+}$ level is almost entirely controlled by the mitochondria. It is difficult to gauge the relevant contribution of terminal mitochondria to the control of terminal $Ca^{2+}$, but it is reasonable to suppose that these structures are subcellular traps for $Ca^{2+}$. In squid axon, at least, mitochondria probably take up $Ca^{2+}$ which may be released into the cytoplasm in the absence of ATP (Blaustein and Hodgkin, 1969).

In the giant synapse of squid, it has been shown that intraterminal injections of $Ca^{2+}$ enhance spontaneous (Miledi, 1973) and depress evoked release (Kusano, 1970). By analogy with the proposed effects of Li or ouabain on transmitter release at the amphibian neuromuscular junction (Baker and Crawford, 1975; Crawford, 1975), anaesthetics may also affect $Ca^{2+}$ uptake into mitochondria resulting in changes in terminal $Ca^{2+}$ and therefore in release of transmitter. Recently, relevant studies have been undertaken on the effects of halothane (Rosenberg and Haugaard, 1973) and chlorpromazine and related agents (Tijoe *et al.*, 1971) on $Ca^{2+}$ uptake into isolated rat brain mitochondria. These studies demonstrated that under certain conditions these anaesthetics depressed reversibly $Ca^{2+}$ uptake, and that the ATP-dependent $Ca^{2+}$ uptake was more sensitive to inhibition (Tijoe *et al.*, 1971). In particular, Rosenberg and Haugaard (1973) showed that halothane in concentrations comparable to that found during clinical anaesthesia produced a reduction in mitochondrial $Ca^{2+}$ uptake. Typical data are shown in Fig. VI.8. It can be seen that the

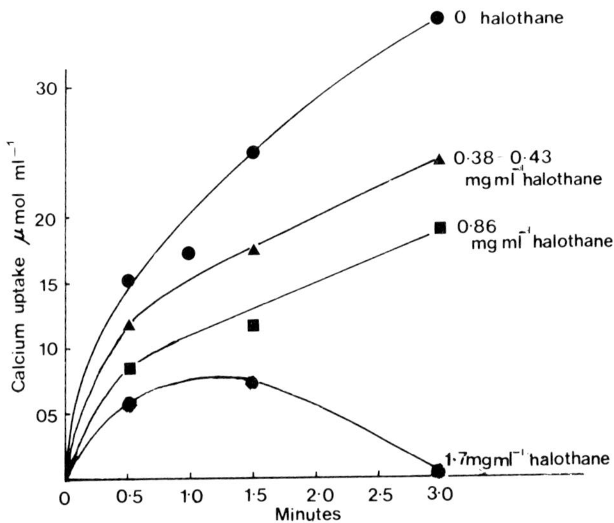


FIG. VI.8. Calcium uptake by brain mitochondria in the presence of glutamate, ATP and inorganic phosphate. Each point represents the mean of the results from six experiments and is the difference between the mean initial calcium concentration and the mean calcium concentration after a given period of incubation either with or without the anaesthetic halothane. (Rosenberg and Haugaard, 1973)

depression of uptake was dose-dependent and comparatively rapid in onset. These authors cite blood levels during anaesthesia of 0·2 mg $ml^{-1}$ for man and 0·5 mg $ml^{-1}$ for dogs. From their data on isolated rat brain mitochondria it is evident that this range of concentrations would slightly depress the $Ca^{2+}$ sequestering capacity of this system during a 2–3 min exposure. At higher concentrations, the $Ca^{2+}$ uptake is further reduced so that it is conceivable that the enhanced spontaneous and reduced evoked release of transmitter suggested by various studies is due to a rise in terminal $Ca^{2+}$, secondary to a depression of mitochondrial function. However, since it has been reported that the uptake of $Ca^{2+}$ into brain mitochondria has different characteristics from the uptake into mitochondria from other systems (Tijoe *et al.*, 1970), it is important that much more data is obtained with relevant mitochondria before too much emphasis is placed on such a mitochondrial role.

Krnjević (1974a) has nevertheless suggested that a reduction in mitochondrial $Ca^{2+}$ sequestering capacity is central to the synaptic effect of anaesthetics. Presynaptically, such a depression should lead to a reduced evoked release but an enhanced spontaneous release, following the rise in terminal $Ca^{2+}$. It might be anticipated that the terminal membrane would hyperpolarise as intracellular $Ca^{2+}$ increased, since $K^+$ conductance of many membrane systems is known to be $Ca^{2+}$-dependent (e.g. Meech, 1974; Goldsmith *et al.*, 1975). Current evidence suggests that pentobarbitone caused depolarisation of the terminal membrane (Nicoll, 1975), although from terminal excitability measurements a hyperpolarisation has been inferred (Krnjević and Morris, 1976). At present some effects, e.g. enhanced spontaneous release, are consistent with preterminal depolarisation, but others such as enhanced evoked release (e.g. Krnjević and Morris, 1976) suggest terminal hyperpolarisation. Consequently, the action of anaesthetics on the preterminal membrane potential are probably complex.

### 2.4 UPTAKE OF TRANSMITTER

Not many studies have been directly concerned with the possibility that anaesthetics might alter the mechanisms underlying removal of the transmitter from its site of action. However, as discussed below, the barbiturates are known to prolong both pre- and postsynaptic inhibition, and this has probably prompted studies on the effects of barbiturates on the uptake processes which are responsible for the termination of the action of GABA (Iversen and Neal, 1968). Such studies revealed that pentobarbitone (2–5 $\times 10^{-3}$ M) inhibited competitively both the high- and low-affinity uptake systems for GABA (Cutler *et al.*, 1974). This effect alone could explain the observed action of barbiturates at GABA-mediated synapses. It should be noted that the inhibition of GABA uptake may only occur at high concentrations of barbiturate (e.g. Jessell and Richards, 1977).

### 2.5 TRANSMITTER SYNTHESIS AND MOBILISATION

Most experimental work seems to discount the possibility that a change in synthesis or mobilisation underlies synaptic depression observed with anaesthetic agents. This conclusion stems from the observation that anaesthetics do not alter the facilitation or post-tetanic potentiation of the EPSP (Løyning *et al.*, 1964; Barker, 1975b; Richards and White, 1975), nor do they affect the change in EPSP amplitude observed as the frequency is varied (Løyning *et al.*, 1964). In Fig. VI.9 is shown the effect of high-

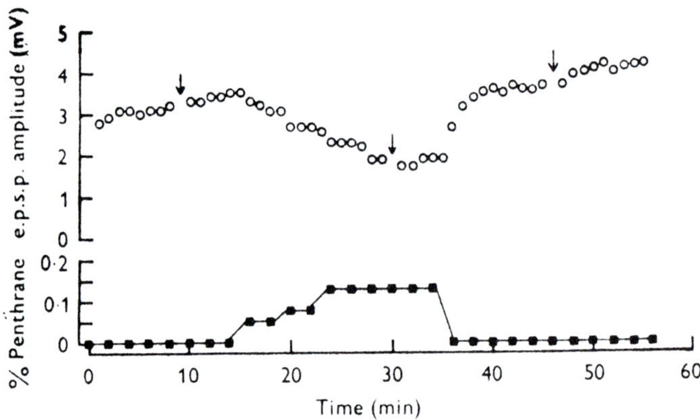


FIG. VI.9. The effect of high-frequency stimulation on the evoked population EPSP recorded in the dentate gyrus before, during and after exposure to methoxyfluorane (Penthrane). At each arrow, stimuli were delivered at 20 Hz for 72 sec. (Richards and White, 1975)

frequency stimulation on the evoked population EPSP of hippocampal neurones before, during and after exposure to methoxyfluorane. In this experiment, high-frequency stimulation produced no change in the amplitude of the EPSP in the absence of the anaesthetic. It is clear that the application of anaesthetic still caused no change in the amplitude of the EPSP following such high-frequency stimulation, showing the anaesthetic did not promote fatigue in these neurones.

It is probable that higher doses of anaesthetics may ultimately interfere with the processes responsible for synthesis and for mobilisation, but most studies have been restricted, for obvious reasons, to dose ranges closest to the clinical values.

## 3 Postsynaptic actions

### 3.1 ALTERED INPUT CHARACTERISTICS

Anaesthetics are capable of producing direct effects on the postsynaptic membrane, in some cases at lower aqueous concentrations than that required to produce presynaptic effects. Somjen (1963) suggested that the primary action of these agents is at the level of the postsynaptic membrane.

Harmel and Malcolm (1958) reported that in the frog spinal cord the dendritic response of the motoneurone to an antidromic ventral root volley was most sensitive to procaine: that is to say, that conduction of the action potential into the dendrites had the lowest safety factor. This was confirmed by Curtis and Phillis (1960), who demonstrated that procaine applied microiontophoretically in the cat spinal cord depressed the action potential response of motoneurones, Renshaw cells and interneurones whether this was elicited by antidromic stimulation, by orthodromic volley, by direct electrical stimulation or by application of chemical excitants. The results from an experiment on a motoneurone are presented in Fig. VI.10. In this study, procaine was passed iontophoretically around a motoneurone from which intracellular recordings were being made (Fig. VI.10G). The spike responses of Fig. VI.10A–F were obtained from a motoneurone which was fired directly (ii) by rectangular current pulses applied intracellularly through the central barrel of the electrode. The size of the current is shown in (i). The antidromic spike potential elicited by stimulation of the ventral root is shown in (iii). As shown, the threshold to direct stimulation was increased reversibly by the application of procaine (B–D) as indicated by the size of the current (i) required to elicit an action potential. Simultaneously, the antidromic invasion of the SD segment was blocked (C and D). In the same study, Curtis and Phillis (1960) reported that the motoneurone spike generated by orthodromic volley often failed, despite the presence of a larger EPSP in the presence of procaine, thus providing strong evidence that with this anaesthetic, synaptic failure is due to a postsynaptic threshold change.

Similar results have been observed for ethanol in the cat spinal cord by Eidelberg and Wooley (1970). These authors showed that in addition to depressing the SD component of the motoneurone action potential, at a concentration of 600 mg $kg^{-1}$, ethanol decreased the resting potential and increased the input resistance of the cell. Depression of excitability of afferent terminals only occurred at the same or higher doses.

Not all anaesthetics, however, exert a uniform action postsynaptically. In the case of diethyl ether and thiopentone it appears that the threshold for generation of an action potential elicited by orthodromic volley is increased

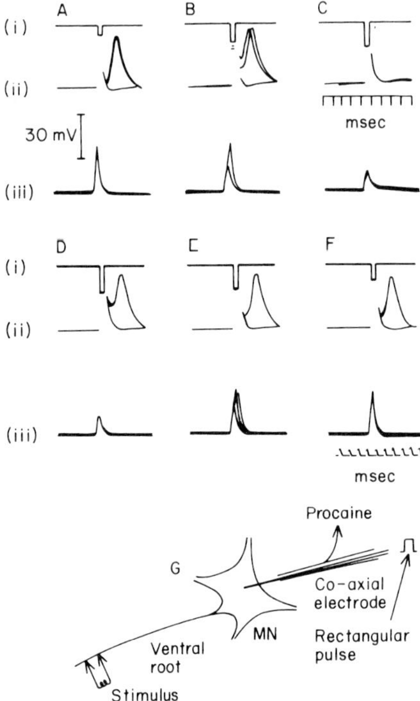


FIG. VI.10. A–F, responses recorded from a cat motoneurone (MN) by means or the inner $K_2SO_4$-containing barrel of a co-axial electrode, as illustrated in G. The outer barrel contained a saturated solution of procaine hydrochloride from which procaine was applied as a cation for 115 sec after the control responses A were recorded. Each record consists of six superimposed responses at a frequency of 1 $sec^{-1}$, the antidromic and directly evoked spikes being recorded 500 msec apart. A, control responses: (i) upper beam, relative magnitude of the current pulse which had a duration of 0·3 msec and evoked a spike in roughly half the tests; (ii) middle beam, spike potential evoked by a rectangular current pulse passed through the intracellular electrode; (iii) lower beam, spike potential evoked by stimulating the ventral root. B–F, as for A except that the pulse size was altered in order to straddle the threshold for direct excitation. B, 90 sec; C, 110 sec after the commencement of the iontophoretic current; D, 25 sec; E, 40 sec; and F, 75 sec after its termination. Time marker, msec for (i) and (ii) below middle beam of C; msec for the antidromic spike (iii) below F. Voltage, 30 mV for (ii) and (iii). (Curtis and Phillis, 1960)

despite no change in the configuration of the antidromic spike (Somjen and Gill, 1963). In Fig. VI.11 are shown tracings of intracellular records taken from a motoneurone in the cat spinal cord. Inhalation of ether produced no change in the antidromic spike potential (lower records), but the synaptically evoked action potential was delayed in onset due to the more slowly rising EPSP, and the threshold was increased as the motoneurone presumably accommodated to the slowly rising synaptic current.

It is worth noting that although the threshold for action potential generation was increased in fifteen of the thirty-one motoneurones tested

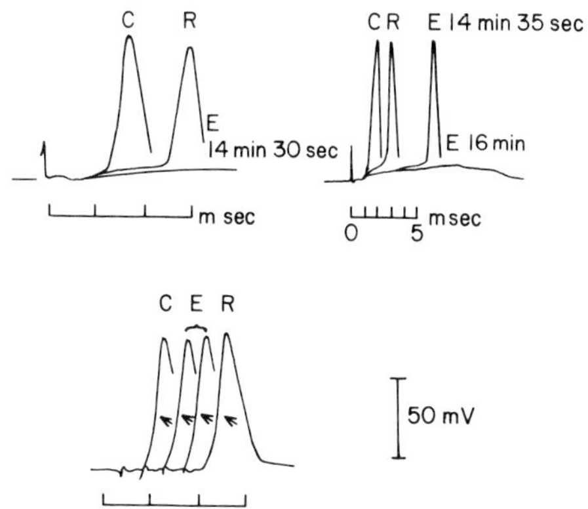


FIG. VI.11. Tracings of intracellular recordings from a motoneurone in $L_7$ segment of a decapitate cat. Above: responses to stimulation of dorsal root recorded on two different sweep speeds; C, unanaesthetised control; E, during ether inhalation, with time elapsed since onset of administration shown in minutes and seconds; R, about 25 min after cessation of ether. Below: antidromically conducted spikes; E, at 14 min 25 sec, and 17 min after start of ether; C and R as before. Arrows mark inflections between "IS" and "SD" spikes. Voltage calibration common to all records. Timebase calibration in msec. (Somjen and Gill, 1963)

in this study, depression of the EPSP produced the final synaptic block. Although Somjen and Gill (1963) observed an increase of threshold with intravenous administration of thiopentone (30–55 mg $kg^{-1}$), it is evident from other work that lower doses of thiamylal sodium, thiopentone or pentobarbitone (10 mg $kg^{-1}$) produced no effect on threshold or input resistance of cat motoneurones, although they did reduce the release of transmitter (Løyning *et al.*, 1964; Weakly, 1969).

The effects of anaesthetics at a postsynaptic site have also been studied at a supraspinal level. In an interesting comparison study, Galindo (1969) examined the action of microiontophoretically applied procaine and pentobarbitone on synaptic transmission in the cuneate nucleus of the decerebrate cat. To distinguish pre- from postsynaptic effects, a relay unit (projecting to the thalamus) in the cuneate was indirectly stimulated orthodromically and antidromically. Extracellular action potentials were recorded from the unit using the centre barrel of a five-barrelled glass microelectrode. Procaine depressed both orthodromic and antidromic spikes, non-specifically, implying that this agent was acting both pre- and postsynaptically. This is shown in Fig. VI.12. On the other hand, in the same experiment, GABA depressed only the orthodromic spike as expected. Other experiments with microiontophoretically applied pentobarbitone suggested that this agent could depress orthodromic activation of the unit

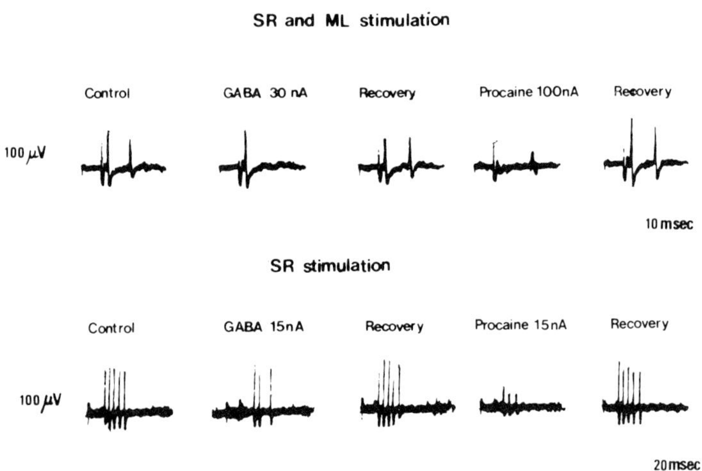


FIG. VI.12. Comparison between the specific postsynaptic blockade induced by γ-aminobutyric acid (GABA) and the non-specific blockade by procaine in the cuneate nucleus of the cat. On the upper row, the same relay unit was simultaneously excited (first deflection marks stimulus artifact) from the superficial radial nerve (SR) and the medial lemniscus (ML). Because of the shorter latency of the latter, the antidromic spike appears first, followed 5 msec later by the orthodromically evoked spike. Recording was made through the central barrel of a five-barrelled micropipette. GABA and procaine were released iontophoretically from two lateral barrels. The former depressed only the orthodromic spike, whereas procaine depressed both. On the lower row a cuneate cell responds to superficial radial nerve stimulation with a burst of five spikes. Both compounds reduce the size of the burst. (Galindo, 1969)

at a lower concentration (i.e. lower releasing current), suggesting a more pronounced presynaptic effect. Halothane given systemically or by superfusing the cuneate produced no postsynaptic effect despite depressing transmission.

Richards and White (1975) demonstrated convincingly that the synaptic depression observed in their studies with halothane and other volatile anaesthetics was not due to an alteration in the threshold of the postsynaptic cell. In the granule cells of the hippocampus there is a strong relationship between the size of the population spike and the population EPSP. Raising the $Ca^{2+}$ and $Mg^{2+}$ concentrations of the c.s.f. increases the threshold for spike generation of these cells. This is shown in Fig. VI.13. However, superfusion of the hippocampal slices utilised in this study with a variety of volatile anaesthetics did not alter the relationship between the spike size and the population EPSP size, suggesting that little or no change in threshold of the cells had occurred (see Fig. VI.13).

The results described above, therefore, seem to some extent consistent, since the local anaesthetics procaine and ethanol can produce primarily postsynaptic depression, the barbiturates depress preferentially at a presynaptic level, and the volatile anaesthetics produce no changes in the input characteristics of the postsynaptic cell. Findings from invertebrate preparations are not in agreement with these results, however, since Chase (1975) found that ethanol, 2–6 per cent, produced depolarisation but a decline in the input resistance of pedal ganglion cells of *Tritonia*, and that postsynaptic changes could not account for the EPSP depression. Also, Chalazonitis (1967) reported that the volatile anaesthetics, ether, chloroform and halothane, produced hyperpolarisation and a marked reduction in the input resistance of *Aplysia* and *Helix* ganglion cells. In the latter study, synaptic block was brought about by an increase in the threshold for spike generation, and the secondary decrease in EPSP amplitude was ascribed to the reduction in input resistance of the postsynaptic cell. Although the concentrations used were higher than those used in the vertebrate studies (e.g. Richards and White, 1975), it is likely that genuine differences in the susceptibility of synaptic processes in invertebrate and vertebrate preparations do exist.

As mentioned previously, Krnjević (1972, 1974a, 1975) has proposed that anaesthetics may produce a synaptic effect via an action on mitochondrial $Ca^{2+}$ sequestering capacity. A depression of $Ca^{2+}$ uptake in the presence of anaesthetics would result in a rise in intracellular $Ca^{2+}$ in the postsynaptic cell. This in turn may lead to an increase in $K^{+}$ permeability of the cell membrane with resultant hyperpolarisation, since it has been shown that in some cells $K^{+}$ permeability is modulated by a rise in intracellular $Ca^{+}$ (Whittam, 1968; Krnjević and Lisiewicz, 1972; Meech, 1974).

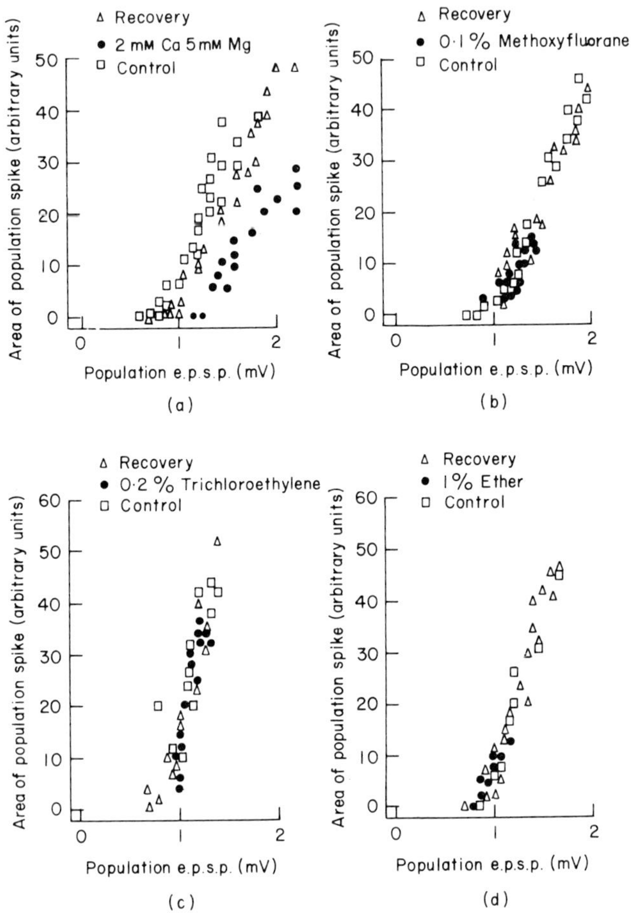


FIG. VI.13. The relationship between the population EPSP and the population spike of hippocampal granule cells and the action of various substances on it. (a) the effect of increasing extracellular $Ca^{2+}$ and $Mg^{2+}$ from 1 mM $Ca^{2+}$, 2 mM $Mg^{2+}$ to 2 mM $Ca^{2+}$, 5 mM $Mg^{2+}$; (b) the action of 0·1 per cent methoxyfluorane; (c) the action of 0·2 per cent trichloroethylene; (d) the action of 1 per cent ether. Elevating $Ca^{2+}$ and $Mg^{2+}$ shifted the relationship to the right, indicating an increase in the threshold of the granule cells. The anaesthetics had no effect on the relationship. (Richards and White, 1975)

Thus, injections of $Ca^{2+}$ ions into invertebrate neurones and cat spinal motoneurones produce hyperpolarisation and an increase of $K^+$ conductance (Krnjević and Lisiewicz, 1972; Meech, 1974). DNP and other metabolic inhibitors produce a similar hyperpolarisation and a reduction in input resistance of cat cortical neurones which is presumed to follow a rise in intracellular $Ca^{2+}$ (Godfraind *et al.*, 1971), since DNP is an inhibitor of both the energy-dependent Ca uptake and the high-affinity binding of $Ca^{2+}$ by mitochondria (Carafolli and Rossi, 1971).

In addition, Catchlove *et al.* (1972) noted that DNP and general anaesthetics had similar effects on cortical neurone firing rate, which is consistent with the suggestion that these agents act through a similar mechanism. Figure VI.14 shows the effects of the intravenous administration of methohexital, or the inhalation of halothane or methoxyfluorane on the firing rate of a cat postcruciate unit activated by acetylcholine. These agents depressed the acetylcholine-induced increase in firing rate, whereas the responses to glutamate were often enhanced. Similar effects were observed with ether, cyclopropane, trichloroethylene and $N_2O$. Since the acetylcholine-induced effects in cortical neurones are associated with a reduction in membrane conductance to $K^+$(Krnjević *et al.*, 1971) it is possible that the depression by anaesthetics of the response to microiontophoretically applied acetylcholine is brought about because anaesthetics and acetylcholine act antagonistically on the mechanisms underlying changes in membrane $K^+$ conductance.

Sodium pentobarbitone (Nembutal) ($10^{-3}$ g $ml^{-1}$) has also been shown to increase the permeability of the postsynaptic membrane to $K^+$ in *Aplysia* ganglion cells (Sato *et al.*, 1967). At the same time, low concentrations ($10^{-5}$ g $ml^{-1}$) block the acetylcholine ($10^{-6}$ g $ml^{-1}$) response in both D and H cells before the membrane is hyperpolarised. This suggests that the primary postsynaptic effect of pentobarbitone involves a binding to the receptor site, and these authors concluded that the hyperpolarisation observed was due to a direct effect of this agent on the acetylcholine receptor and not due to an effect mediated via a rise in intracellular $Ca^{2+}$.

### 3.2 DEPRESSION OF THE EPSP

Intracellular studies have revealed that barbiturates depressed the EPSP amplitude in the vertebrate spinal cord (Somjen and Gill, 1963; Løyning *et al.*, 1964; Weakly, 1969) and that a variety of anaesthetic agents reduced the height of the EPSP recorded from invertebrate cells (Chalazonitis, 1967; Chase, 1975; Barker, 1975a). Extracellular studies have also shown that volatile anaesthetics reduced the height of the population EPSP recorded from mammalian olfactory cortical cells (Richards *et al.*, 1975)

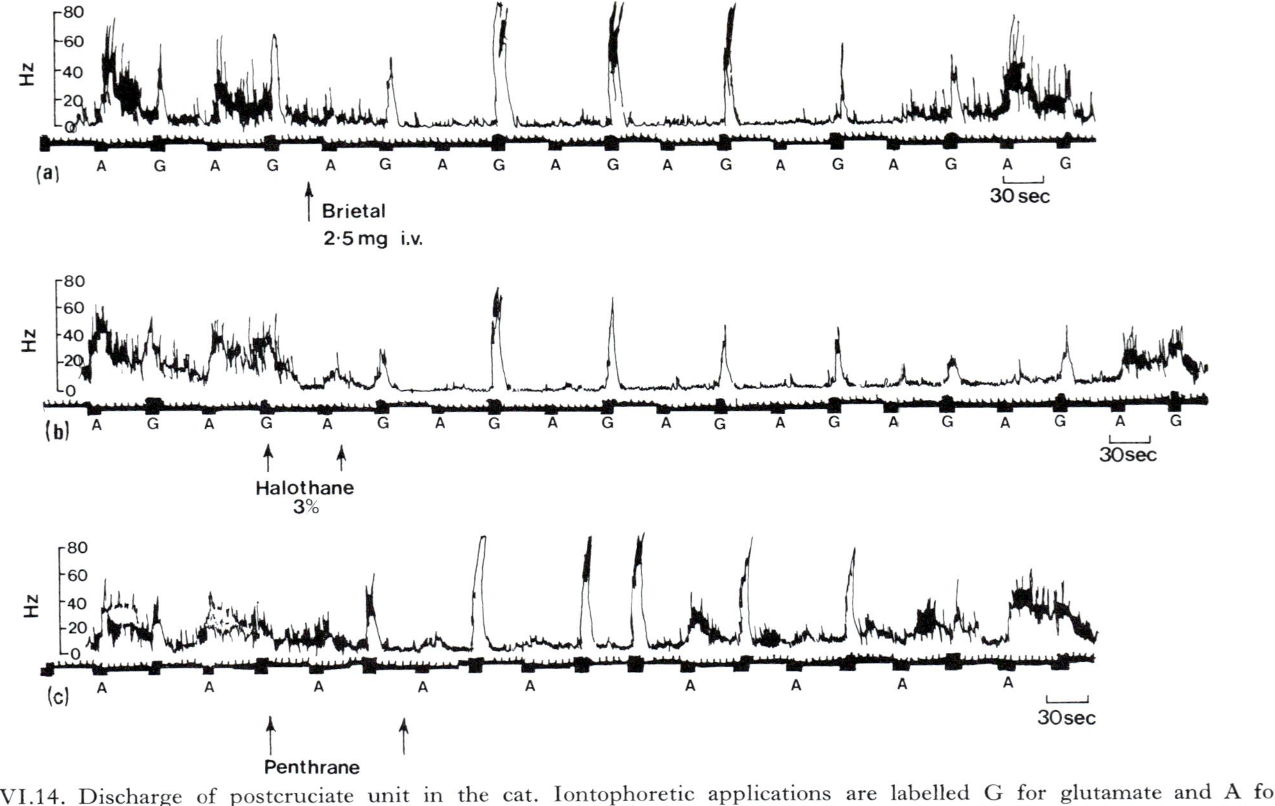


FIG. VI.14. Discharge of postcruciate unit in the cat. Iontophoretic applications are labelled G for glutamate and A for acetylcholine (both 100 nA). Traces show effects of giving methohexital (Brietal) intravenously (a) and halothane (b) or methoxyfluorane (Penthrane) (c) by inhalation. The Penthrane concentration was not known precisely, but was estimated as 2 per cent. (Catchlove *et al.*, 1972)

and the granule cells of the dentate gyrus (Richards and White, 1975). With the local anaesthetic procaine, however, synaptic block was clearly associated with no reduction in the EPSP amplitude (Curtis and Phillis, 1960).

Definitive work in the cat spinal cord has shown that thiamylal (a short-acting barbiturate) reduced the EPSP amplitude at concentrations which have no effect on the antidromic spike. This is shown in Fig. VI.15. In A is shown the EPSP recorded intracellularly from a motoneurone of a cat lightly anaesthetised with sodium pentobarbitone. In B is shown the antidromic spike and in C and D the hyperpolarising current (bottom trace) required to block the IS–SD and axon–IS components of the spike (top trace). These currents were presumably proportional to the firing threshold of these components. The top row represents control (arrow *a* in E), the middle and bottom rows refer to potentials recorded at the time indicated by *b* and *c* in E. From A and E it is evident that an intravenous injection of thiamylal $Na^+$ (10 mg $kg^{-1}$) depressed reversibly the height of the EPSP without any detectable change of the antidromic spike (B) or the firing threshold (C and D). Further, the time constant of decay of the EPSP was not affected, implying that the passive electrical characteristics of the motoneurone membrane remained unchanged.

Weakly (1969) produced convincing evidence that the EPSP reduction by pentobarbitone or thiopentone (10 mg $kg^{-1}$) in the cat motoneurone was due to a reduction in the quantal content and not the average amplitude of the "unitary" EPSP. Hence, at the concentration of thiamylal used by Løyning *et al.* (1964) (i.e. 10 mg $kg^{-1}$ in Fig. IV.15) it is likely that the depression observed was due to a reduction in transmitter release and not due to postsynaptic factors.

Excitatory transmission in the guinea pig olfactory cortex and the dentate gyrus of the hippocampus is depressed by low concentrations of volatile anaesthetics, and good evidence exists that this depression is also due to a presynaptic effect (Richards, 1974; Richards and White, 1975; Richards *et al.*, 1975). These experiments were conducted on isolated thin slices of brain tissue whose anatomical organisation is well defined and from which stable field potentials may be recorded with extracellular electrodes. The results showed that the population EPSP elicited from both these preparations in response to stimulation of an afferent input, that is either the lateral olfactory tract (olfactory cortex) or the perforant path fibres (dentate gyrus), was depressed by comparatively low concentrations of anaesthetics. The data obtained are shown in Table VI.2. It can be seen that the anaesthetics tested produced a 50 per cent depression of the population EPSP and, as a result, depression of the resulting population spike at doses close to the minimum alveolar concentrations (MAC) values. In addition, the perforant path fibre EPSP was more susceptible to depression

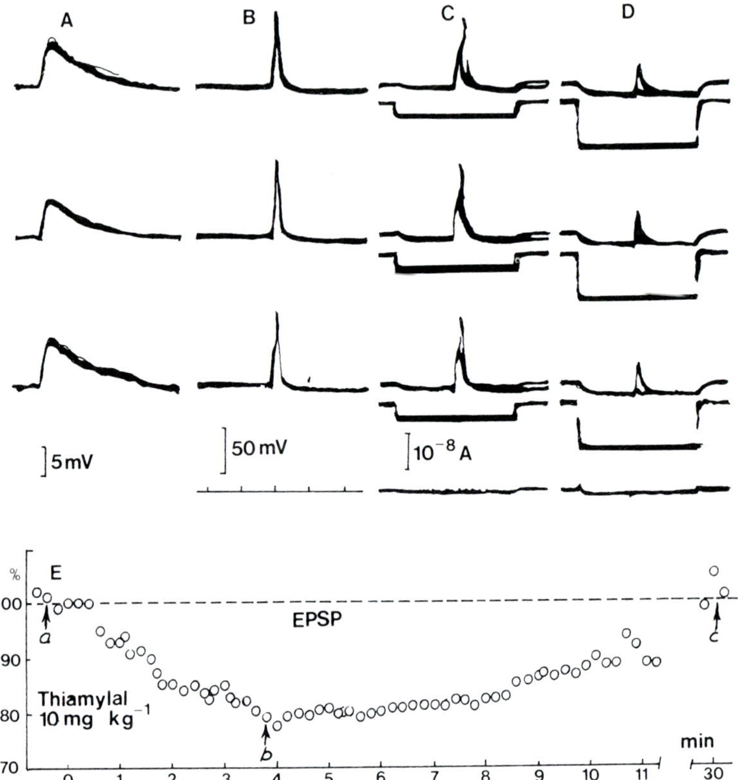


FIG. VI.15. Effects of thiamylal on the monosynaptic EPSP and the antidromic spike potential in a flexor hallucis and digitorum longus motoneurone of a cat. A shows the monosynaptic EPSP and B the antidromic spike potential. C and D show block of the conduction of the antidromic spike potential in the IS–SD and axon–IS components, respectively, by passing a hyperpolarising current through the intracellularly located microelectrode. Lower traces in C and D show recordings of the current applied. In E the EPSP sizes are plotted before and after an injection of 10 mg kg$^{-1}$ thiamylal. Top row potentials in A–D were recorded before the injection (arrow *a* in E), the potentials in the second and third row are recorded after the injection at the times indicated by arrows *b* and *c* in E. The bottom row in C and D shows the potential change produced by the same current just after the microelectrode was withdrawn from the cell. The time scale gives msec for A and 5 msec for B–D. The voltage calibration represents 5 mV for A and 50 mV for the upper traces in B–D. The current calibration represents $10^{-8}$ A for the lower traces in C and D. (Løyning *et al.*, 1964)

## TABLE VI.2

A comparison of the depression of the synaptic transmission produced by volatile anaesthetics in the dentate gyrus and olfactory cortex (Richards and White, 1975)

| Anaesthetic | MAC (dog) (%) | Dentate gyrus: $ED_{50}$ for depression of pop EPSP (v/v%) | Dentate gyrus: % Depression of pop spike at $ED_{50}$ for EPSP depression | Dentate gyrus: No. of expts. | Olfactory cortex: $ED_{50}$ for depression of pop EPSP (v/v%) | Olfactory cortex: % Depression of pop spike at $ED_{50}$ for EPSP depression | Olfactory cortex: No. of expts. |
|---|---|---|---|---|---|---|---|
| Ether | 3 | 1·5–2·5 | 60–100 | 11 | 3·0–4·0 | 30–60 | 7 |
| Halothane | 0·9 | 0·3–0·4 | 90–100 | 12 | 0·8–1·0 | 40–60 | 11 |
| Methoxyfluorane | 0·2 | 0·1–0·2 | 75–100 | 15 | 0·2–0·3 | 30–50 | 9 |
| Trichloroethylene | 0·2–0·3[a] | 0·2–0·5 | 80–100 | 17 | 0·35–0·5 | 40–50 | 5 |

[a]Calculated value.

than the lateral olfactory tract EPSP. Richards and White (1975) have suggested that this may be due to the fact that the perforant path fibres release less transmitter anyway.

A survey of the action of anaesthetic agents on synaptic transmission in invertebrate preparations has led to the interesting possibility that the EPSP depression may be a function of the kind of conductance change elicited by the transmitter–receptor complex at the postsynaptic membrane. Barker (1975a, 1975b) has shown that the anaesthetic agents pentobarbitone, chloroform, chloralose, urethane and ethanol depress excitatory transmission which is mediated postsynaptically by a rise in $Na^+$ permeability alone, or a rise in $Na^+$–$K^+$ or $Na^+$–$Ca^{2+}$ or $Na^+$–$K^+$–$Ca^{+2}$ permeability. Depolarisation of the postsynaptic membrane which is not mediated by a rise in $Na^+$ permeability, and all kinds of hyperpolarisation, are not affected, according to Barker's view. The depression of excitatory transmission is thus unrelated to the kind of the transmitter substance and to the kind of synaptic activity. Hence, the depolarising phase of a biphasic postsynaptic potential recorded from ganglion cells of *Helix* or *Aplysia* was as sensitive to depression by pentobarbitone (0·2 mM) as was the EPSP recorded from the lobster nerve muscle junction. Such a selective action of anaesthetics on the $Na^+$ permeability mechanism might provide a clue to the nature of the processes underlying receptor-coupled $Na^+$ permeability activation.

### 3.3 ALTERATION IN RECEPTOR SENSITIVITY

Many studies have concluded that a variety of anaesthetics may produce their characteristic synaptic effect by an alteration in postsynaptic receptor sensitivity to the endogenous transmitter molecule. It has been proposed that such changes in sensitivity to acetylcholine can occur at membrane concentrations one-hundredth that required to produce fluidity changes (Changeux *et al.*, 1975). One perhaps should be cautious about the interpretations resulting from microiontophoretic application of alleged transmitter substances, or analogues, since it is conceivable that the anaesthetic molecule may interfere not only with the receptor–transmitter binding process, but also may affect the rate at which the applied transmitter (a) is deactivated by enzymatic degradation, or (b) is removed from its site of action by an active uptake process. In this respect it is worth noting that pentobarbitone depresses the uptake of GABA, probably therefore potentiating its action at postsynaptic sites (Cutler *et al.*, 1974). However, many investigations have reported a depression of postsynaptic receptor activity to applied test molecules, hence it would be necessary to argue that anaesthetics augment degradation or uptake processes. At present, there is little evidence to support this.

Anaesthetics depress postsynaptic responses to acetylcholine and its analogue carbachol at the amphibian neuromuscular junction (Thesleff, 1956; Karis *et al.*, 1966, 1967), to acetylcholine on autonomic ganglion cells (Matthews and Quilliam, 1964), to glutamate on spinal neurones (Ransom and Barker, 1975), to glutamate and acetylcholine on cuneate cells (Galindo, 1969), and to acetylcholine and excitant amino acids on cortical neurones (Crawford, 1970; Richards *et al.*, 1975). Halothane may depress the sensitivity of vascular smooth muscle to noradrenaline (Cristoforo and Brody, 1968) and some volatile anaesthetics also appear to alter the sensitivity of the guinea pig ileum to acetylcholine (Speden, 1965).

In contrast, ethanol potentiates the action of carbachol and acetylcholine at the amphibian neuromuscular junction (Sachdev *et al.*, 1963), cyclopropane enhances the contraction induced by noradrenaline of the denervated nictitating membrane (Gravenstein *et al.*, 1960), and general anaesthetics often enhance the effects of microiontophoretically applied glutamate on cortical cells (Krnjević, 1974a).

In Fig. VI.16 is shown the depressant effect of sodium pentobarbitone (0·05 mg $ml^{-1}$) on the change in membrane potential of the frog end-plate induced by microiontophoretic application of acetylcholine. In these experiments, conducted by Thesleff (1956), the change in membrane potential was reversibly reduced despite no change in membrane resistance of the muscle membrane.

In later studies it has been shown that both diethyl ether (5 per cent) and halothane (1·1 per cent) also reduced the depolarising response of the muscle end-plate to carbachol and acetylcholine, this effect again being

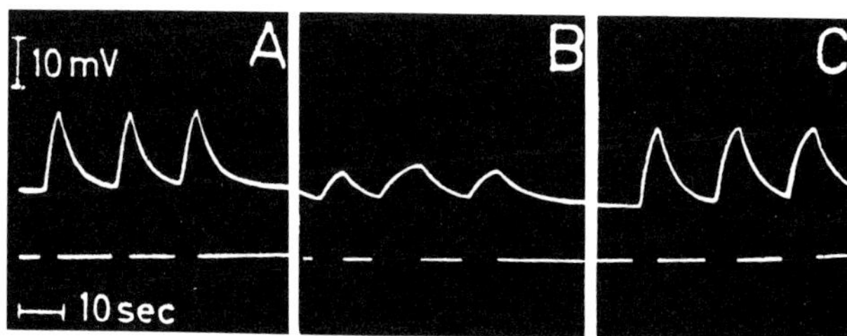


FIG. VI.16. The effect on resting membrane potential of acetylcholine released close to the end-plate region of a frog single muscle fibre by electrophoresis from a micropipette. (A) before and (B) after the addition of sodium pentobarbitone in a concentration of 0·05 mg $ml^{-1}$; (C) after washing with Ringer's fluid. The position of the micropipette remained unchanged during the experiment. The duration of the current passing through the micropipette is marked by the interruptions in the lower line. The resting membrane potential was 85 mV. (Thesleff, 1956)

attributable to a change in receptor sensitivity, since it was not accompanied by any corresponding change in input characteristics of the muscle membrane (Karis *et al.*, 1966, 1967). The neuromuscular block resulting from application of these agents did not resemble a depolarising block, nor that produced by *d*-tubocurarine, since it was not antagonised by endrophonium or succinylcholine (Karis *et al.*, 1966). The results could clearly not be explained in terms of an altered acetylcholinesterase activity, since carbachol is resistant to hydrolysis by this enzyme. These authors thus concluded that the primary action of the agents tested was on the processes underlying the "switching on" of the permeability mechanisms activated by the receptor–transmitter binding. It was not, however, possible to determine from this study whether the anaesthetics produced receptor depression rather than blockage of the ion channels normally opened by receptor activation.

The potentiation by ethanol of acetylcholine or carbachol-induced contractures seen by many authors (e.g. Sachdev *et al.*, 1963) is more than likely a reflection of the action of this agent on the end-plate currents underlying the end-plate potential change. Subsequent work has demonstrated that short-chain alcohols enhanced the muscle end-plate potential, an effect ascribed to a prolongation of the end-plate current (Gage, 1976). This effect is described more fully below.

Not many studies of the relative potency of anaesthetics at nerve-muscle junctions have been performed. Speden (1965) reported that for the guinea pig ileum response to acetylcholine, trichlorethylene was most effective, followed by ether, which in turn was more potent in producing receptor depression than chloroform, methoxyfluorane and halothane. Thus, in this system potency did not correlate well with the hydrophobicity of the agent. In the stimulated, perfused cat superior cervical ganglion, Matthews and Quilliam (1964) observed differences between the effects of various central depressant agents on the retraction of the nictitating membrane elicited by acetylcholine injection into the perfusion stream. Hence, paraldehyde was less potent than methylpentynol (a sedative) or its carbamate, and the latter produced initially some enhancement of ganglionic transmission possibly induced by postsynaptic receptor stimulation.

There is the suggestion that in some preparations, at least, pentobarbitone may bind to the postsynaptic receptor normally occupied by GABA or acetylcholine. Sato *et al.* (1967) noted that high concentrations of pentobarbitone mimicked the effects of application of acetylcholine to ganglion cells of *Aplysia*. In the frog spinal cord, Nicoll (1975) has shown that pentobarbitone ($4 \times 10^{-4}$ M) depolarised primary afferent terminals, an effect which was blocked by the GABA antagonists picrotoxin and bicuculline. Finally, in cultured mammalian spinal cord neurones, pentobarbitone prolonged the conductance change induced by GABA and these two agents produced qualitatively similar conductance changes (Ransom and Barker, 1975). It is, however, difficult once again to distinguish effects produced by "turning on" of the conductance mechanism normally engaged by receptor activation and altered conformational change of the receptor by binding of the anaesthetic agent.

The literature on the effects of anaesthetics on the chemical sensitivity of supraspinal neurones is confusing. In the cat cuneate nucleus, microiontophoretic release of procaine and pentobarbitone depressed the excitatory effects of microiontophoretically applied glutamate, but halothane (1–2 per cent administered by inhalation) had no effect (Galindo, 1969). However, 2 per cent halothane potentiated the effect of GABA and glycine (applied microiontophoretically). In slices of the guinea pig prepiriform cortex, perfusion with pentobarbitone, methoxyfluorane (0·2 per cent) and trichloroethylene depressed the glutamate-induced activity of cortical cells, while again halothane (1 per cent) had no effect (Richards, 1974).

In a systematic study, Crawford (1970) examined the effects of a wide range of anaesthetics (administered systemically) on the frequency of spontaneous and drug-induced firing of single pericruciate cortical neurones in "cerveau isole" cats. This author concluded that although the barbiturates were effective depressants of firing induced by acetylcholine, or excitant amino acids, "analgesic" levels of nitrous oxide, trichloroethylene and halothane had little effect; halothane at a concentration above 1·5 per cent depressed and, contrary to Richard's work, methoxyfluorane (0·2–1·5 per cent) produced no depression of chemical sensitivity. In contrast, Catchlove *et al.* (1972) reported that a range of inhalation anaesthetics, including halothane, trichloroethylene, nitrous oxide and methoxyfluorane, depressed acetylcholine-induced responses of cat pericruciate neurones but often actually enhanced the response to glutamate.

The different results obtained in these studies make clear-cut interpretations difficult. The test excitant amino acid employed in Crawford's study was DL-homocysteic acid, and in the study of Galindo (1969) and Catchlove *et al.* (1972) glutamate was used. It is possible that anaesthetics depress the uptake of glutamate and hence augment the effect of its application, whereas the removal of DL-homocysteic acid from its site of action may not be so sensitive to depression. (DL-homocysteic acid does not occur naturally in mammals.) This consideration may explain the difference between the results of Catchlove *et al.* (1972) and Crawford (1970) with respect to the excitant amino acid data. However, the disagreement between the experiments of Richards (1974) with glutamate and the results of Catchlove *et al.* (1972) cannot be explained in this fashion.

The resultant effect produced by an anaesthetic agent in all of these experiments is unlikely to be due solely to a direct action of the agent on the postsynaptic receptor sensitivity of the cell whose activity is being monitored. More often it probably results from depression, or enhancement, of a synaptic input which is activated by application of either the test amino acid or acetylcholine. Richards (1974) took the precaution of bathing the olfactory cortical slices used in his study in c.s.f. containing 10 mM $Mg^{2+}$. This procedure removed the background synaptic input into the cells, the activity of which was being monitored. It is of interest, therefore, to note that in this study, as in previous work, halothane—a very hydrophobic agent—had no depressant effect on glutamate responses.

A major stumbling block with all of these experiments is that they provide no real answer to the question of how the membrane receptor behaves in response to the natural transmitter in the presence of an anaesthetic. It could also be argued that just as glutamate may be a non-specific postsynaptic excitant, so too the anaesthetics are non-specific, postsynaptic depressants or excitants, and that this action is not related to any physiological action of these agents. Nevertheless, such a general action of anaes

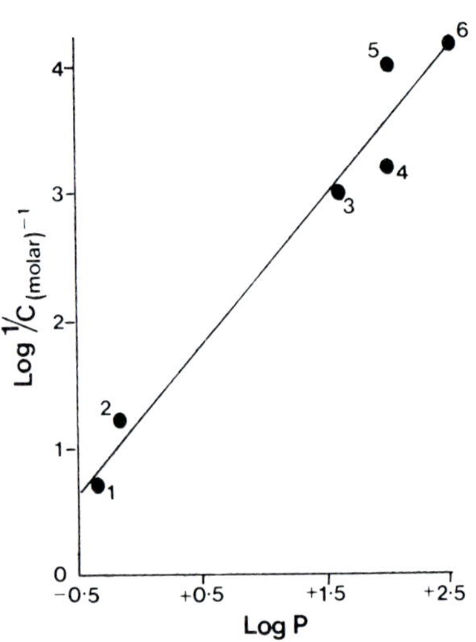


FIG. VI.17. Plot of the relative ability of anaesthetics to depress postsynap excitation in a snail neurone as a function of hydrophobicity (log P) of the com pounds. 1 = ethanol, 2 = urethane, 3 = chloralose, 4 = chloroform, 5 = pent barbitone, 6 = diphenylhydantoin (an anticonvulsant). (Barker, 1975b)

thetics on receptor activity, or the conductance mechanisms activated by receptor–test agent interaction, may provide information about the nature of the processes underlying these phenomena.

Barker (1975b) has proposed that depression of receptor activity in *Helix* neurones is correlated with the hydrophobicity of the agent. Data supporting this hypothesis are shown in Fig. VI.17. On the ordinate is plotted the log of the reciprocal of the concentration required to produce a 50 per cent depression of the acetylcholine-induced depolarisation, and on the abscissa the log of the partition coefficient of the agents tested. Statistical analysis of the data reveals that activity of the agent is significantly correlated with hydrophobicity, suggesting that this physical characteristic of the anaesthetic molecule may determine its activity in this preparation. Thus, the agent could attach itself to the hydrophobic portion of the receptor protein, or alter the lipid environment of the receptor, or perhaps become attached to a hydrophobic portion of a membrane channel, hence affecting its permeability characteristics. The latter suggestion has been put forward as part explanation for the relative potencies of quaternary ammonium compounds in blocking $K^+$ channels (Armstrong, 1975). Unfortunately, the lack of depressant activity exhibited by halothane and trichloroethylene in certain vertebrate studies does not lend support to the notion that hydrophobicity is important in determining potency at all postsynaptic receptor sites.

### 3.4 ANALYSIS OF END-PLATE STUDIES

Intracellular recordings at the neuromuscular junction demonstrate clearly that the amplitude and time course of end-plate electrical events are altered by anaesthetics, the actual effect being dependent on both the nature of the anaesthetic and its concentration. Thus, the short-chain alcohols (up to pentanol) reduced the rate of decay of the miniature end-plate current (m.e.p.c.), whereas hexanol produced a biphasic decay of this current, and longer-chain alcohols increased the rate of decay (Gage *et al.*, 1974, 1975).

The local anaesthetics can produce m.e.p.c.s with an initial fast followed by a slower rate of decay (Furukawa, 1957; Maeno, 1966; Steinbach, 1968a; Kordaš, 1970), although differences in action do exist between agents (Maeno *et al.*, 1971).

The barbiturates and the inhalational anaesthetics ether and halothane also modify the time course of end-plate currents. Adams (1974) reported that amylobarbitone and thiopentone (5 mM) increased the rate of decay, and Quastel and Linder (1975) observed that pentobarbitone ($<1{\cdot}0$ mM) produced biphasic decays similar to those observed with local anaesthetics. Low concentrations of ether and halothane shortened the m.e.p.c. (Gage and Hamill, 1975), while higher concentrations of diethyl ether lengthened the m.e.p.c. (Karis *et al.*, 1966; Quastel and Linder, 1975), and higher concentrations of halothane produced biphasic m.e.p.c.s similar to that produced by procaine (Gage and Hamill, 1975).

An elegant analysis of the effects of alcohols on the m.e.p.c. at amphibian end-plates has been carried out recently by Gage *et al.* (1975). The alcohols, ethanol to hexanol, were found to increase the amplitude and duration of the miniature end-plate potential. This was due to a specific prolongation of the decay phase of the m.e.p.c. which is normally exponential. The time course of the current follows the end-plate conductance change and, since the rate of decay is much slower than the growth, the amplitude and time constant of decay determines the amount of charge transferred and hence the amplitude of the postsynaptic miniature end-plate potential. The intriguing question is whether the decay of this current is determined by the rate of dissociation of the acetylcholine-receptor complex or the rate of conformational change of the receptor from its active to inactive form. The interaction of the receptor with acetylcholine may be written in the simple form

$$nT + R \underset{k_{-1}}{\overset{k_1}{\rightleftharpoons}} TnR \underset{\beta}{\overset{\alpha}{\rightleftharpoons}} TnR^*$$

where $T$ denotes acetylcholine, $R$ the receptor, $TR$ and $TR^*$ are the "closed" and "open" conformation of the complexes. $n$ molecules of acetylcholine are required to open the channel, and $k_1$, $k_{-1}$, $\alpha$ and $\beta$ are rate constants. This model has been discussed by Anderson and Stevens (1973). Gage *et al.* (1975) showed that the time constant of decay of the m.e.p.c. increased up to sevenfold in 0·5 M ethanol and that the potency of the alcohols increased as the chain length of the alcohol increased. Similar increases have been observed by Quastel and Linder (1975) at a mammalian end-plate. Figure VI.18 shows the dramatic effect that exposure to a high concentration of ethanol (0·5 M) produced on the magnitude and time course of the miniature end-plate potential (m.e.p.p.) of amphibian fibres.

Proof that such changes were not due to a change in the length of time that acetylcholine exists in the synaptic cleft (i.e. an anticholinesterase effect) was provided by the fact that ethanol also increased the depolarisation and conductance change produced by carbachol (Quastel and Linder, 1975), and also that although anticholinesterase drugs prolong the m.e.p.c. they do not alter the frequency spectra of acetylcholine noise (Gage *et al.*, 1975; Quastel and Linder, 1975). The spectral density of e.p.c. fluctuations basically characterises these fluctuations in terms of the amplitude of each frequency component. Thus, if the e.p.c. fluctuations are rapid, then a

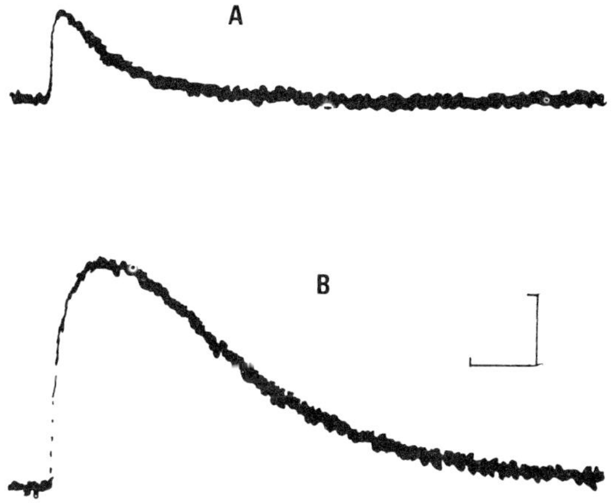


FIG. VI.18. The increase in amplitude and prolongation of the time course oe miniature end-plate potentials caused by ethanol. The m.e.p.p. was recorded at th. amphibian neuromuscular junction in a standard solution (A). Another m.e.p.p (B) was recorded from the same fibre after 10 min exposure to 0·5 M ethanol· Calibrations: vertical, 0·5 mV; horizontal, 10 msec. (Gage *et al.*, 1975)

larger number of spectral estimates will appear in the high-frequency range, whereas if the fluctuations are slow, spectral estimates will occur primarily in the low-frequency range. Descriptions of acetylcholine noise using power spectra have previously provided information about the conductance and kinetics of closing of end-plate channels (Anderson and Stevens, 1973). Present evidence shows that ethanol, unlike anticholinesterase agents, shifts the power spectra to lower frequencies, consistent with the hypothesis that ethanol is slowing the m.e.p.c. decay by lengthening the open time of the end-plate channels.

Since the time constant of decay of synaptic current is voltage-sensitive, it has been proposed that the reaction underlying the decay is accompanied by a change in dipole moment (Anderson and Stevens, 1973). Since the decay is normally exponential, it is not unreasonable to suggest that it is controlled by the rate of a first order reaction which could involve a protein (receptor) conformation change. If the reacting molecules have dipole moments, then it is to be expected that the dielectric constant of the environment will affect the reaction rate, and this is the key to the analysis of Gage *et al.* (1975). Assuming an initial dielectric constant value for membrane lipid of 3, and homogeneous mixing of lipid and alcohol, Gage *et al.* (1975) calculated, using the dielectric constant of the alcohol and its partition coefficient, first the weight fraction of alcohol in the membrane and therefore the dielectric constant of the resultant alcohol–membrane mixture. It was then possible to predict the effect of such a change in dielectric constant on the time constant of decay of m.e.p.c. Their analysis predicted an exponential relationship between the aqueous concentration of alcohol and time constant of decay of the current as observed experimentally. Such an increase in dielectric constant might also explain the prolongation of m.e.p.c.s observed in the presence of high concentrations of ether (Karis *et al.*, 1966; Quastel and Linder, 1975), although a rigorous test of this possibility has not been performed.

When dealing with the action of anaesthetic agents on the time course of decay of end-plate electrical events, a second consideration must be taken into account. It is likely that conformational changes in protein molecules within the membrane are dependent on the fluidity of the environmental lipid (Keynes, 1972) and this is likely to be increased by a variety of anaesthetic agents (see Seeman, 1972). A fluidity change may therefore be responsible for the increase in decay rate of m.e.p.c. observed with longer-chain alcohols, ether and halothane (Gage *et al.*, 1974; Gage and Hamill, 1975), and could be responsible for the alteration in the decay rate of m.e.p.c.s seen with pentobarbitone (Quastel and Linder, 1975) or the local anaesthetics (Maeno, 1966; Steinbach, 1968a).

Procaine and other local anaesthetics produce m.e.p.c.s with an initial rapid, followed by a later slower, phase. As Steinbach (1968a) has pointed out, the molecular properties of xylocaine and derivatives producing this effect at the frog neuromuscular junction are very similar to acetylcholine, so that these agents could be binding to the acetylcholine receptor. At the same time, similar effects have been observed with structurally unrelated compounds, such as hexanol (Gage *et al.*, 1974), halothane (Gage and Hamill, 1975) and pentobarbitone (Quastel and Linder, 1975). It is probably therefore not necessary at present to invoke steric considerations.

One explanation of the two-phase decay of m.e.p.c. has been offered by Maeno (1966), who suggested that procaine depressed sodium conductance ($g$Na) early during the end-plate conductance change and left a residual $g$Na which decayed much more slowly. Potassium conductance ($g$K), assessed by measuring the short-circuiting effect of the end-plate potential on the overshoot of the muscle action potential, was however unaffected. In later voltage-clamp experiments conducted by Maeno *et al.* (1971), the e.p.c. was measured at the equilibrium potential for $K^+$ ($E_K$) and at the equilibrium potential for $Na^+$ ($E_{Na}$). In the presence of procaine, the e.p.c. recorded at the $E_K$ value (representing $\Delta g$Na during the e.p.p.) was altered differently from the e.p.c. recorded at the $E_{Na}$ value (i.e. $\Delta g$K during the e.p.p.). However, lidocaine and derivatives affected similarly $\Delta g$Na and $\Delta g$K during the end-plate potential.

Steinbach (1968b) has proposed an alternative kinetic model to explain the biphasic decay of e.p.c.s recorded in the presence of lidocaine. In this scheme, receptor–acetylcholine binding occurs first and the rapid decay of the e.p.c. is due to both the conformational relaxation of the complex and the subsequent formation of a partially active acetylcholine–receptor–anaesthetic complex. Dissociation of this second complex governs the decay rate of the slow phase of the e.p.c. As pointed out by Gage (1976), there are difficulties with this model, not the least of which is the finding that derivatives of lidocaine may produce triphasic decay rates.

Although at present the picture emerging from these end-plate studies is a complicated one, the changes in kinetics of the acetylcholine–receptor reaction induced by anaesthetics may yet provide some further insight into the membrane processes underlying this reaction sequence.

## 4 Postsynaptic inhibition

General anaesthetics not only depress the process of synaptic excitation in the vertebrate central and peripheral nervous system, but there is also good evidence that anaesthetics, in particular the barbiturates, enhance the process of postsynaptic inhibition. The prolongation of postsynaptic inhibition by barbiturates may be responsible for the discrepancy between the duration of the recurrent inhibition of the monosynaptic reflex in the decerebrate cat (45 msec) and the duration of the intracellular IPSP (~200 msec) recorded under barbiturate anaesthesia (Larson and Major, 1970). The balance between facilitation by the parallel fibres of the antidromic invasion of Purkyně cells and inhibition of this invasion by the basket cells was also altered in favour of inhibition as anaesthesia was deepened in the cat (Eccles *et al.*, 1971). This, too, may be due to enhancement of inhibitory processes by anaesthetics.

Evidence that the barbiturates enhance postsynaptic inhibition has been provided in the cat hippocampus (Nicoll *et al.*, 1975), in the rabbit olfactory bulb (Nicoll, 1972) and in mouse spinal neurones (Ransom and Barker, 1975). As shown by Nicoll (1972), in work on the dendrodendritic inhibitory pathway of the olfactory bulb, the prolongation of postsynaptic inhibition may be produced at concentrations which do not depress excitatory transmission. Whereas large doses of pentobarbitone (>40 mg kg$^{-1}$) were required to block the antidromic invasion of the mitral cell dendrites, or to reduce the mitral cell synaptic excitation of granule cells, lower doses of pentobarbitone (3 mg kg$^{-1}$) prolonged the inhibition of mitral cells by granule cell processes. In this study, hexobarbitone, halothane, urethane, chloralose and chloral hydrate also prolonged postsynaptic inhibition. Ethanol was not effective, however.

Since evidence has been presented that GABA is the transmitter involved in the inhibition (Nicoll, 1971), it is possible that the potentiation observed in this investigation was due to a depression of GABA uptake (Cutler *et al.*, 1974) or, alternatively, a decrease in the rate of relaxation of the GABA-induced conductance change (Ransom and Barker, 1975). An effect on GABA metabolism is perhaps ruled out by the observation of Nicoll (1972) that amino-oxyacetic acid, in doses which are known to inhibit GABA catabolism by inhibiting GABA-α-keto-glutaric transaminase (Kelly *et al.*, 1971), did not augment postsynaptic inhibition.

The effect of an intravenous injection of pentobarbitone on the characteristics of the IPSP recorded intracellularly from a cat hippocampal neurone is shown in Fig. VI.19.

This figure shows the effect of sequential injection of pentobarbitone on

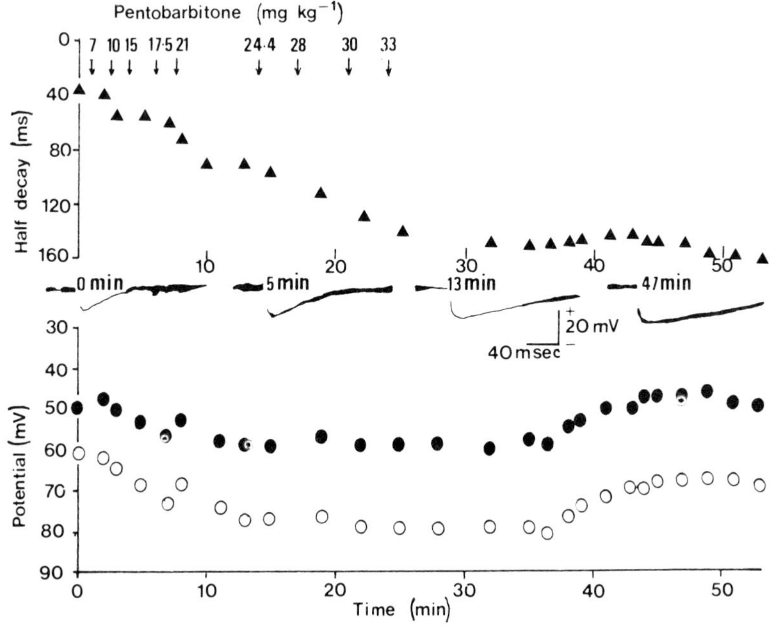


FIG. VI.19. Prolongation by pentobarbitone of the IPSP recorded in a cat hippocampal neurone. Top graph shows the half decay time of the IPSP (▲); bottom graph the membrane potential (●), and IPSP peak (○). Sample records taken at the times illustrated are shown between the graphs. The cumulative dose is shown at the top of the figure. (Nicoll *et al.*, 1975)

the half decay time (▲), the resting potential (●) and the amplitude of the IPSP (difference between ○ and ●). Sample records taken at the times illustrated are shown between the graphs. The time constant of decay of the IPSP increased some fourfold, while the amplitude of the IPSP doubled. Thiopentone and thiamylal also prolonged the IPSP. At the concentration tested, the barbiturates produced no change in resting potential. If GABA is the transmitter involved in postsynaptic inhibition in the hippocampus as would seem to be the current belief (see Krnjević, 1974b), then the enhancement may be due to the same factors discussed above. Alternatively, an effect on GABA uptake may only be relevant if high concentrations of barbiturates are employed (Jessell and Richards, 1977).

Studies with procaine have furnished some interesting results with respect to the relative resistance to depression of IPSPs recorded from the cat motoneurone (Curtis and Phillis, 1960), and to the relative susceptibilities of post- and presynaptic inhibition in the fish brain (Furukawa *et al.*, 1964). Figure VI.20 shows that microiontophoretic application of procaine to a cat motoneurone actually augmented the EPSP (C–E), had no effect on the IPSP (G–I) and depressed excitatory transmission (D) and antidromic invasion (H). In this study, the threshold behaviour of the postsynaptic cell was therefore most vulnerable to procaine.

## 5 Presynaptic inhibition

It is known that some anaesthetics increase the duration of presynaptic inhibition in the mammalian (Eccles *et al.*, 1963) and amphibian spinal cord (Eccles and Malcolm, 1946; Schmidt, 1963; Nicoll, 1975), although the intensity of the inhibition can be reduced if depression of excitatory transmission occurs along the interneuronal pathway involved (Schmidt, 1963). The effect of agents has not been examined directly by intracellular recording of the presynaptic depolarisation, but more usually the intensity and time course of this form of inhibition has been investigated in the spinal cord by measuring the dorsal root potential (Eccles *et al.*, 1962a) or by examining the depression of a spinal monosynaptic reflex (m.s.r.) induced by a conditioning volley to other group I afferents (Eccles *et al.*, 1962b). Hence, in decerebrate cats Eccles *et al.* (1963) looked at the effects of Nembutal (20 mg $kg^{-1}$) on the inhibitory action of four posterior biceps plus semitendinosus (PBST) volleys on a monosynaptic testing reflex of a gastrocnemius-soleus (GS) motoneurone (Fig. VI.21).

The presynaptic inhibitory action of four PBST volleys was tested by recording the m.s.r. evoked to two GS volleys before (○) and after (●) an intravenous injection of Nembutal. By varying the interval between the

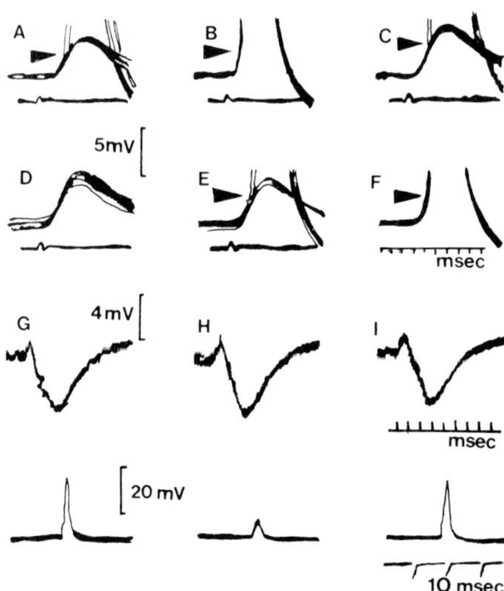


FIG. VI.20. A–F, monosynaptic excitatory postsynaptic potentials recorded from a cat motoneurone in response to orthodromic stimulation, the volley arriving at the dorsal surface of the spinal cord indicated in the lower beam. Each figure consists of 15–20 superimposed records at a frequency of 5/sec. A, B, control responses with maximal stimulation in B, while in A the EPSP just straddled the threshold of the cell. C–F, as for B, but during and after the application of procaine to the cell by a current of 200 nA for 75 sec; C, 48 sec; D, 72 sec; E, 95 sec and F, 135 sec after this current commenced. The horizontal arrows indicate the size of the threshold EPSP. G–I, potentials similarly recorded from a second motoneurone. The inhibitory potential (upper beam) was evoked by stimulation of the peroneal nerve whereas the antidromic spike potential (lower beam) was produced by stimulation of the segmental ventral root. G, control response; H, 60 sec after a current of 330 nA had been passing procaine from the outer barrel of the electrode; I, 70 sec after this current ceased. (Curtis and Phillis, 1960)

conditioning PBST volleys and the test two GS volleys, the time course of the inhibition could be measured and an inhibition curve constructed as shown in Fig. VI.21. Minutes after the injection the inhibition was increased and prolonged (●), and after 3 hr the inhibition curve had returned almost to the control shape (+). A second injection of Nembutal again increased the intensity and duration of the inhibition (×). The amplitude and duration of the dorsal root potentials were also increased in these experiments, so that the more effective inhibition in the presence of Nembutal was attributed to a larger presynaptic depolarisation.

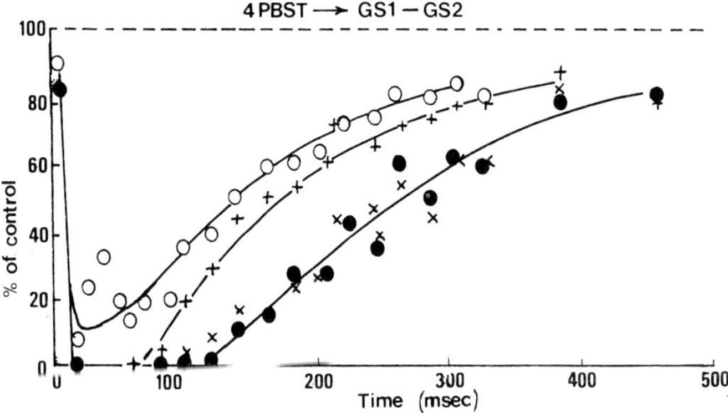


FIG. VI.21. Effect of Nembutal (sodium pentobarbitone) on presynaptic inhibition of a monosynaptic reflex in the cat. The presynaptic inhibitory action of 4 PBST volleys (maximum for Group I and at 300/sec) was tested by monosynaptic reflexes evoked to two GS volleys (see text) at 1·5 msec interval and recorded monophasically in the S1 ventral root. The first GS volley itself evoked no reflex, but merely facilitated the reflex evoked by the second volley that was maximum for Group I. The sizes of the test reflexes were calculated as percentages of the control reflex and plotted against the testing intervals (first PBST volley to second GS volley) to give the inhibitory curves. At each testing interval there were several superimposed traces and the means were plotted. ○ (control), initial curve in decerebrated unanaesthetised preparation; ●, after Nembutal 20 mg $kg^{-1}$; +, 3 hr later; ×, after second injection of Nembutal 20 mg $kg^{-1}$. (Eccles *et al.*, 1963)

In the toad spinal cord, chloralose and the barbiturates thiamylal and phenobarbitone also increased the intensity and duration of presynaptic inhibition (Schmidt, 1963), and Nicoll (1975) has demonstrated that in the frog spinal cord some barbiturates lengthened presynaptic inhibition but also depolarised the primary afferent terminals directly. This is shown in Fig. VI.22. Pentobarbitone at a bath concentration of $4 \times 10^{-4}$ M increased the excitability of primary afferent terminals (A1) and this was accompanied by a depolarisation of the neighbouring dorsal root recorded simultaneously across a sucrose gap (A2). The potentials between the graphs show sample antidromic spikes at the times indicated. $4 \times 10^{-3}$ M pentobarbitone also caused depolarisation of the root (B2), but caused an eventual reduction in terminal excitability (B1). This direct depolarising action of pentobarbitone could also therefore lead to a reduction in transmission at the first central synapse in the spinal cord, as does the specific presynaptic inhibitory process.

Since GABA is probably the synaptic transmitter involved in producing presynaptic inhibition in the spinal cord (Eccles *et al.*, 1963; Nicoll, 1975),

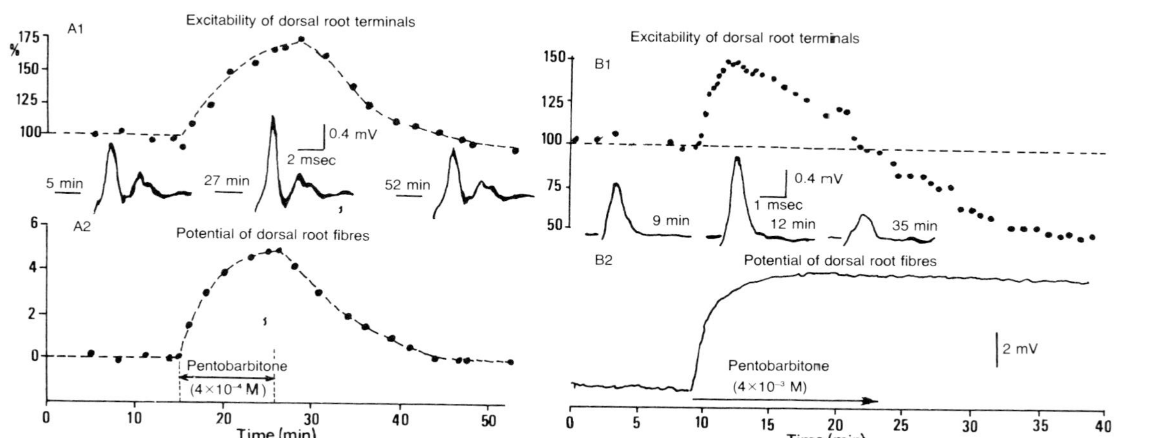


FIG. VI.22. Effect of pentobarbitone on the membrane potential and excitability of primary afferents of the isolated frog spinal cord. (A) Addition of $4 \times 10^{-4}$ M pentobarbitone reversibly increases the excitability of dorsal root terminals (A1), the excitability increase having the same time course as the depolarisation of the neighbouring dorsal root recorded simultaneously across a sucrose gap (A2). The inserts show samples of the antidromic volley at the times indicated. (B) Addition of $4 \times 10^{-3}$ M pentobarbitone causes an initial increase in excitability (B1), followed by a decrease below the control values. B2 shows a pen recording of the depolarisation recorded from the neighbouring dorsal root, which is maintained throughout the application. (Nicoll, 1975)

it is reasonable to presume that the prolongation of presynaptic inhibition with barbiturates is due at least in part to their effect on the uptake of GABA (Cutler *et al.*, 1974).

It is feasible that increased release of GABA could account for the effects seen with barbiturates. However, CHEB (a convulsant barbiturate) can increase transmitter release (Thomson and Turkanis, 1973) but does not affect presynaptic inhibition (Nicoll, 1975). Also, pentobarbitone, but not phenobarbitone, alters the efflux of GABA from cortical slices (Cutler and Dudzinski, 1974; Cutler *et al.*, 1974) but both enhance presynaptic inhibition (Schmidt, 1963), and the effects of pentobarbitone are complex in that this agent depressed spontaneous efflux and the efflux evoked by high $K^+$ (Cutler and Dudzinski, 1974; Cutler *et al.*, 1974) but enhanced electrically induced efflux (Cutler and Dudzinski, 1974). In view of these differences among this family of agents, it is worth checking the credentials of the barbiturates before assuming that all of these agents enhance presynaptic inhibition by a similar action on the membrane uptake of GABA. In addition, the intriguing possibility remains that these agents may prolong the conductance change induced by GABA across the primary afferent terminal membrane, just as they prolong the conductance change underlying postsynaptic inhibition (Ransom and Barker, 1975).

## 6 Anaesthetics–pressure interactions

One of the hitherto unexplored areas of anaesthetic action at a synaptic level is the interplay between anaesthetics and hydrostatic pressure. It has been argued that pressure reversal of any effect produced by an anaesthetic may be one necessary criterion in determining whether this action underlies the anaesthetic state (see Kendig and Cohen, 1975), since it has been shown that high pressure can abolish the gross signs of anaesthesia (Johnson and Flagler, 1950). What then is the action of pressure on an anaesthetic-induced synaptic effect?

One preliminary study shows that 0·5 mM halothane depressed synaptic transmission through the rat superior cervical ganglion, and surprisingly transmission was depressed further by 137 atm of helium pressure (Kendig and Cohen, 1975). Thus far, intracellular work has demonstrated that excitatory synaptic transmission is depressed by pressure *per se* at the crustacean neuromuscular junction (Campenot, 1975) and by He pressure at the squid stellate ganglion (Henderson, 1976), but the interaction between pressure and anaesthetics remains unstudied.

In experiments with *Helix* neurones it has been observed that excitatory transmission is often depressed by hydrostatic pressure, but also in preliminary experiments, IPSP activity may be abolished selectively by low

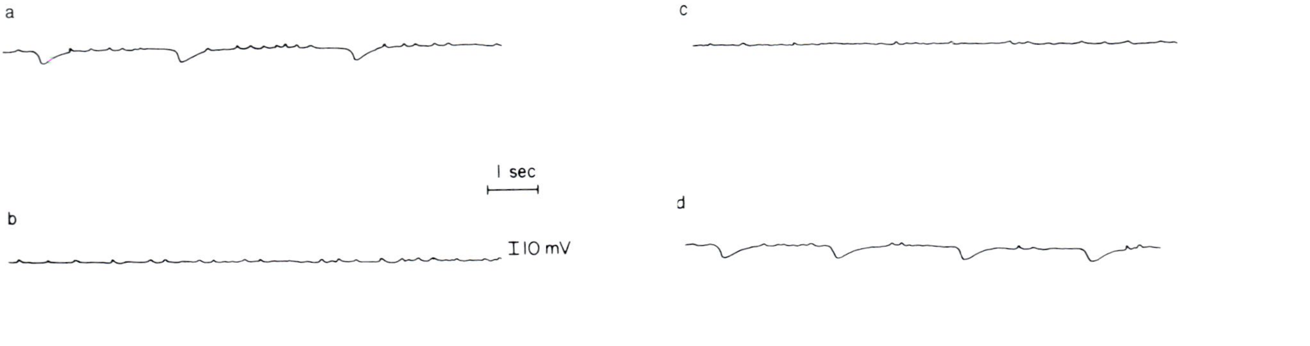


FIG. VI.23. The effects of hydrostatic pressure (hydraulic compression to 100 atm) on spontaneous synaptic activity of a ganglion cell of *Helix*. (a) control; (b) 100 atm, 10 min post-compression; (c) 100 atm, 40 min post-compression; (d) recovery, 5 min. (Wann *et al.*, unpublished observations)

to moderate pressures (Fig. VI.23). In these same experiments, Nembutal (0·5 mM) increased IPSP bombardment of these cells, although the effects of pressure on this enhanced activity has not been examined. There may be an interesting interplay between pressure and Nembutal in this system which remains to be investigated.

A study of the interaction between pressure and anaesthetics at a synaptic level could lead to characterisation of molecular volume changes underlying an anaesthetic effect, and therefore might provide a clue to its molecular basis. It would be necessary, however, to examine the pressure susceptibility of a discrete anaesthetic-induced effect using a simple synaptic system.

## 7 Conclusions

It is evident that anaesthetics exert wide-ranging effects on the process of synaptic transmission. Their primary action is dependent on the nature and concentration of the agent tested, and also depends on the type of synapse under investigation. Thus the barbiturates and the inhalational anaesthetics depress excitatory transmission by interfering with the presynaptic release mechanisms whereas the local anaesthetic procaine (for example) may produce predominantly postsynaptic depression. None of the anaesthetics display the specificity in affecting a synaptic process that the classical pharmacological agents possess (e.g. *d*-tubocurarine).

Many agents (e.g. the inhalation anaesthetics) may affect specific receptor-coupled cation conductance mechanisms, and depression of the $Na^+$ conductance system and activation of the $K^+$ conductance system have both been observed experimentally. In invertebrate preparations, receptor depression by anaesthetics can be correlated with the hydrophobicity of the agent.

The barbiturates also enhance the processes of postsynaptic and presynaptic inhibition. This action may be related in part to their depressant effect on the GABA transport system, although the GABA receptor may prove to be an equally important target for anaesthetics.

The actions of anaesthetics on the process of synaptic transmission are thus as diverse as the structure of the agents themselves. An appreciation of the complex effects produced by anaesthetics at a simple synaptic level must, however, be a stepping stone to a better understanding of their actions on an integrated central nervous system.

## References

Adams, P. R. (1974). The mechanism by which amylobarbitone and thiopentone block the end-plate response to nicotinic antagonists. *J. Physiol.* **241**, 41–42P.

ALLISON, A. C. (1974). The effects of inhalational anaesthetics on proteins. *In* "Molecular Mechanisms in General Anaesthesia" (M. J. Halsey, R. A. Millar, and S. A. Sutton, eds), pp. 164–181. Churchill-Livingstone, London.

ALLISON, A. C. and NUNN, J. F. (1968). Effects of general anaesthetics on microtubules—a possible mechanism of anaesthesia. *The Lancet*, **2**, 1326–1329.

ANDERSON, C. R. and STEVENS, C. F. (1973). Voltage clamp analysis of acetylcholine produced end-plate current fluctuations at frog neuromuscular junction. *J. Physiol.* **235**, 655–691.

ARMSTRONG, C. M. (1975). Potassium pores of nerve and muscle membranes. *In* "Membranes" (G. Eisenman, ed.), vol. 3, pp. 325–358. Dekker, New York.

BAKER, P. F. and CRAWFORD, A. C. (1975). A note on the mechanism by which inhibitors of the sodium pump accelerate spontaneous release of transmitter from motor nerve terminals. *J. Physiol.* **247**, 209–226.

BANKS, P. and TILL, R. (1975). A correlation between the effects of anti-mitotic drugs on microtubule assembly *in vitro* and the inhibition of axonal transport in noradrenergic neurones. *J. Physiol.* **252**, 283–294.

BARKER, J. L. (1975a). Inhibitory and excitatory effects of CNS depressants on invertebrate synapses. *Brain Research*, **93**, 77–90.

BARKER, J. L. (1975b). Selective depression of postsynaptic excitation by general anaesthetics. *In* "Molecular Mechanisms of Anaesthesia" (B. R. Fink, ed.), vol. 1, pp. 135–153. Raven Press, New York.

BLAUSTEIN, M. P. and HODGKIN, A. L. (1969). The effect of cyanide on the efflux of calcium from squid axons. *J. Physiol.* **200**, 497–527.

BORLE, A. E. (1973). Calcium metabolism at the cellular level. *Fed. Proc.* **32**, 1944–1950.

BURGEN, A. S. V. (1971). Inhalation anaesthesia: mode of action. *In* "General Anaesthesia" (T. C. Gray and J. F. Nunn, eds), vol. 1, pp. 428–438. Butterworth, London.

CAMPENOT, R. B. (1975). The effects of high hydrostatic pressure on transmission at the crustacean neuromuscular junction. *Comp. Biochem. Physiol.* **52B**, 133–140.

CARAFOLI, E. and ROSSI, C. S. (1971). Calcium transport in mitochondria. *In* "Advances in Cytopharmacology" (F. Clementi and B. Cecarelli, eds), vol. 1, pp. 209–227. Raven Press, New York.

CATCHLOVE, R. F. H., KRNJEVIĆ, K. and MARETIĆ, H. (1972). Similarity between effects of general anaesthetics and dinitrophenol on cortical neurones. *Can. J. Physiol. Pharmacol.* **50**, 1111–1113.

CHALAZONITIS, N. (1967). Effects of anaesthetics on neural mechanisms—selective actions of volatile anaesthetics on synaptic transmission and autorhythmicity in single identifiable neurons. *Anesthesiol.* **28**, 111–123.

CHANGEUX, J. P., BENEDETTI, L., BOURGEOIS, J-P., BRISSON, A., CARTAUD, J., DEVAUX, P., GRÜNHAGEN, H., MOREAU, M., POPOT, J-L., SOBEL, A. and WEBER, M. (1975). Some structural properties of the cholinergic receptor protein in its membrane environment relevant to its function as a pharmacological receptor. *Symp. Quant. Biol.* **XL**, 211–230.

CHASE, R. (1975). The suppression of excitatory synaptic responses by ethyl alcohol in the nudibranch mollusc, *Tritonia Diomedia*. *Comp. Biochem. Physiol.* **50C**, 37–40.

CRAWFORD, A. C. (1975). Lithium ions and the release of transmitter at the frog neuromuscular junction. *J. Physiol.* **246**, 109–142.

CRAWFORD, J. M. (1970). Anaesthetic agents and the chemical sensitivity of cortical neurones. *Neuropharmacol.* **9**, 31–46.

CRISTOFORO, M. F. and BRODY, M. J. (1968). The effects of halothane and cyclopropane on skeletal muscle vessels and baroreceptor reflexes. *Anesthesiol.* **29**, 36–43.

CURTIS, D. R. and PHILLIS, J. W. (1960). The action of procaine and atropine on spinal neurones. *J. Physiol.* **153**, 17–34.

CUTLER, R. W. P. and DUDZINSKI, D. S. (1974). Effect of pentobarbital on uptake and release of [$^3$H] GABA and [$^{14}$C] glutamate by brain slices. *Brain Research*, **67**, 546–548.

CUTLER, R. W. P., MARKOWITZ, D. and DUDZINSKI, D. S. (1974). The effect of barbiturates on [$^3$H] GABA transport in rat cerebral cortex slices. *Brain Research*, **81**, 189–197.

DREYER, C., BISCHOFF, D. and GÖTHERT, M. (1974). Effects of methoxyflurane anaesthesia on adrenal medullary catecholamine secretion. *Anesthesiol.* **41**, 18–26.

ECCLES, J. C. and MALCOLM, J. L. (1946). Dorsal root potentials of the spinal cord. *J. Neurophys.* **9**, 139–160.

ECCLES, J. C., FABER, D. S. and TÁBOŘÍKOVÁ (1971). The action of a parallel fibre volley on the antidromic invasion of Purkyně cells of cat cerebellum. *Brain Research*, **25**, 335–356.

ECCLES, J. C., MAGNI, F. and WILLIS, W. D. (1962a). Depolarisation of central terminals of Group I afferent fibres from muscle. *J. Physiol.* **160**, 62–93.

ECCLES, J. C., SCHMIDT, R. F. and WILLIS, W. D. (1962b). Presynaptic inhibition of the spinal monosynaptic reflex pathway. *J. Physiol.* **161**, 282–297.

ECCLES, J. C., SCHMIDT, R. and WILLIS, W. D. (1963). Pharmacological studies on presynaptic inhibition. *J. Physiol.* **168**, 500–530.

EIDELBERG, E. and WOOLEY, D. F. (1970). Effects of ethyl alcohol on spinal cord neurones. *Arch. Int. Pharmacodyn.* **185**, 388–396.

FURUKAWA, J. (1957). Properties of the procaine end plate potential. *Jap. J. Physiol.* **7**, 199–212.

FURUKAWA, T., FUKAMI, Y. and ASADA, Y. (1964). Effects of strychnine and procaine on collateral inhibition of the Mauthner cell. *Jap. J. Physiol.* **14**, 386–399.

GAGE, P. W. (1965). Effect of cardiac glycosides on neuromuscular transmission. *Nature*, **205**, 84–85.

GAGE, P. W. (1976). Generation of end-plate potentials. *Physiol. Rev.* **56**, 1, 177–247.

GAGE, P. W. and HAMILL, O. (1975). General anaesthetics: synaptic depression consistent with increased membrane fluidity. *Neuroscience* (*Lett.*) **1**, 61–65.

GAGE, P. W., MCBURNEY, R. N. and SCHNEIDER, G. T. (1975). Effects of some aliphatic alcohols on the conductance change caused by a quantum of acetylcholine at the toad end-plate. *J. Physiol.* **244**, 409–429.

GAGE, P. W., MCBURNEY, R. N. and VAN HELDEN, D. (1974). End plate currents are shortened by octanol; possible role of membrane lipid. *Life Sci.* **14**, 2277–2283.

GALINDO, A. (1969). Effects of procaine, pentobarbital and halothane on synaptic transmission in the central nervous system. *J. Pharm. Exp. Therap.* **169**, 185–195.

GERGIS, S. D., SOKOLL, M. D., CRONNELLY, R., DRETCHEN, K. L. and LONG, J. P. (1975). Changes of acetylcholine release at frog's neuromuscular junction by drugs used during anaesthesia. *In* "Molecular Mechanisms of Anaesthesia" (B. R. Fink, ed.), vol. 1, pp. 181–190. Raven Press, New York.

GODFRAIND, J. M., KAWAMURA, H., KRNJEVIĆ, K. and PUMAIN, R. (1971). Actions of dinitrophenol and some other metabolic inhibitors on cortical neurones. *J. Physiol.* **215**, 199–222.

GOLDSMITH, M. W., HARPER, A. A. and WANN, K. T. (1975). Some effects of 2,4-dinitrophenol (2,4-DNP) on the electrical properties of the frog skeletal muscle membrane. *J. Physiol.* **251**, 59–60P.

GRAVENSTEIN, J. S., SHERMAN, E. T. and ANDERSEN, T. W. (1960). Cyclopropane-epinephrine interaction on the nictitating membrane of the spinal cat. *J. Pharm. Exp. Therap.* **129**, 428–432.

HARMEL, M. and MALCOLM, J. L. (1958). The site of action of procaine on the isolated spinal cord of the frog. *J. Physiol.* **140**, 213–219.

HENDERSON, J. V. (1976). Squid giant synapse at high pressure. *Fed. Proc.* **35**, 237.

HINKLEY, R. E. and GREEN, L. S. (1971). Effects of halothane and colchicine on microtubules and electrical activity of rabbit vagus nerves. *J. Neurobiol.* **2**, 97–105.

IVERSEN, L. L. and NEAL, M. J. (1968). The uptake of [$^3$H] GABA by slices of rat cerebral cortex. *J. Neurochem.* **15**, 1141–1149.

JESSELL, J. M. and RICHARDS, C. D. (1977). Barbiturate potentiation of hippocampal i.p.s.ps is not mediated by blockade of GABA uptake. *J. Physiol.* **269**, 42-44 P.

JOHNSON, F. H. and FLAGLER, E. A. (1950). Hydrostatic pressure reversal of narcosis in tadpoles. *Science*, **112**, 91–92.

KARIS, J. H., GISSEN, A. J. and NASTUK, W. L. (1966). Mode of action of diethyl ether in blocking neuromuscular transmission. *Anesthesiol.* **27**, 42–51.

KARIS, J. H., GISSEN, A. J. and NASTUK, W. L. (1967). The effect of volatile anaesthetic agents on neuromuscular transmission. *Anesthesiol.* **28**, 128–134.

KELLY, J. S., RENAUD, L. P. and GOTTESFELD, Z. (1971). The effects on inhibition of AOAA evoked increase in the cortical GABA level. *Experientia*, **27**, 1114.

KENDIG, J. J. and COHEN, E. N. (1975). Neural sites of pressure-anaesthesia interactions. *In* "Molecular Mechanisms of Anaesthesia" (B. R. Fink, ed.), vol. 1, pp. 421–427. Raven Press, New York.

KENNEDY, R. D., FINK, B. R. and BYERS, M. R. (1972). The effect of halothane on rapid axonal transport in the rabbit vagus. *Anesthesiol.* **36**, 433–443.

KEYNES, R. D. (1972). Excitable membranes. *Nature*, **239**, 29–32.

KORDAŠ, M. (1970). The effect of procaine on neuromuscular transmission. *J. Physiol.* **209**, 689–699.

KRNJEVIĆ, K. (1972). Excitable membranes and anaesthetics. *In* "Cellular Biology and Toxicity of Anaesthetics" (B. R. Fink, ed.), pp. 3–9. Williams and Wilkins, Baltimore.

KRNJEVIĆ, K. (1974a). Central actions of general anaesthetics. *In* "Molecular Mechanisms in General Anaesthesia" (M. J. Halsey, R. A. Millar, and J. A. Sutton, eds), pp. 65–89. Churchill-Livingstone, London.

KRNJEVIĆ, K. (1974b). Chemical nature of synaptic transmission in vertebrates. *Physiol. Rev.* **54**, 418–540.

KRNJEVIĆ, K. (1975). Is general anaesthesia induced by neural asphyxia. *In* "Molecular Mechanisms of Anaesthesia" (B. R. Fink, ed.), vol. 1, pp. 93–98. Raven Press, New York.

KRNJEVIĆ, K. and LISIEWICZ, A. (1972). Injections of calcium ions into spinal motoneurones. *J. Physiol.* **225**, 363–390.

KRNJEVIĆ, K. and MORRIS, M. E. (1976). Input-output relation of transmission through cuneate nucleus. *J. Physiol.* **257**, 791–815.

KRNJEVIĆ, K., PUMAIN, R. and RENAUD, L. (1971). The mechanism of excitation by acetylcholine in the cerebral cortex. *J. Physiol.* **215**, 247–268.

KUSANO, K. (1970). Influence of ionic environment on the relationship between pre- and postsynaptic potentials. *J. Neurobiol.* **1**, 435–457.

LARRABEE, M. G. and POSTERNAK, J. M. (1952). Selective action of anesthetics on synapses and axons in mammalian sympathetic ganglia. *J. Neurophys.* **15**, 91–114.

LARSON, M. D. and MAJOR, M. A. (1970). The effect of hexobarbital on the duration of the recurrent IPSP in cat motoneurons. *Brain Research*, **21**, 309–311.

LØYNING, Y., OSHIMA, T. and YOKOTA, T. (1964). Site of action of thiamylal sodium on the monosynaptic spinal reflex pathway in cats. *J. Neurophys.* **27**, 408–428.

MAENO, T. (1966). Analysis of sodium and potassium conductances in the procaine end-plate potential. *J. Physiol.* **183**, 592–606.

MAENO, T., EDWARDS, C. and HASHIMURA, S. (1971). Difference in effects on end-plate potentials between procaine and lidocaine as revealed by voltage clamp experiments. *J. Neurophys.* **34**, 32–46.

MATTHEWS, E. K. and QUILLIAM, J. P. (1964). Effects of central depressant drugs upon acetylcholine release. *Brit. J. Pharmacol.* **22**, 415–440.

MEECH, R. W. (1974). The sensitivity of *Helix Aspersa* Neurones to injected calcium ions. *J. Physiol.* **237**, 259–277.

MILEDI, R. (1973). Transmitter release induced by injection of calcium ions into nerve terminals. *Proc. R. Soc. Lond. B*, **183**, 421–425.

NICOLL, R. A. (1971). Pharmacological evidence for GABA as the transmitter in granule cell inhibition in the olfactory bulb. *Brain Research*, **35**, 137–149.

NICOLL, R. A. (1972). The effects of anaesthetics on synaptic excitation and inhibition in the olfactory bulb. *J. Physiol.* **223**, 803–814.

NICOLL, R. A. (1975). Presynaptic action of barbiturates in the frog spinal cord. *Proc. Nat. Acad. Sci. U.S.A.* **72**, 1460–1463.

NICOLL, R. A., ECCLES, J. C., OSHIMA, T. and RUBIA, F. (1975). Prolongation of hippocampal inhibitory postsynaptic potentials by barbiturates. *Nature*, **258**, 625–627.

POISNER, A. M. and BERNSTEIN, J. (1971). A possible role of microtubules in catecholamine release from the adrenal medulla: effect of colchicine, vinca alkaloids and deuterium oxide. *J. Pharm. Exp. Therap.* **177**, 102–108.

PRICE, H. L. and PRICE, M. L. (1967). Relative ganglion-blocking potencies of cyclopropane, halothane and nitrous oxide, and the interaction of nitrous oxide with halothane. *Anesthesiol.* **28**, 349–353.

QUASTEL, D. M. J. and LINDER, T. M. (1975). Pre- and postsynaptic actions of central depressants at the mammalian neuromuscular junction. *In* "Molecular Mechanisms of Anaesthesia" (B. R. Fink, ed.), vol. 1, pp. 157-165. Raven Press, New York.

QUASTEL, D. M. J., HACKETT, J. T. and OKAMOTO, K. (1972). Presynaptic action of central depressant drugs: inhibition of depolarisation-secretion coupling. *Can. J. Physiol. Pharmacol.* **50**, 279–284.

RANSOM, B. R. and BARKER, J. L. (1975). Pentobarbital modulates transmitter effects on mouse spinal neurones grown in tissue culture. *Nature*, **254**, 703–705.

RICHARDS, C. D. (1974). The action of general anaesthetics on synaptic transmission within the central nervous system. *In* "Molecular Mechanisms of General Anaesthesia" (M. J. Halsey, R. A. Millar and J. A. Sutton, eds), pp. 90–109. Churchill-Livingstone, London.

Richards, C. D. and White, A. E. (1975). The actions of volatile anaesthetics on synaptic transmission in the dentate gyrus. *J. Physiol.* **252**, 241–257.

Richards, C. D., Russell, W. J. and Smaje, J. C. (1975). The action of ether and methoxyflurane on synaptic transmission in isolated preparations of the mammalian cortex. *J. Physiol.* **248**, 121–142.

Rosenberg, H. and Haugaard, N. (1973). The effects of halothane on metabolism and calcium uptake in mitochondria of the rat liver and brain. *Anesthesiol.* **39**, 44–53.

Sachdev, K. S., Panjwani, M. H. and Joseph, A. D. (1963). Potentiation of the response to acetylcholine on the frog's rectus abdominus by ethyl alcohol. *Arch. Int. Pharmacodyn.* **145**, 36–43.

Sato, M., Austin, G. M. and Yai, H. (1967). Increase in permeability of the postsynaptic membrane to potassium produced by "Nembutal". *Nature*, **215**, 1506–1508.

Schmidt, R. F. (1963). Pharmacological studies on the primary afferent depolarisation of the toad spinal cord. *Pflügers Archiv.* **277**, 325–346.

Schmitt, F. D. (1968). Fibrous proteins—neuronal organelles. *Proc. Nat. Acad. Sci. U.S.A.* **60**, 1092–1101.

Seeman, P. (1972). The membrane actions of anaesthetics and tranquilisers. *Pharmacol. Rev.* **24**, 583–655.

Sherrington, C. S. (1906). "The Integrative Action of the Nervous System". Yale University Press, New Haven, Connecticut.

Smith, D. S. (1971). On the significance of cross bridges between microtubules and synaptic vesicles. *Phil. Trans. Roy. Soc. Lond.* **261**, 395–405.

Somjen, G. (1963). Effects of ether and thiopental on spinal presynaptic terminals. *J. Pharm. Exp. Therap.* **140**, 396–402.

Somjen, G. and Gill, M. (1963). The mechanism of the blockade of synaptic transmission in the mammalian spinal cord by diethyl ether and by thiopental. *J. Pharm. Exp. Therap.* **140**, 19–30.

Speden, R. N. (1965). The effect of some volatile anaesthetics on the transmurally stimulated guinea pig ileum. *Brit. J. Pharmacol.* **25**, 104–118.

Steinbach, A. B. (1968a). Alteration by xylocaine (lidocaine) and its derivatives of the time course of the end-plate potential. *J. Gen. Physiol.* **52**, 144–161.

Steinbach, A. B. (1968b). A kinetic model for the action of xylocaine on receptors for acetylcholine. *J. Gen. Physiol.* **52**, 162–180.

Thesleff, S. (1956). The effect of anaesthetic agents on skeletal muscle membrane. *Acta Physiol. Scand.* **37**, 335–349.

Thomson, T. D. and Turkanis, S. A. (1973). Barbiturate-induced transmitter release at a frog neuromuscular junction. *Brit. J. Pharmacol.* **48**, 48–58.

Tijoe, S., Bianchi, C. P. and Haugaard, N. (1970). The function of ATP in $Ca^{2+}$ uptake by rat brain mitochondria. *Biochim. Biophys. Acta*, **216**, 270–273.

Tijoe, S., Haugaard, N. and Bianchi, C. P. (1971). The effects of psychoactive agents on calcium uptake by preparations of rat brain mitochondria. *J. Neurochem.* **18**, 2171–2178.

Wall, P. D. (1967). The mechanisms of general anaesthesia. *Anesthesiol.* **28**, 46–53.

Weakly, J. N. (1969). Effect of barbiturates on "quantal" synaptic transmission in spinal motoneurones. *J. Physiol.* **204**, 63–77.

Weisenberg, R. C. (1972). Microtubule formation *in vitro* in solutions containing low calcium concentrations. *Science*, **177**, 1104–1105.

Whittam, R. (1968). Control of membrane permeability to potassium in red blood cells. *Nature*, **219**, 610.

CHAPTER VII

# Mechanical, transport and thermal properties



## 1 Introduction

In the integrated animal, the outstanding effect of anaesthetics and inert gases is to reversibly depress consciousness, but these agents also have important non-narcotic effects in respiration, transport phenomena in tissues, and in thermal balance.

When human subjects breathe a gas mixture of oxygen and the inert gas sulphur hexafluoride, their ventilatory performance is altered by the high density of the mixture. Their breathing closely resembles that seen in subjects exposed to high pressure, as in diving. The effects of breathing dense gas mixtures are instructive to respiratory physiologists and studied by diving physiologists for safety's sake. Experiments in which subjects breath unusual gas mixtures also demonstrate how individual gases permeate the body at rates which are influenced by their physical properties. These studies have both theoretical and practical importance in anaesthesia and diving physiology. In the case of subjects breathing a helium-oxygen mixture at high pressure, several important phenomena in respiratory and circulatory physiology are revealed and, because of the high heat capacity of helium, the subjects also exhibit an extraordinary heat loss, thereby providing interesting problems in thermal regulation.

So, in this section, the mechanical, transport and thermal effects of anaesthetics and inert gases are outlined. The treatment is introductory and the intention is to provide a framework around which the reader can build an understanding of the non-narcotic effects which these agents have in the integrated organism.

## 2 Mechanical effects of inert gases in ventilation

### 2.1 INTRODUCTION

Pulmonary ventilation is a process which involves the physics of gases and the physiological and anatomical complexities of the lung, and because of its great importance in medicine, pulmonary physiology is an intensively studied subject. Developments in human diving which require humans to ventilate their lungs with highly compressed, and therefore dense, gas mixtures have stimulated enquiries into the effects of gas density on pulmonary ventilation. This discussion will distinguish two different density effects, (a) a gross airway resistance effect and (b) effects which are apparent in the small airways of the lung.

Ventilation involves muscular work to move gas in and out of the lungs and to overcome the resistance of the tissues to movement. Inert gases have been used to alter the physical properties of gas mixtures which may be safely breathed in order to identify the properties of gases which influence ventilation. The chief property of interest is the density of the gas mixture, because the turbulent flow which occurs to varying degrees in the airways is sensitive to gas density. Turbulent flow begins to develop in airways when the product of velocity, density and airway diameter divided by viscosity (i.e. the Reynolds number) exceeds approximately

1000. Increasing the density of the gas mixture clearly increases the likelihood of flow being turbulent.

What predictions can be made about pulmonary ventilation when the density of the respired gas mixture is increased? More effort will be required to maintan normal ventilation. Further, the effect of gas density on ventilatory effort should be constant, irrespective of the gas which is used to achieve the particular density. Finally, the diffusive end stages of pulmonary ventilation by which gases mix in the alveoli may be expected to alter in accordance with Graham's law of diffusion.

## 2.2 THE WORK OF BREATHING

An excellent example of the way in which gas density affects the effort of breathing is provided by the experiment of Glauser *et al.* (1967). Human subjects were required to breathe gas mixtures in a quiet, basal fashion and were then caused to hyperventilate under the influence of 7 per cent $CO_2$ in the breathing mixture. Their oxygen consumption was simultaneously measured in each case. The oxygen cost of quiet ventilation with air was some 80 ml $O_2$ STPD $min^{-1}$ less than hyperventilating with $CO_2$-rich air. The dense gas mixture comprising 7 per cent $CO_2$, 73 per cent $SF_6$ and 20 per cent $O_2$ required an additional 184 ml $O_2$ STPD $min^{-1}$ to achieve the same high level of ventilation as with the $CO_2$-rich air (Table VII.1).

TABLE VII.1

Oxygen consumption of human subjects breathing gas mixtures of different density, $\dot{V}$ is 36 litre $min^{-1}$. (modified from Glauser *et al.*, 1967)

| Density at 20°C, g $litre^{-1}$ (see also Table VII.2) | | Oxygen consumption $\dot{V}O_2$ | Oxygen cost of additional ventilation $\frac{\dot{V}O_2}{\dot{V}E}$ |
|---|---|---|---|
| 1. (7% $CO_2$ in air) | 1·248 | 353 ml STPD $min^{-1}$ | 2·7 ml STPD/litre BTPS |
| 2. (7% $CO_2$, 20% $O_2$, 73% $SF_6$) | 4·885 | 537 ml STPD $min^{-1}$ | 8·8 ml STDP/litre BTPS |
| | | Difference 184 ml STPD $min^{-1}$ | |

Thus, at a ventilation rate of 36 litre $min^{-1}$, increasing the gas density fourfold from 1·248 g $litre^{-1}$ ($CO_2$-rich air) to 4·885 g $litre^{-1}$ (7 per cent $CO_2$, 73 per cent $SF_6$, 20 per cent $O_2$) causes an extra oxygen consumption of 184 ml $O_2$ $min^{-1}$. In the design of the experiment care was taken to maintain ventilation constant but it is quite possible that with the difference in gas density some increase in muscular and tissue displacement took place, thereby introducing a small addition to the elastic resistance to breathing. This is probably a minor effect and the data undoubtedly provide an overall metabolic measurement of the resistance produced by a high-density gas mixture. Generally it is found that human ventilation is able to cope with very considerable increases in gas density.

Other measurements of the work of breathing dense gases have produced similar results. Salzano *et al.* (1971) showed the extra oxygen consumption associated with breathing a gas mixture some 4·4-fold the density of air was approximately 150 ml $O_2$ STPD $min^{-1}$ ($\dot{V} = 25$ litre $min^{-1}$). In this case, the greater than normal gas density was provided by confining the subjects in a hyperbaric chamber pressurised with a helium–oxygen (heliox) mixture to 31·3 atm.

The practical needs of the deep sea diver have stimulated numerous measurements of pulmonary function at high pressure. It is a curious fact of physics that compressing a gas increases its density proportionately, but exerts negligible effect on its viscosity. Accordingly, hyperbaric conditions can be conveniently mimicked at atmospheric pressure by using inert gases such as $SF_6$ whose viscosity is normal but whose density is high. (Table VII.2).

## 2.3 EFFECT OF GAS DENSITY ON AIRWAY RESISTANCE

The rate of gas flow achieved during inspiration or expiration for a given pressure difference between the alveoli and the mouth provides a measure of the resistance to flow generated by the gas in question. Mixtures of different densities were studied by Maio and Farhi (1967) using a special spirometer to measure flow and an oesophageal balloon to obtain a measure of pleural pressure. Pressure-flow curves for different gases of the same density superimpose, as in the case of air at 2 atm pressure and a $SF_6$–$O_2$ mixture at 0·5 atm pressure (Fig. VII.1). The increased airway resistance encountered with high gas densities, as in Fig. VII.1, is also apparent during measurements of dynamic lung compliance (Forkert *et al.*, 1975) and of the maximal flow rate achieved around the middle of expiration. Figure VII.2 shows the effect which relative gas density exerts on the latter parameter, halving it when density is twice normal or more. Up to a fivefold increase in density there is a marked decline in flow rate, but at higher densities the fall off in flow rate with increased density is much smaller. When this was first observed (Maio and Farhi, 1967) it was thought that at high expiratory flow rates the respiratory muscles would have insufficient time to exert maximal force, but when high gas density increases airway resistance and gas flow is much slower, then the muscles would be able to counteract the increased resistance by exerting their full force. The recent measurements carried out by Vorosmarti *et al.* (1975)

TABLE VII.2

Densities of inert gases and their mixtures

| A. | Pure gas, 1 atm absolute | g litre$^{-1}$, 0°C |
|---|---|---|
| | He | 0·178 |
| | $H_2$ | 0·089 |
| | Ne | 0·900 |
| | $N_2$ | 1·251 |
| | Ar | 1·784 |
| | Kr | 3·708 |
| | Xe | 5·851 |
| | $SF_6$ | 6·602 |
| | $N_2O$ | 1·978 |
| | $O_2$ | 1·429 |
| | Air | 1·293 |

B. Mixtures used by Glauser *et al.* (1967) (see also Table VII.1)

| | g litre$^{-1}$, 20°C, 1 atm absolute | density:viscosity ratio |
|---|---|---|
| Air | 1·205 | 6·617 |
| 20% $O_2$, 80% $SF_6$ | 5·188 | 32·244 |
| 7% $CO_2$, 93% Air | 1·248 | 6·988 |
| 7% $CO_2$, 20% $O_2$, 73% $SF_6$ | 4·885 | 29·997 |

C. Mixtures used by Vorosmarti *et al.* (1975) (see also Figs. VII.2 and VII.3)

| | Density relative to air | Pressure (atm absolute) |
|---|---|---|
| $He-O_2$ | 0·43 | 1 |
| $N_2-O_2$ | 1·0 | 1 |
| $SF_6-O_2$ | 3·75 | 1 |
| $Ne-O_2$ | 3·2 | 5·7 |
| $Ne-O_2$ | 5·1 | 10·1 |
| $Ne-O_2$ | 3·2 | 19·2 |
| $Ne-O_2$ | 9·5 | 19·2 |
| $Ne-O_2$ | 3·9 | 23·7 |
| $Ne-O_2$ | 4·25 | 26·0 |
| $Ne-O_2$ | 15·0 | 26·0 |

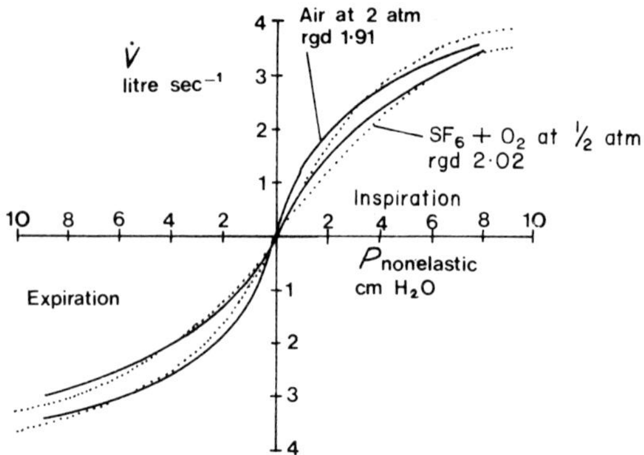


FIG. VII.1. Effect of gas density on airway resistance. Airflow, litre sec$^{-1}$, is plotted on the vertical scale against pleural pressure, cm/$H_2O$. Continuous lines relate to air breathed by human subjects at 2 atm, relative gas density 1·91. Dotted lines relate to $SF_6 + O_2$ breathed at 0·5 atm, relative gas density 2·02. Each pair of lines encloses the mean ± the standard deviation. (Maio and Farhi, 1967)

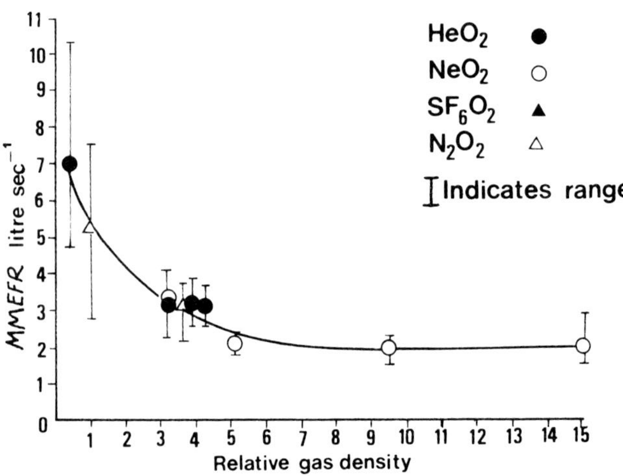


FIG. VII.2. Effect of gas density on maximal breathing capacity. Midmaximal expiratory flow rate in litre sec$^{-1}$ is plotted against relative gas density, which was varied by selecting appropriate gas mixtures and pressures. Data for human subjects quietly breathing. See also Table VII.2. (Vorosmarti *et al.*, 1975)

seen in Fig. VII.2 show that midmaximal expiratory flow becomes constant over the relative density range 5–15, the explanation for which is unlikely to lie in the velocity of muscle contraction, as originally proposed. Under these extraordinary conditions, we see a limiting airway resistance in the ventilatory system (Mead *et al.*, 1967). This may be due to airway collapse or increased turbulence, or both, increasing with gas density.

Airway resistance has been directly measured both during quiet breathing and with subjects hyperventilating with a range of gas densities. Figures VII.3(a) and (b) shows that the relationship between gas density

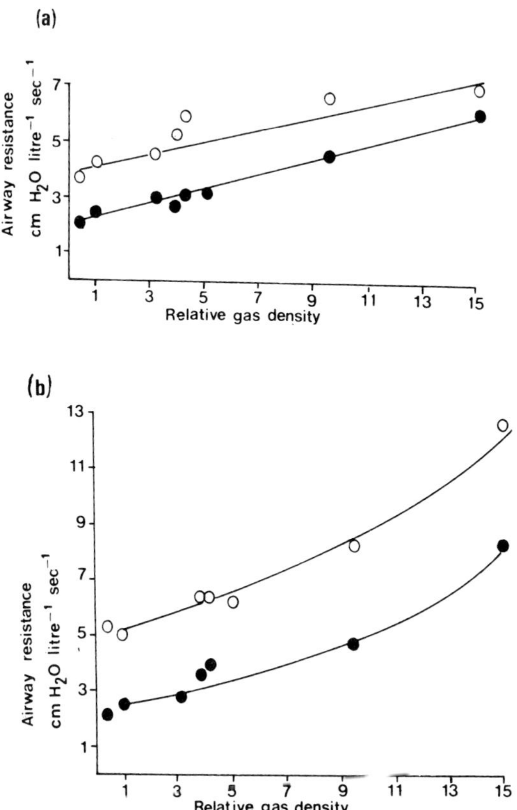


FIG. VII.3. Effect of gas density on inspiratory and expiratory airway resistance. (a) Quiet breathing: ● = inspiration, ○ = expiration. (b) Hyperventilation: ● = inspiration, ○ = expiration. See also Table VII.2. (Vorosmarti *et al.*, 1975)

and airway resistance is linear in quiet breathing, but signs of an upswing in resistance at higher gas densities may be seen during hyperventilation, probably due to an increase in the extent of turbulent flow.

It is of some interest to note the maximum density which pulmonary ventilation can cope with. Test mammals breathing helium-oxygen at pressures up to 250 atm with a relative gas density of 35 appear in reasonable health. Mice with their lungs filled with saline (relative density 900) also ventilate surprisingly well but cannot eliminate $CO_2$ sufficiently rapidly. Humans breathing helium-oxygen and subjected to 50 and 60 atm pressure (relative gas density 8–9) or hyperbaric neon-based mixtures (relative gas density 15) are able to ventilate sufficiently to sustain moderate exercise (Morrison and Florio, 1971; Varene *et al.*, 1974; Vorosmarti *et al.*, 1975; Petersen and Wright, 1976).

## 2.4 REGIONAL VENTILATION AND PERFUSION

It is generally held that bulk gas flow does not extend to the alveoli, and a diffusive mixing process is ultimately responsible for replenishing alveolar gas. According to Graham's law the rates of diffusion of gases are inversely proportional to the square root of their relative density.

The diffusion of oxygen and other gases within the alveoli has been shown to depend on the density (diffusivity) of the gas present. Evidence for this comes from some experiments with humans (Georg *et al.*, 1965) and from work with dogs. In the latter case, the animals were anaesthetised, paralysed, artificially ventilated, and at intervals forced to hold their breath. During the forced breath-hold a declining arterial $pO_2$ was recorded to provide a measure of the rate at which oxygen diffused from alveoli to blood. The dogs respired either helium-oxygen or air, and it was found that the rate of oxygen diffusion from alveoli to blood during a breath-hold decreased with increased gas density (Johnson and van Liew, 1974).

The alveolar–arterial (A–a) $pO_2$ difference provides a measure of the efficiency with which alveolar ventilation takes place. It has been used as a criterion in examining density-dependent effects in alveolar gas exchange by Saltzman *et al.* (1971). Human subjects breathed gases whose density ranged twenty-fold, from helium-oxygen at atmospheric pressure to a nitrogen–oxygen mixture at 7 atm, and were found to have a constant A–a $pO_2$ difference. Presumably in this study the rate-determining step in the overall diffusion of oxygen into the lung was unaffected by alveolar gas density. It is possible that a favourable change in the ventilation:perfusion ratio occurred with increased gas density, although the subjects were lying down.

Regional ventilation may be observed directly in subjects breathing radioactive $^{133}Xe$ and by means of this elegant technique Wood and Bryan (1971) have shown that an increase in gas density, brought about by $SF_6$, favours ventilation in the upper portion of the lung. This finding may appear paradoxical, but in the upright human lung airway diameters tend to be large in the upper parts due to the gravitational pull of the lung tissue, and presumably during ventilation of the dense gas mixture the larger airway diameters in the upper part of the lung more than compensate for their reduced compliance.

The clinical use of inert gases to vary gas density seems to be rare, the only case known to the authors being the use of helium-oxygen for the relief of excessive airway resistance (Eversole, 1938; Lambertsen, 1971).

## 3 The distribution of inert gases and anaesthetics in the body

### 3.1 INTRODUCTION

Gases pervade the tissues by diffusion and perfusion. Of long-standing interest to physiologists is the rate at which the steady state balance between the inspired gas partial pressure and tissue partial pressure is approached.

In this section the ways in which the physical properties of anaesthetics and inert gases influence their distribution throughout the organism will be outlined. Kety pointed out in an important review published in 1951 how they serve as instructive examples of gas exchange in general.

"The relative magnitudes and time relationships of these various processes are not obvious in the steady state of metabolic exchange. They become apparent, however, when a new molecular species is introduced into the atmosphere which the organism breathes and, if the new substance happens to be an inert gas, its behaviour in the organism may be explained and predicted on the basis of relatively simple physical laws."

The application of "relatively simple physical laws" attains severe mathematical complexity in this field, which the present treatment will avoid.

Inert gases and anaesthetics with a wide range of physical properties are available to demonstrate some of the salient features of gas transfer processes. We may distinguish three stages within the overall distribution process: (1) pulmonary equilibration, in which alveolar gas attains a balance with inspired gas, (2) the diffusion of alveolar gas into the circulation, and (3) the diffusion of dissolved gas from capillary blood into the tissues. These three constitute an interacting sequence whose relationships can be conveniently studied in a test animal or human subject breathing an experimental gas mixture. During such a "wash-in" experiment the approach to the steady state partial pressures in the three compartments mentioned reveals a variety of factors and complications. The reverse "wash-out" experiment does not always yield a mirror image of a "wash-in" experiment. Practical interest in "wash-in" and "wash-out" kinetics lies chiefly in anaesthesia and the decompression procedures used in human diving. For an admirable treatment of gas exchanges in anaesthesia see Eger (1974) and for a discussion of decompression phenomena see the relevant chapters in Bennett and Elliot (1975) and Hills (1977).

## 3.2 FACTORS INFLUENCING THE TRANSPORT OF DISSOLVED GASES IN THE BODY

The partial pressure of water vapour in the mammalian respiratory tract is usually higher than that of the inspired gas, so humidification is the first change undergone by newly inspired gas. Immediately following inspiration of the anaesthetic or inert gas, equilibration in the alveoli proceeds at a rate which depends on (1) the inspired concentration, the rate of alveolar ventilation and the size of the functional residual capacity, and (2) the solubility of the anaesthetic or inert gas in the blood. In the case of a poorly soluble gas ($N_2O$) it is easy to see that $P_A$ will rise quickly. Conversely, in the case of a highly soluble gas $P_A$ will rise much less quickly. Figure VII.4 from Salanitre *et al.* (1967) shows experimental data to illustrate a variety of cases. The gases $N_2O$, $C_2H_4$ and $C_3H_6$, all of which have a low blood solubility, show an initially rapid rate of rise in their $P_A$s in contrast to the more soluble agents. In the latter group the alveolar steady state partial pressure is only attained after 1 hr or even longer. On that time scale it is necessary to take the equilibration with the whole body into account. It is worth comparing the rates of alveolar equilibration of cyclopropane and $N_2O$. These gases have similar solubilities in blood and lean tissue but cyclopropane is the more soluble in fat. When cyclopropane is inspired its equilibration in the alveoli is slower than that of $N_2O$ because of its much greater simultaneous equilibration with body fat (Fig. VII.4, Table VII.3). Highly fat-soluble substances such as chloroform virtually disappear into the fatty tissues of the body, and blood returning to the lungs contains little of it, giving rise to a steep inward partial pressure gradient. Thus the rate at which inert gases and anaesthetics enter the pulmonary blood is influenced initially by only one physical property of the gas, its solubility in blood and body tissues and by the physiological parameters perfusion and partial pressure gradient.

These factors also operate in a "wash-out" experiment, a particularly dramatic case of which may be seen in patients recovering from $N_2O$ anaesthesia. During anaesthesia a large quantity of $N_2O$ accumulates in the body. When the patient starts to breathe air after surgery the steep rise in the alveolar concentration of $N_2O$ is sufficient to depress the oxygen partial pressure in the lung by as much as 15 per cent of the normal level, producing symptoms of mild anoxia. The condition is called "diffusion anoxia" (Fink, 1955; Rackow *et al.*, 1961).

What factors influence the rate of equilibration between blood and tissues? First, there is the solubility of the gas in the given tissue. Secondly,

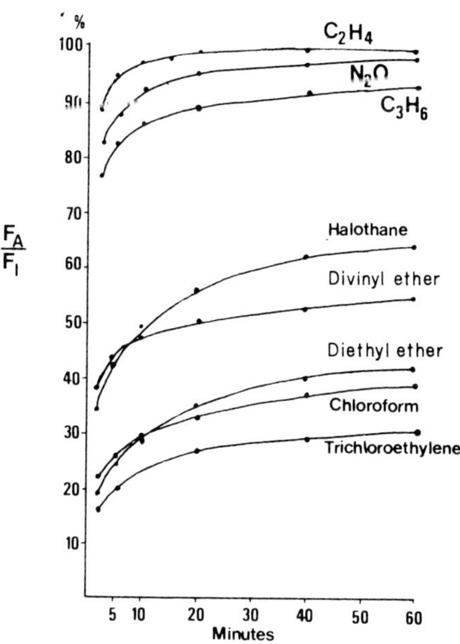


FIG. VII.4. Attainment of the steady state alveolar concentration of anaesthetics and inert gases. The ratio of end-tidal (i.e. alveolar) concentration of anaesthetics to the inspired concentration is plotted against time after the onset of breathing the gas mixture. The figures are for human subjects quietly breathing in the sitting position. (Salanitre *et al.*, 1967)

there is the rate of perfusion of the tissue and very likely also the rate of diffusion of the gas in bulk water or its rate of diffusion through cell membranes. And thirdly there is the partial pressure difference between arterial blood and the tissue. The simple and illuminating approach of Eger (1974), following earlier workers, is suited to our purpose, and we will start with the Fick principle to describe the steady state transfer of gas from lung to blood.

TABLE VII.3

Physical properties of anaesthetics and inert gases which influence their distribution in the body (from Bennett and Miller, 1973)

| | Ostwald solubility coefficients at 37°C[a] | | | |
|---|---|---|---|---|
| | Human blood–gas | Water–gas | Olive oil–gas | Molecular weight |
| He | 0·0098 | 0·0095 | 0·0175 | 4·0 |
| $H_2$ | 0·019 | 0·019 | 0·068 | 2·016 |
| Ne | 0·0109 | 0·011 | 0·023 | 20·18 |
| $N_2$ | 0·014 | 0·014 | 0·074 | 28·01 |
| Ar | 0·0293 | 0·03 | 0·154 | 39·95 |
| Kr | 0·6 | 0·05 | 0·49 | 83·8 |
| Xe | 0·14 | 0·09 | 1·9 | 131·3 |
| Ethylene | 0·14 | 0·093 | 1·28 | 28·05 |
| $N_2O$ | 0·47 | 0·47 | 1·4 | 44·01 |
| Cyclopropane | 0·45 | 0·21 | 11·4 | 42·08 |
| $CF_4$ | — | 0·0048 | 0·072 | 88 |
| $SF_6$ | — | 0·0052 | 0·28 | 146 |
| Fluoroxene | 1·4 | 0·86 | 47·8 | 126 |
| Diethyl ether | 13·0 | 13·0 | 64·3 | 74·12 |
| Halothane | 2·4 | 0·84 | 224 | 197 |
| Chloroform | 8·6 | 4·3 | 265 | 119·38 |
| Methoxyfluorane | 12·8 | 4·6 | 960 | 164·98 |

[a] Each number is the ratio of concentration in the solvent: concentration in gas phase with which it is in equilibrium

The Fick principle states that the quantity of a substance delivered to, or removed from, an organ or tissue may be derived from the difference between the arterial and venous concentrations. In the case of the lung

$$U_L = \dot{Q}(C_a - C_v) \tag{1}$$

$U_L$ = uptake from the lung, $C_a$ = arterial concentration, $C_v$ = venous concentration and $\dot{Q}$ = blood flow. From this well-known premise it is now possible to confirm what has already been asserted, that in the transfer of gas from lung to blood three factors operate, the solubility of the gas, the rate of perfusion and the arterial–venous difference in partial pressure.

By definition the blood–gas partition coefficient is

$$\lambda_B = C_a/C_A \tag{2}$$

where $C_a$ = arterial concentration and $C_A$ = alveolar concentration. This may be changed to partial pressure terms in which $P$ is barometric pressure.

$$C_a - \frac{\lambda_B P_A}{P}. \tag{3}$$

Also

$$C_V = \frac{\lambda_B P_V}{P}$$

Substitution into equation (1) gives

$$U_L = \lambda_B \dot{Q} \frac{(P_A - P_V)}{P} \tag{4}$$

which restates that the rate of gas uptake from the lung is given by the blood–gas partition coefficient $\lambda_B$ (i.e. solubility), perfusion ($Q$ ml lung$^{-1}$) and the arterial–venous partial pressure difference. Equation (4) emphasises the importance of gas solubility, demonstrated in Fig. VII.4.

By similar reasoning Eger argues that gas transfer from the blood to the tissues is given by

$$U_T = \lambda_B \dot{Q} \frac{(P_a - P_t)}{P} \tag{5}$$

in which the terms have the corresponding meaning to those in equation (4).

Tissues differ in their degree of perfusion and their capacity to dissolve gases, hence the arterial–venous differences which arise across them differ. Different tissues take up a newly inspired gas at characteristic rates. Brain is well perfused and equilibrates with anaesthetics much faster than does cartilage. Fat deposits are generally fairly well perfused but they only attain their steady state with an anaesthetic slowly because they have a very large dissolving capacity. Figure VII.5 shows an example of an organ, the brain of a dog, reaching its steady state value with arterial blood shortly after the animal began to inspire nitrous oxide. Figure VII.6 shows the body "wash-in" for $N_2O$ in human subjects. Interestingly, the "wash-out" curve is remarkably similar, demonstrating, in this case, that the equilibration process can work in reverse, as the simple treatment given above implies. The smooth curves in Fig. VII.6 mask many different equilibration rates which different tissues possess. Most workers in this field find it necessary to distinguish only a few tissue types, primarily on the criteria of perfusion and gas solubility. Inspection of the curves in Fig. VII.6 immediately suggests a "slow" and a "fast" compartment.

### 3.3 THE RELATIVE ROLES OF DIFFUSION AND PERFUSION IN GAS EXCHANGE IN THE LUNGS

The distinction between a diffusion-limited or a perfusion-limited process is made clear in Fig. VII.7. Curve A shows the gas partial pressure gradient

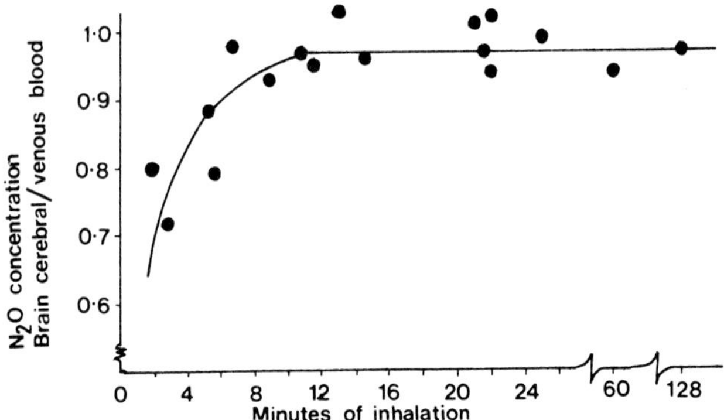


FIG. VII.5. The rate of equilibration of nitrous oxide between the brain and cerebral venous blood of the dog. The ratio of the brain:blood nitrous oxide concentration is plotted on the vertical scale against time in minutes. (Kety *et al.*, 1948)

from alveolus to blood in conditions in which perfusion limits the rate of transfer. That is, diffusion through the alveolar wall is very rapid. In B, diffusion through the wall is very slow and perfusion is relatively fast. Thus diffusion in the wall is the rate-limiting factor in B.

In a previous section the diffusive movement of gas within alveoli was mentioned. Here we consider the transfer of gas from alveoli to the pulmonary blood. Scheid and Piiper (1975) argue that it is possible to distinguish diffusion-limited and perfusion-limited gas transfer processes and to relate them to the physical properties of the gases. The diffusion of gas into capillary blood through the alveolar wall may be written, in their terminology:

$$D = d\beta_M \frac{F}{X} \tag{6}$$

where $D$ = diffusive conductance, $d$ = diffusion coefficient in the alveolar wall, $\beta_M$ = solubility of gas in the alveolar wall, $F$ = area of alveolar wall and $X$ = thickness of alveolar wall.

As the blood moves past an individual alveolus the partial pressure of gas in the blood approaches the level in the alveolus. In the steady state, the overall driving partial pressure may be given by

$$\frac{P_A - P_C}{P_A - P_V}$$

i.e. the *ratio* of the difference in partial pressure between alveolar and out-

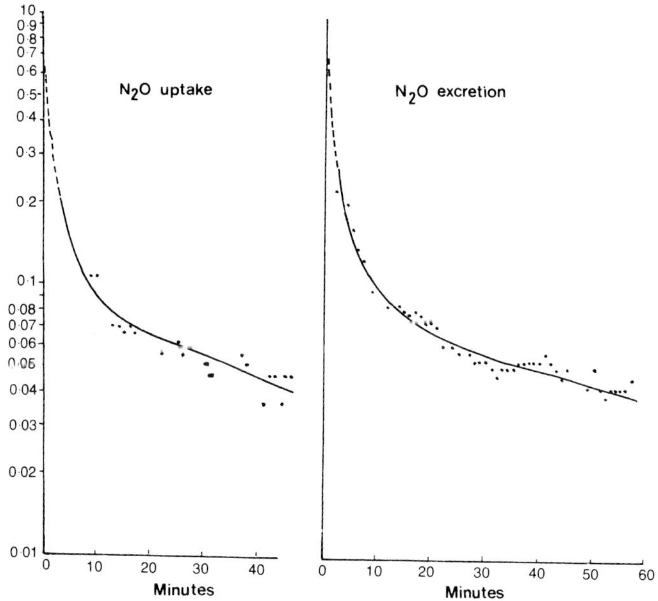


FIG. VII.6. The pulmonary uptake and elimination of nitrous oxide during a sub-anaesthetic exposure. Uptake or elimination is plotted on the vertical scale in relative units against time from the start of the experiment. (Salanitre *et al.*, 1962)

going capillary blood to alveolar to incoming venous blood (Scheid and Piiper, 1975).

While the driving partial pressure is obviously important, so too is the perfusion of the alveoli, and Scheid and Piiper use the term convective conductance to quantify it. Convective conductance is the product of $\dot{Q}$, blood flow, and the solubility of the gas in blood, $\beta_b$. The driving partial pressure and diffusive and convective conductances are related in the following manner (see Scheid and Piiper, 1975, Fig. 2)

$$\frac{P_A - P_c}{P_A - P_V} = \exp - \frac{D}{\beta_b \dot{Q}} \tag{7}$$

There are two physical properties of gases which influence their transfer across the alveolar wall, $\beta$b and $D$, the latter being defined in equation (6). A gas which is highly soluble in blood, such as carbon monoxide, will move across the alveolar wall at a rate largely limited by the term $D$, that is in a diffusion limited way. In contrast, inert gases and anaesthetics have a

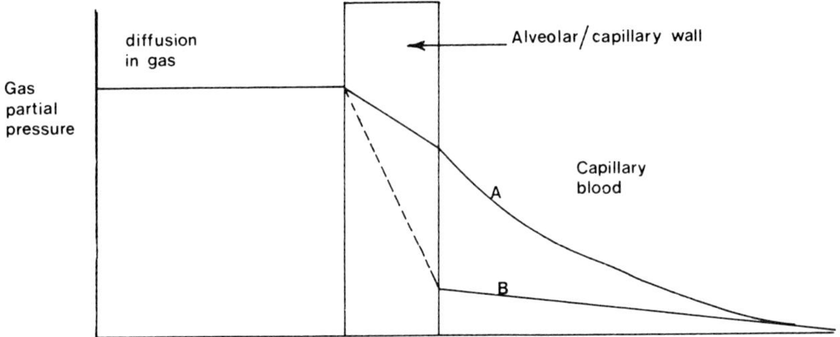


FIG. VII.7. Diagram to show the partial pressure profiles of perfusion-limited (A) and diffusion-limited (B) transport across the lung. See text.

lower blood solubility, the transfer of a small amount of these gases into the blood causes a marked partial pressure increase, and hence their removal by perfusion determines the overall rate of transfer. Thus all factors being constant these gases generally equilibrate with the pulmonary circulation at a rate determined by perfusion (Fig VII.7, slope A).

### 3.4 THE RELATIVE ROLES OF DIFFUSION AND PERFUSION IN GAS TRANSFER IN TISSUES

This problem demands a detailed explanation of the processes involved in gas exchange in heterogeneous tissue. The distinction between diffusion-limited and perfusion-limited processes is obvious from Fig. VII.8. In A the blood gas partial pressure is maximal at the capillary wall and diffusion thereafter is the slow and the rate-determining process in gas transfer. Increasing the velocity of blood flow in the capillary will not affect the rate at which dissolved gas enters the tissue. In B the partial pressure of the dissolved gas is less than maximal at the capillary wall, due to prior diffusion into the tissue. This is the state of affairs which is regarded by many as the normal physiological condition, but in recent years this has been disputed by several workers. Quite apart from the theoretical interest in this question there are at least two physiological techniques which depend on distinguishing between diffusion- and perfusion-limited transfer processes. They are the use of inert gases for measuring the perfusion of an organ, and the technique of decompression in diving, which is a particularly critical "wash-out" experiment.

It is relatively simple to postulate differences between diffusion-limited and perfusion-limited systems, but not at all easy to obtain the necessary,

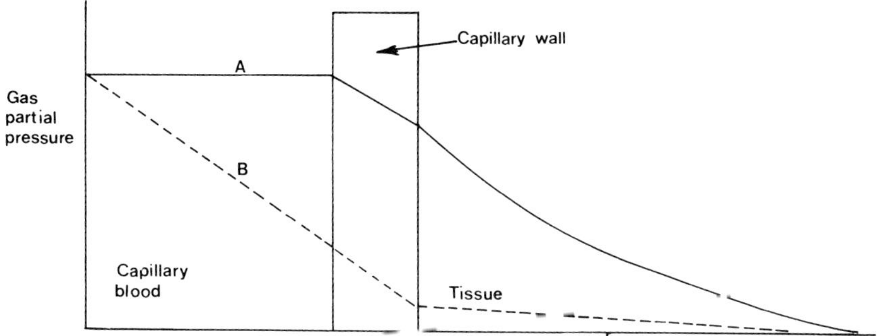


FIG. VII.8. Diagram to show the partial pressure profile of diffusion-limited (A) and perfusion-limited (B) transport across the capillary wall. See text.

decisive measurements to test actual cases (Hills, 1970). Diffusion-limited gas transfer, for example, will lead to a close similarity between arterial and venous levels during wash-in. Similarly, one may argue that increasing the velocity of blood flow should increase the rate of gas exchange in a perfusion-limited system, but such data as exist are unequivocal (Hills, 1967, 1970). Finally, the time taken for a tissue to equilibrate by a perfusion-limited process should be unaffected by the diffusion coefficient of the gas and only influenced by its partition coefficient, but no data which can separate these two possibilities seem to exist. The situation is confusing, the physiological problem complex, and specialists in the field are divided in their views on the roles of diffusion and perfusion.

### 3.5 USE OF INERT GASES TO MEASURE PERFUSION

Inert gases have been widely used as tracers to measure the perfusion of an organ. The Fick principle is the basis of the method. Equation (1) rearranged gives

$$\text{blood flow} - \frac{M \times t}{(a - v)} \tag{8}$$

where $M$ = quantity of tracer substance removed from or added to an organ, $t$ = time and $a - v$ = arterial–venous concentration difference. The tracer must be reasonably soluble in blood and a freely diffusible substance which does not react with the tissues. Its purpose is to move into or from the bloodstream at a rate determined by perfusion, and not by diffusion. The tracer should also be easily assayed. Suitable inert gases include $C_2H_2$, $N_2O$, Xe, Kr and trifluoroidomethane ($CF_3I$). These gases are intermediate in their solubility (Table VII.3); a gas such as nitrogen is inconveniently insoluble for the purpose of measuring perfusion although $H_2$ has been used. The more soluble agents such as ether equilibrate too slowly with tissues of interest. Xe, Kr and $CF_3I$ may be used in the radioactive form and monitored *in vivo*.

The measurement of blood flow in the brain determined with $N_2O$ provides an example to illustrate the method. As we have seen in Fig. VII.5 the brain of a dog attains a steady state with $N_2O$ in about 10 min. The equilibration process can be followed by measuring the $pN_2O$ in an artery and the jugular vein, and when these two are equal it may be assumed that the brain is saturated with $N_2O$. The time taken to equilibrate is required in the equation below, from Kety and Schmidt (1945).

$$\text{blood flow} = \frac{100\,\dot{Q}t}{\int_{T_1}^{T_2} (C_\mathrm{a} - C_\mathrm{v})\mathrm{d}t} \tag{9}$$

where $\dot{Q}t$ is the calculated quantity of nitrous oxide taken up by 100 gm of brain in time $t$, i.e. at the steady state. By dividing the quantity of $N_2O$ thus obtained by the integral of the arterial–venous concentration differences over the appropriate time interval, blood flow in units of volume per mass brain per time is obtained. Similarly, the "wash-out" of brain $N_2O$ can be used to obtain blood flow.

Many developments of this simple procedure which was originally introduced by Kety and Schmidt in 1945 have made blood flow measurements complicated. A prime assumption in the method is that perfusion limits the transfer of the diffusible gas and this has not been unequivocally established. An interesting comparison between blood flow rates in the cat brain determined with the radioactive inert gas $CF_3I$ and another indicator, $^{14}C$ antipyrine, showed differences because equilibration of the latter was partly determined by its diffusion through the capillary wall. This does not prove that the inert gas is distributed in an entirely perfusion-limited way but it suggests that diffusion is only a minor factor (Eckman *et al.*, 1975).

### 3.6 INERT GASES IN DECOMPRESSION

The tissues of a diver breathing inert gases at high pressure acquire a high concentration of dissolved gas. Diving thus provides an extreme form of "wash-in" experiment, and during the ascent to the surface an equally extreme form of gas "wash-out" occurs. During decompression (i.e. ascent) the hydrostatic pressure experienced by the diver may decrease more rapidly than the partial pressure of gas in the tissues, and the ensuing state of supersaturation leads to the formation of gas bubbles. These may eventually rupture tissue, occlude blood vessels and exert haematological effects, but the early symptoms of decompression sickness is usually a pain in a joint, probably caused by mild mechanical pressure on a nerve.

The ideal solution to the practical problem of decompression will provide for the fastest possible, symptom-free ascent. The decompression schedules which are used in practical diving work fairly well in preventing the short-term symptoms of decompression sickness, but they could be improved and their basis is a happy (or unhappy, depending on your outlook) mixture of practical experience and applied physiology.

Let us here consider the different decompression or "wash-out" properties of the principle diving gases, helium and nitrogen.

### 3.6.1 *Bubble formation*

Bubbles of gas emerge from solution when the partial pressure of the dissolved gas exceeds the ambient hydrostatic pressure. In many practical situations pre-existing gas nuclei favour the growth of visible bubbles in supersaturating conditions. In experiments with solvents supersaturated with gases it is possible to eliminate gas nuclei and then measure the extent of the supersaturation which leads to bubble formation.

Enormous supersaturation pressures are required to cause bubbles to form within water held in capillaries. Hemmingsen (1975) found the minimum supersaturated partial pressures required to induce spontaneous bubbles of argon and nitrogen in water were 140 atm, and for the less soluble gas helium the equivalent value was 300 atm. The process is really one of cavitation, which Hemmingsen regards as a spontaneous and therefore rather variable event, influenced by the concentration of the dissolved gas, the properties of the solvent and the material enclosing the water. The presence of gas hydrates seem to exert no effect on bubble formation.

The improbability of a gas bubble forming in a solution from which gas micronuclei have been removed is dramatically illustrated by Hemmingsen. Water in equilibrium with nitrogen at 200 atm contains 700 water molecules for every 1 nitrogen molecule. Cavitation is really the problem of removing the water molecules since the nearest neighbour distance between dissolved nitrogen molecules (above) is $28 \times 10^{-8}$ cm and in the gas phase of a bubble at 200 atm it would be $6 \times 10^{-8}$ cm.

What has all this to do with the formation of bubbles during the decompression of a diver? The magnitude of the supersaturation pressures suggest that completely different conditions exists in tissues which bubble much more readily than water. Numerous methods have been used to detect decompression bubbles *in vivo*, but it is one thing to demonstrate the appearance of bubbles in a convenient tissue, it is quite another to correlate painful symptoms with the particular bubble whose expansion is causing pain, and the precise conditions which lead to the formation of the critical bubbles in the first place. However, from measurements made on convenient models and tissues, there has come the idea that bubbles form spontaneously with little supersaturation (Fig. VII.9). The metastable condition demonstrable in pure water does not exist *in vivo*.

### 3.6.2 *Helium and nitrogen in decompression*

In decompression practice in diving a safe decompression ratio is used as if a metastable condition exists in tissues, but this is probably a consequence of the symptoms used as a criterion for bubble formation. The existence of a safe decompression ratio was discovered by high-altitude physiologists who learned that men could safely experience a halving of their ambient pressure without experiencing decompression symptoms. Recent data show that divers saturated with air at 1·8 ATA may return abruptly to normal pressure with safety. The "no-stop" decompression threshold pressure for helium-oxygen is higher (2·1 ATA) and for argon–oxygen it is lower (Zaltzman and Zinov'eva, 1965; Hempleman, 1975). These pressures correlate with the tissue-solubility of the inert gases in a way which is readily understood and, as already mentioned, they are two orders of magnitude smaller than comparable data for water. Using a super-saturation ratio of between 1·6 and 1·0 Bühlman *et al.* (1967) found humans attained a steady state with inspired helium 2·6 times faster than with $N_2$. The rates of diffusion of these two gases differ by this factor, suggesting that tissue gas exchange is limited by a diffusion-limited process, which is plausible in the case of cartilaginous tissue in joints which was probably the critical tissues in this case. However, the oil-water partition coefficients of the two gases also differ by a factor of 2·95, which suggests that equilibration times are merely approximately determined by the quantity of gas dissolved, pointing to no particular limitation in the rate of equilibration.

Other differences between helium and nitrogen are seen as a consequence of the "early bubbles" which may be demonstrated in decompression experiments with test animals. In decompression research the great need to correlate physical conditions with physiological symptoms often makes it necessary to use the behaviour or death of animals as the criterion for dangerous bubble formation.

Consider the significance of early bubbles on the efflux of gas from a tissue undergoing decompression.

A step decompression which elicits no bubbles in a previously saturated animal will generate an inert gas partial pressure gradient between the tissues and the capillary blood. Any further decompression will steepen the partial pressure gradient up to the point at which bubbles form. The effect of bubbles is to drain the tissue of gas and reduce the partial pressure gradient driving gas into the circulating blood. Further decompression only enhances bubble growth, which proceeds at a rate determined by the diffusion coefficient and the solubility of the gas in question.

The experience of the early diving physiologists was consistent with the idea that bubbles capable of causing symptoms only formed when the (large) decompression ratio was exceeded. Experiments carried out first by Hempleman and then by others show that this idea is false. Bubbles seem to form rather easily (as in Fig. VII.9) and the art of safe and rapid decompression lies in handling this condition. The way in which helium and nitrogen exert different effects on early bubble growth is demonstrated in the following experiments (Hempleman, 1960; Griffiths *et al.*, 1971).

A test mouse is given a conditioning hyperbaric exposure, decompressed and then after a selected time interval is given a second exposure to pressure and finally decompressed. The subsequent survival of the animal is scored. Figure VII.10(A) illustrates the procedure. The decompression step used is a fairly severe one. The susceptibility of the animals after the conditioning exposure was found to increase with the time interval between exposures. For example, Fig. VII.10(B) shows that when the interval between the conditioning and the second compression in nitrogen is 5 min, 50 per cent of the animals die after the second exposure. When the interval is either very short or much longer (20 min), only 20 per cent or so die after the second exposure. These results are consistent with the idea that the first decompression caused bubbles to form. Once formed the bubbles grow during the interval, and during the second compression more gas dissolves in the tissues to subsequently feed the growth of bubbles which occurs after the second decompression. Bubbles inhibit the safe efflux of the gas from the animal as a whole during the interval and after the second decompression. When the interval between compressions is very short there is little bubble growth and fewer die after the second decompression. When the interval between compressions is much longer than 5 min, bubble growth reaches its limit and there is sufficient time for gas to be eliminated from the tissue to protect the animals from the effects of a second compression.

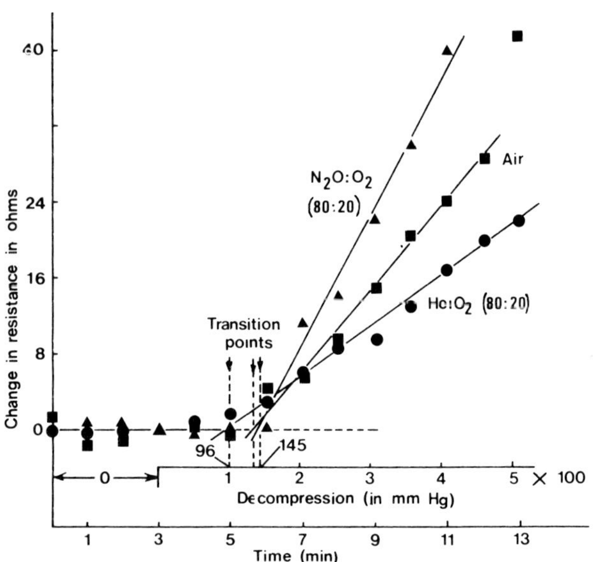


FIG. VII.9. Appearance of gas bubbles *in vivo* following a small decompression step. An electrical method is used to determine the growth of bubbles in rats' tails. Decompression begins at time 3 min and after 2 min the pressure has decreased to 100 mm Hg. The slopes relate to the solubilities of the different gas mixtures breathed by the rats before and during decompression. (Hills, 1971)

According to this argument the upper curve in Fig. VII.10(A) reflects the balance between nitrogen diffusing into extravascular bubbles and into the bloodstream. Helium (lower curve Fig. VII.10(B), in contrast, produces effects consistent with its more rapid diffusion, which is related to its low molecular weight. These results also provide some evidence for bubble formation taking place in tissues and not in the blood, because in the latter case perfusion would probably dictate the rate of bubble growth, masking the differences between the bulk diffusion of helium and nitrogen. "Wash-out" curves during decompression are thus liable to be influenced by the presence of bubbles which are not present during the "wash-in" phase. Bubbles may also occlude capillaries and impair perfusion. For these reasons hyperbaric "wash-in" and "wash-out" curves may not be symmetrical.

The differential rates of uptake found between helium and nitrogen and their differential solubilities can be put to good use to accelerate safe

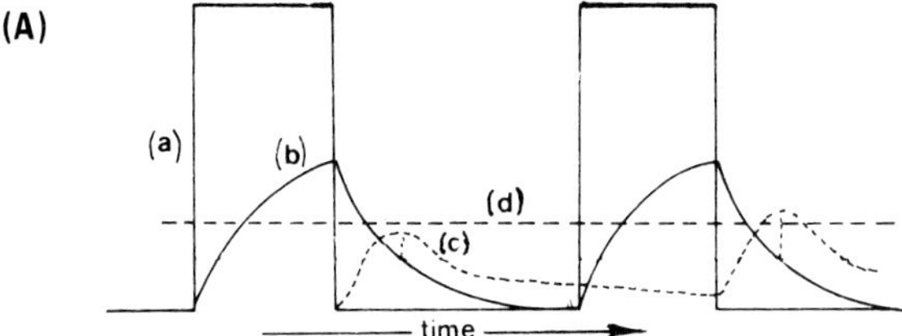


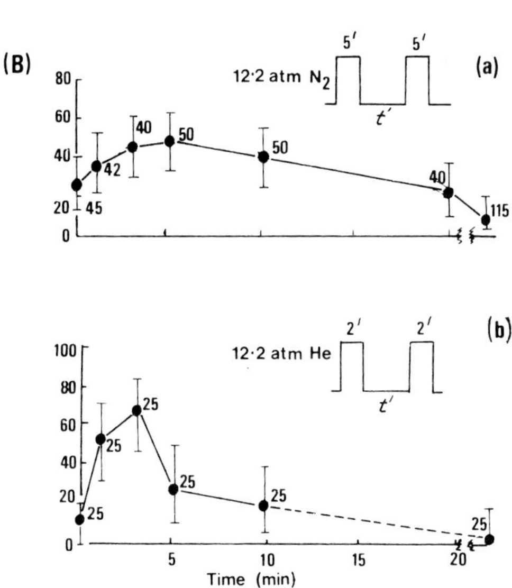


FIG. VII.10. Evidence for "early bubbles" in decompression. (A) Pressure profile of exposure (*a*). The partial pressure of the inert gas dissolved in the tissue is shown by (*b*), and the quantity of gas in a bubble by (*c*). The dotted line (*d*) indicates the threshold for symptoms caused by the growth of bubbles. (B) The relationship between time spent at 1 atm oxygen between two short exposures to pressure and percentage mortality after the second exposure. Vertical scale, percentage mortality. Insets show pressure profile. (a) Two 5 min exposures to 12·2 atm $N_2$ plus oxygen. (b) Two 2 min exposures to 12·2 atm He plus oxygen. (Griffiths *et al.*, 1971)

decompression. In principle, it is possible to minimise the amount of gas which dissolves in a diver's tissues during a brief dive by using a "slow" gas, i.e. $N_2$ rather than He. Of more practical value is the use of $N_2$ during decompression after a helium dive, especially after a saturation dive. By changing the diver's breathing mixture from a helium to a nitrogen mixture at constant hydrostatic pressure, it is possible to obtain a very steep partial pressure gradient of helium between the tissues and capillary blood (Keller and Bühlman, 1965). Under these conditions nitrogen washes in more slowly than helium washes out. This procedure is called "mixed gas decompression".

Another example of the possible use of a slowly equilibrating inert gas is the suggestion that free escape from submarines be accomplished using $CF_4$ in the breathing mixture (Table VII.3) The normal free escape method involves the subject breathing air trapped in a hood covering the head and shoulders while he floats rapidly up to the surface. The time course of the escape allows negligible "wash-in" of gas during the short period of exposure to pressure, thus precluding decompression problems. Gait and Miller (1975) argue that $CF_4$ will equilibrate much more slowly than the nitrogen in air. From simulated escapes with test mammals they suggest that men should be able to undertake free escape from twice the present depth limit if $CF^4$ were used, thus extending the technique to 400 m depth.

### 3.6.3 *Isobaric counterdiffuson*

Changing the composition of the gas mixtures breathed by divers while at high pressure can bring about the extraordinary phenomenon of bubbles forming and growing in the tissues which experience no change in ambient pressure.

During the course of some simulated dives a number of subjects experienced itching skin, which developed into a rash of small subcutaneous blisters (Lambertsen and Idicula, 1975). A number of vestibular problems were also encountered. The symptoms appeared when the subjects breathed a nitrogen- or neon-based gas mixture from a breathing apparatus while still confined in a helium environment at constant pressure. The partial pressure gradients set up by changing the inspired gas mixture are shown in Fig. VII.11. Evidently, the outward diffusion of nitrogen from the cutaneous capillaries and the steady state partial pressure of helium in the skin combine to generate a supersaturated area where bubbles form. Isobaric supersaturation might be capable of forming bubbles in deep tissues when, for example, a tissue equilibrated with helium at high pressure is perfused with blood equilibrated with a high partial pressure of nitrogen. Simple model experiments demonstrate this process (Graves *et al.*, 1973;

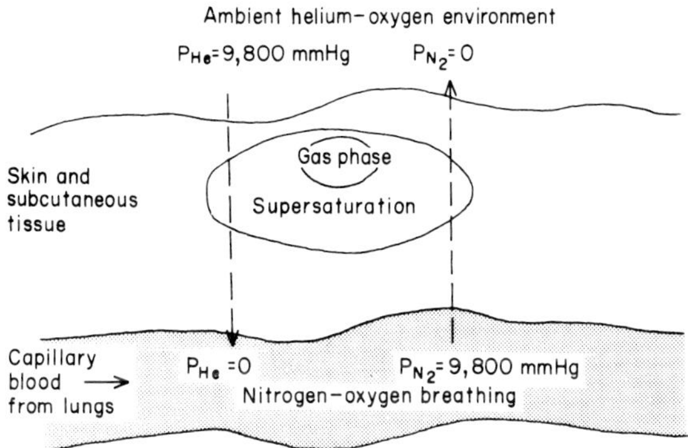


FIG. VII.11. Isobaric counterdiffusion produces dermal gas lesions (Lambertsen and Idicula, 1975)

Strauss and Kunkle, 1974). For example, if gelatin slabs are equilibrated with nitrogen and then decompressed to induce bubble formation, it may be shown that the replacement of nitrogen with helium at constant high pressure causes the bubbles to grow in size. Figure VII.12 illustrates the principle behind the mixed gas decompression technique.

### 3.6.4 *Osmotic effects of dissolved gases*

Dissolved gases exert an osmotic effect, like any other solute. For example, Kylstra *et al.* (1968), using a polyurethane–polyether semipermeable membrane, obtained an osmotic response from an aqueous $N_2O$ solution. Hills (1971) has shown that dissolved $N_2$ exerts an osmotic pressure across a membrane of cartilage mounted in a conventional osmometer. The physiological problem, however, is to demonstrate, or refute, that osmotic effects arise from the large gas concentration gradients encountered in anaesthesia or in hyperbaric conditions. By means of a high-speed technique, erythrocytes have been shown to respond by a slight shrinkage when exposed to $N_2O$ dissolved in isotonic saline. The period of shrinkage lasted 1 sec and normal cell volume was restored in the following 2 sec, but it is of no significance in practical circumstances (Longmuir and Grace, 1969).

In an attempt to elicit osmotic effects from gases *in vivo*, Hills (1972) set up a concentration gradient of $N_2O$ across the lungs of rabbits. One of the animals' lungs was ventilated with a mixture of $N_2O$:$O_2$ in the proportions of 80:20, while the other lung was simultaneously ventilated

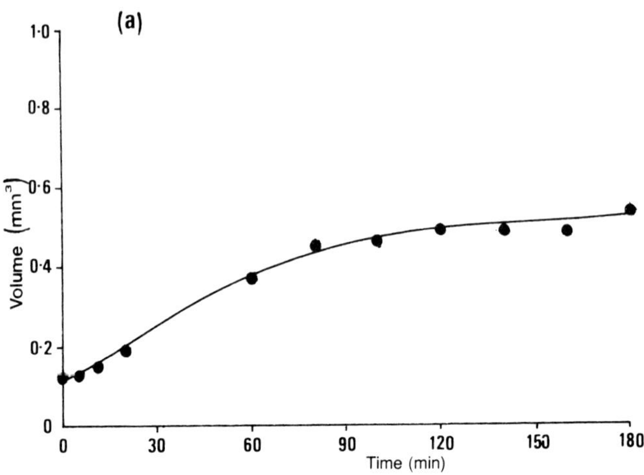


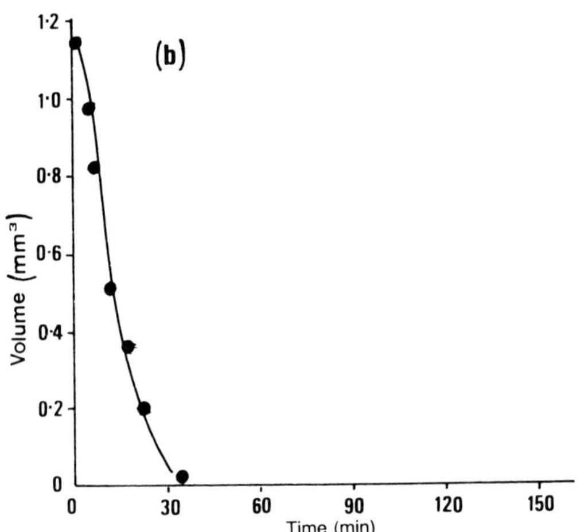


FIG. VII.12. Formation of gas bubbles in gelatin by isobaric counter diffusion. (a) The growth in bubble volume with time when the ambient nitrogen is replaced by helium. (b) The decline in bubble volume with time when helium is replaced by nitrogen. (Strauss and Kunkle, 1974)

with air. After $2\frac{1}{2}$ hr the animals were sacrificed and the amount of accumulated water in each lung was measured. The lung exposed to nitrous oxide ventilation contained significantly more water than the control lung, which may be explained by the presence of a high concentration of solute ($N_2O$) in the alveolar fluid of the experimental lung. Although $N_2O$ equilibrates quickly with the pulmonary circulation, Hills argues that the transient period of gas equilibration provides sufficient time for some osmotic flow from the blood to the alveoli. The possibility that $N_2O$ increases the water permeability of the treated lung more than the control also needs to be examined.

The idea that dissolved gases like $N_2O$ or $N_2$ can affect tissue water by disturbing the balance between the colloidal osmotic pressure of the blood proteins and blood pressure is an intriguing one. Clearly, it could only occur during "wash-in" or "wash-out" conditions. Hills (1972) suggests that $N_2O$ may exert an osmotic pressure of a magnitude comparable to that exerted by blood proteins, taking the reflexion coefficient of $N_2O$ as 0·05 (i.e. the leakiness of the semipermeable membrane to $N_2O$ is considerable). Halsey and Eger (1973) concluded from first principles that $N_2O$ would exert a transcapillary osmotic pressure no greater than that exerted by blood proteins and probably much smaller. In a direct attempt to measure a change in the water balance between blood and tissues in patients undergoing $N_2O$ anaesthesia they measured haematocrit levels at time intervals during the "wash-in" and "wash-out" periods. No changes could be detected, demonstrating the insignificance of the osmotic effects in this particular case.

### 3.7 CONCLUSIONS

The physical properties of inert gases and anaesthetic influence their manner of distribution in the body in a number of ways which are well understood. For example, the rate of change of alveolar partial pressure is closely related to the solubility of the newly inspired gas in blood and tissues. In many tissues it is difficult to identify the rate-determining gas property, as in the case of the rates at which helium and nitrogen equilibrate with the body. Kety's (1951) assertion ". . . if the new substance happens to be an inert gas its behaviour in the organism may be explained and predicted on the basis of relatively simple physical laws" is generally confirmed, but much of the original challenge of that statement remains.

## 4 Inert gases and thermal balance in mammals

In aviation, and space flight and in diving it is frequently necessary for men to be exposed to artificial atmospheres. The sensitivity of humans in particular and animals in general to apparently small changes in their inspired or ambient gas mixture is rather surprising and has stimulated a wide variety of investigations. This section deals with the effects of the inert gas helium on the thermal balance of mammals. Although we will be concerned with the flux of heat and not of molecules, Kety's (1951) statement (p. 251) on the scientific value of inert gases applies here also.

### 4.1 THE USES OF HELIUM IN SPACE FLIGHT AND IN DIVING

The atmosphere within a manned space vehicle has to meet both physiological and engineering requirements.

Oxygen is required at a partial pressure of about 0·2 atm to support respiration. A lower $pO_2$ may cause hypoxia and higher $pO_2$ leads to damage of the lung tissue. The occupants of the vehicle give off a little water vapour and carbon dioxide and thus the total pressure of $pH_2O$, $pCO_2$ and $pO_2$ is inconveniently low to constitute the total atmospheric pressure within the vehicle. It would require the occupants to undergo a decompression stage before entering the vehicle. The question therefore arises, what should the total internal pressure be, and what gas should be used to supplement the oxygen, carbon dioxide and water vapour to achieve it?

Mechanical criteria dictate an upper limit to the internal pressure of the vehicle of about half normal atmospheric pressure (Roth, 1966). A greater internal pressure would necessitate much stronger and heavier walls to contain the differential pressure which exists when the vehicle is in space. A vehicle able to contain a pressure of 0·5 atm has walls no heavier than is required for other structural reasons.

The choice of gas to be used is also influenced by engineering considerations. Both the electronic equipment and the humans within the vehicle emit heat, which has to be lost to space by radiation. There is a distinct advantage in using a gas mixture which transfers heat well. It will help electronic components to lose heat to the atmosphere within the vehicle, thereby prolonging their life, and it will help transfer heat to the vehicle, wall from which it is lost to space. Helium is the gas of choice to meet this requirement and accordingly some space vehicles have contained a helium–oxygen gas mixture at a total pressure of about 0·5 atm. The engineering advantage of helium is physiologically a disadvantage, because efficient heat transfer by a helium-oxygen mixture upsets the body's normal thermal balance. It will be shown that this is easily rectified by raising the temperature of the space vehicle interior to levels which would be uncomfortably hot for a man in air.

In human diving, helium is used as a diluent for oxygen in the breathing mixture and serves as a hydraulic fluid of low density which the pulmonary system can cope with at quite high pressures (p. 250). As was mentioned in Chapter II, helium's physical properties, particularly its solubility in hydrophobic solvents, are such that it exerts no narcotic or anaesthetic effects in the 1–60 atm range with which diving physiology is currently occupied. At much higher partial pressures it does exert a physiological effect in cells like other weak narcotic gases, and there is some evidence that it can perturb certain biochemical reactions at low partial pressures (Chapter III.)

## 4.2 THERMAL BALANCE IN NORMAL CONDITIONS

Before considering the problems of thermal balance in helium environments, either at reduced pressure or at high pressure, it is necessary to outline the normal steady state thermal balance which exists between man, or any warm blooded animal, and the environment. Heat is lost to the environment by radiation ($H_R$), convection ($H_C$), conduction ($H_D$) and through the evaporation of water ($H_V$). At equlibrium the rate of heat loss is matched by the rate at which it is produced metabolically. Accordingly, we may write:

$$\text{total heat produced} = H_R + H_C + H_D + H_V. \tag{10}$$

If the body produces heat faster or slower than it is lost to the environment, the body accordingly warms or cools, and we may introduce an additional term ($H_S$) to account for the storage of heat or ($-H_S$) for the loss of heat.

The four separate mechanisms by which heat exchanges between the body and its environment, radiation, conduction, convection and vaporisation, are convenient abstractions but they do not correspond to anatomical sites. Heat transfer actually occurs across only two different surfaces, the skin and the walls of pulmonary-oral-nasal airways. The different heat transfer pathways are variously affected by changes in the gaseous environment.

Radiant heat transfer occurs simultaneously in two directions: emission of heat from the skin and absorption of radiant heat from the environment. The net heat transfer equation is:

$$H_R = S_0\, \varepsilon_1\, \varepsilon_2\, (T^4 - T_0{}^4)A \tag{11}$$

where $H_R$ = radiant heat transfer g cal$^{-1}$ sec$^{-1}$; $S_0$ = Stefan–Boltzmann constant $5{\cdot}75 \times 10^{-8}$ W m$^{-2}$ °K$^4$; $T$ and $T_0$ × temperature of body and surfaces in the environment, respectively, in °K; $\varepsilon_1$ and $\varepsilon_2$ = emissivity of the surface of the body and environment (Hardy, 1949). $A$ = radiant surface area (75 per cent of the surface area of the body is normally assumed to radiate heat to the environment).

Radiant heat transfer is unaffected by the properties of the gaseous environment although confinement within a space vehicle or hyperbaric chamber obviously introduces other changes which influence radiant heat balance.

Thermal conduction may be treated in a manner analogous to electrical conduction. The following equation is equivalent to Ohm's law:

$$H_D = \frac{T_2 - T_1}{d/KA} \qquad (12)$$

where:

$H_D$ = flow of heat
$K$ = thermal conductivity of conductor
$A$ = area of conductor
$d$ = thickness of conductor
($K$, $A$ and $d$: resistance to heat flow)
$T_2$ and $T_1$ = temperature of body and conductor.

Static air trapped in clothing normally provides important insulation, as air is a poor heat conductor. Substituting helium for the nitrogen in air greatly reduces the insulation which clothing provides because helium has a high thermal conductivity.

For the body to evaporate off a given quantity of water a specific amount of energy is required. That quantity is not quite so simply calculated as one might expect. The greatest part of the heat required to evaporate body water is given by the product of the quantity of water vaporised and the latent heat of vaporisation of water at the appropriate temperature. There are two additional factors however: heat is required to expand the saturated water vapour to the particular level of humidity which prevails at the time and heat is also absorbed when saturated water vapour at skin temperature cools to the environmental temperature. For a more detailed treatment of these phenomena see Hardy (1949). It is obvious that helium will not affect the amount of heat involved in the vaporising processes but it may indirectly affect the rate at which evaporation proceeds by altering ventilation or convection at the skin surface.

Convective heat loss is an important mechanism of heat transfer in the present context, and the most complicated of all heat transfer mechanisms. Figure VII.13 shows the temperature profile of air adjacent to the body, with temperature plotted on the vertical axis. We may first distinguish an unstirred layer of air which adheres to the body surface and through which heat moves by conduction. Beyond that the air is moving and carries away heat at a rate largely determined by the velocity of the air flow. When the velocity is high, as in forced convection, the unstirred layer will be small. In still air the unstirred layer may extend for several millimetres from the body surface, but some convection always occurs because of density differences between warmed and unwarmed air. This is called free convection. It is helpful to regard the convective heat transfer as the combination of (1) the fluid flow over the body surface and (2) the amount of heat which the fluid can absorb during its proximity to the body. Figure

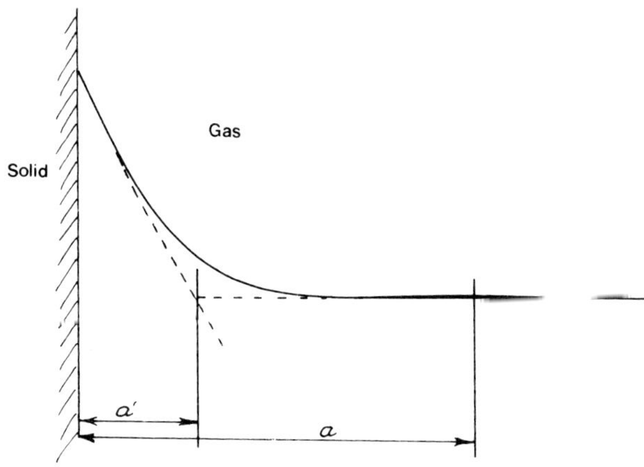


FIG. VII.13. Temperature gradient from a warm body to the ambient gaseous environment. Temperature is indicated on the vertical scale, and distance from the body (left) on the horizontal scale. Within the boundary layer ($a'$) the air is static, the temperature gradient is approximately linear and due to conduction. Beyond $a'$ the air is stirred and the temperature gradient terminates. (Birkebak, 1966)

VII.13 shows that part of convective heat transfer is a conduction process in the boundary layer. But the total heat transfer by convection ($H_c$) depends on the temperature and velocity of the convecting fluid, the surface area in contact with the fluid and on the physical properties of the fluid. The relevant physical properties are: thermal conductivity ($K$), density ($\rho$), viscosity ($\eta$), and specific heat ($C_p$). The following equation incorporates these properties and fits experimental data in an approximate fashion (Epperson *et al.*, 1966).

$$H_c = a\frac{K}{D}\left(\frac{DV\rho}{\mu}\right)^{\frac{1}{2}}\left(\frac{C_p\mu}{\kappa}\right)^{\frac{1}{3}} \qquad (13)$$

where $D$ is a constant for the characteristic shape of the body, $a$ is a constant, depending on the units employed, $v$, velocity and other terms have been defined above.

Inert gases affect convective heat transfer in a manner determined by their physical properties, which differ considerably. Helium, for example, in addition to possessing a high thermal conductivity also differs from nitrogen (and air) in other respects. The way in which these properties

TABLE VII.4

Some physical properties of gases at 0°C and 1 atm (Radford, 1964)

| | Thermal conductivity, $\mu$ cal cm $^{-1}$sec $^{-1}$ | Viscosity micropoises | Specific heat, $C_p$ (cal $g^{-1}$, °K) | Density (g $litre^{-1}$) |
|---|---|---|---|---|
| He | 352·0 | 188·7 | 0·745 | 0·1785 |
| Ne | 108·7 | 298·1 | 0·148 | 0·900 |
| A | 39·7 | 210·4 | 0·0745 | 1·784 |
| Kr | 21·2 | 232·7 | 0·0356 | 3·708 |
| Xe | 12·4 | 210·1 | 0·0227 | 5·851 |
| $H_2$ | 416·0 | 85·0 | 2·411 | 0·0899 |
| $N_2$ | 58·0 | 167·4 | 0·177 | 1·251 |
| $O_2$ | 58·5 | 192·6 | 0·157 | 1·429 |
| $CO_2$ | 34·0 | 138·0 | 0·154 | 1·977 |
| $N_2O$ | 36·8 | 136·2 | 0·159 | 1·978 |
| Air | 58·0 | 170·8 | — | 1·293 |

alter at high pressure is discussed below. Table VII.4 summarises the physical properties of gases which are important in heat transfer.

Finally, it is worth restating the full thermal balance equation to emphasise that any one heat transfer pathway can be calculated if all the others are known

$$\text{metabolic heat} = H_R + H_C + H_D + H_V \pm H_S. \qquad (14)$$

As we shall see, this is an approach frequently used in thermal balance studies particularly to measure the most complex of pathways, $H_C$.

### 4.3 THERMAL BALANCE IN A HELIUM–OXYGEN ENVIRONMENT AT NORMAL AND SUBNORMAL PRESSURES

The engineering advantages of using helium in a manned space vehicle prompt the question, what is the effect of helium on man's thermal balance? Several workers have made measurements of the thermal status of resting and exercising subjects confined in an experimental chamber filled with helium (Epperson *et al.*, 1966; Fox *et al.*, 1966; Bowers and Fox, 1967). In the first study, the chamber was large enough to accommodate four subjects for two weeks. The approach was to measure all the terms in the thermal balance equation (14) except $H_C$, which was obtained by calculation. Three gas mixtures were studied; (1) air at normal pressure, (2) helium–oxygen at normal pressure with a $pO_2$ of 0·2 atm and (3) helium at a total pressure of 0·5 atm, of which the $pO_2$ was 0·2 atm. Mixture (3) is referred to as "low-pressure helium" The experimental measurements were carried out over a 150 min period, comprising 40 min

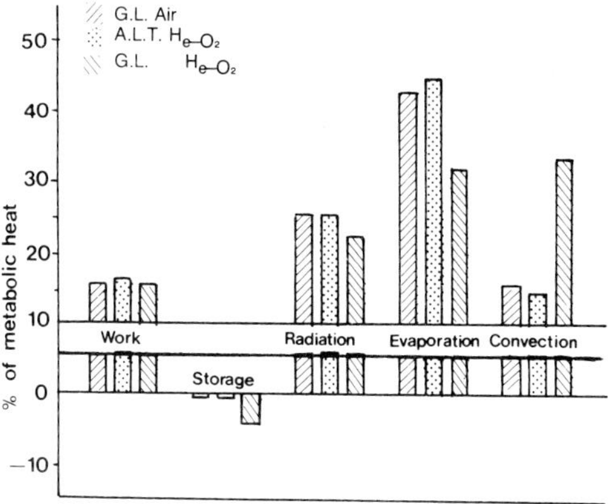


FIG. VII.14. Heat exchange in human subjects in air and helium-oxygen. Air was at normal atmospheric pressure; helium-oxygen was either at low pressure (205·5 mm Hg + $pO_2$ of 165·4 mm Hg) or at atmospheric pressure (pHe 579·3mm Hg + $pO_2$ 159 mm Hg). (Epperson *et al.*, 1966)

rest followed by a 1 hr period of moderate work on a bicycle ergometer and concluded by 50 min rest period.

Some of the terms in equation (14) are more easily obtained than others. $H_D$ (conduction) was negligible, $H_M$ (the metabolic heat production) was calculated from oxygen consumption, and *RQ* data, a standard procedure. Stored heat, $H_S$, was obtained from the product of body temperature and the specific heat of the body tissues. $H_R$ was calculated from equation (11) on the assumption that the subjects' effective radiating area was three-quarters of the body surface area and the emissivity of the skin was 1. Calculation of the evaporative heat loss term, $H_V$, assumed an effective latent heat of evaporation of water of 0·626 kcal $g^{-1}$. Measurements were made at 20 min intervals throughout the test period but the overall picture is best obtained by comparing the amount of heat exchanged by different pathways over the whole test period. This comparison is shown in Fig. VII.14, in which heat is expressed as a percentage of metabolic heat. Air and low-pressure helium produced similar results. The significant feature is the rather large convective heat loss in helium at normal pressure, which is associated with reduced evaporative and radiant heat losses. These three parameters are, of course, interrelated. The increased heat loss from the body surface might be expected to cause a reduction in skin temperature, which it does (Fig. VII.15). A reduced skin temperature will reduce both radiant and evaporative heat losses. The twofold increase in convective heat loss in helium compared to air is caused partly by the high thermal

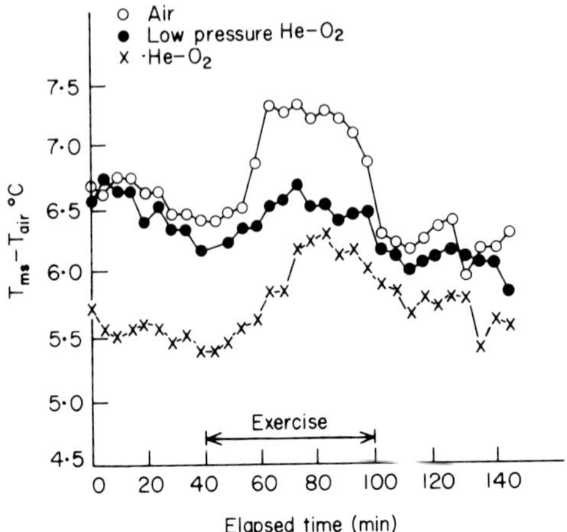


FIG. VII.15. The temperature gradient for mean skin temperature to air or helium-oxygen. The data were obtained from the subjects used in the experiment described in the text and Fig. VII.14. (Epperson *et al.*, 1966)

conductivity of helium, but as one would expect from equation (13) the relationship between convective heat loss and any one physical property is not simple. The thermal conductivity of a gas mixture is not a linear additive property of its constituents. Helium's thermal conductivity is six times, and that of the helium–oxygen mixture is four times, that of air. What property of helium, or its mixtures, causes the actual convective heat loss to be only twice that of the value in air? The low density of helium which is partly responsible, see equation (17). Furthermore, low skin temperature itself also reduces convective heat losses. Note in Fig. VII.14 that the convective loss in low-pressure helium–oxygen is very similar to that found in air at normal pressure. It will be seen in a later section that hyperbaric helium is a very potent convective heat loss agent.

The reduced evaporative heat loss in helium which is apparent in Fig. VII.14 has also been detected in experiments carried out by Fox *et al.* (1966) and provides a good example of an indirect effect of helium on a heat transfer pathway.

The work of Bowers and Fox (1967) has confirmed the overall affect of helium in man using a slightly different procedure to compute convective heat loss. These authors concluded that changing from air to a helium mixture increased the convective transfer of heat by 55 per cent, and they produced some evidence to show that helium exerted no direct effect on the metabolic rate of their experimental subjects.

## 4.4 COMFORT TEMPERATURES IN HELIUM-OXYGEN ENVIRONMENTS

The ambient temperature which gives a comfortable skin temperature in air strikes the person in a helium environment as cool and eventually unpleasantly cold. If men are to occupy space vehicles and undersea habitats in comfort it is necessary to control their ambient temperature to compensate for the presence of helium in the atmosphere. It is particularly important to have the correct comfort temperature during sleep.

Designers of space vehicles and sub-sea habitats need to know the range of temperature to provide and also the constancy with which the temperature has to be controlled. To provide such data, Bonura *et al.* (1967) used a space vehicle simulator in which a crew member, lying on a bunk, adjusted the temperature of the helium–oxygen gas mixture circulating within the vehicle to a comfortable level. The rate of temperature change was 0·5°C min$^{-1}$. Experiments were run at different circulating rates and at ambient pressures of $\frac{1}{3}$, $\frac{1}{2}$ and $\frac{2}{3}$ atm, all with a normal $pO_2$. Recordings of the gas temperature showed a zig-zag pattern with the limits providing a measure of comfort temperature. The rate of gas flow did not alter the comfort temperature, which was primarily influenced by the helium content and density of the gas. Figure VII.16 summarises

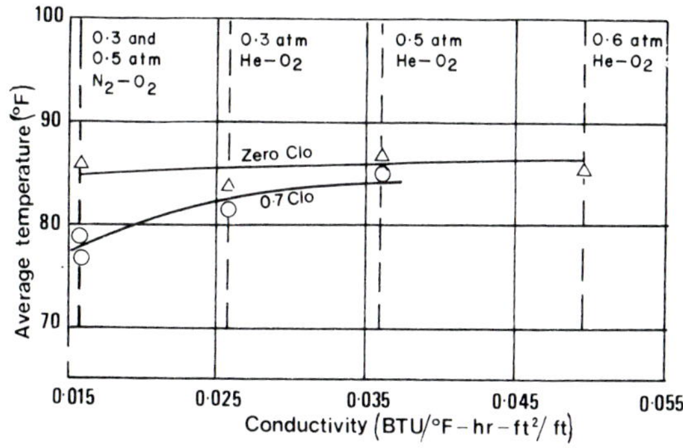


FIG. VII.16. Effect of helium on thermal comfort temperatures. Average comfort temperature selected in different gas mixtures. See text. (Bonura *et al.*, 1967)

the results. For a man resting without the insulation of clothing, the mean comfort temperature varied little with thermal conductivity. But with light clothing there is a marked increase in the comfort temperature, from 25°C in air to 29·4°C in helium, because helium reduces the amount of insulation provided by the clothing.

Without clothing the skin to air temperature differences are normally between 3 and 7°C, but the skin to helium–oxygen temperature gradient is less than 2°C in comparable conditions. Comfort temperature is mainly an index of skin temperature.

Another feature of a helium environment is its marked capacity for reducing the range of the thermal comfort zone, a property well known by the occupants of undersea habitats or hyperbaric chambers who readily sense small changes in their ambient temperature. Helium's high thermal conductivity and convective capacity cause a small temperature change to exert a disproportionate effect on the body's thermal balance. Webb (1970) aptly describes the situation as close "thermal coupling" to the environment.

### 4.5 THERMAL BALANCE IN HELIUM–OXYGEN AT HIGH PRESSURE

Increasing the pressure of the gaseous environment increases the convective heat loss by increasing density and hence heat capacity. Helium–oxygen is a gas mixture in which humans can be safely pressurised to at least 61 atm, and test mammals may be pressurised to over 200 atm without short-term injury. Thus, hyperbaric experiments which simulate deep diving conditions can provide extreme conditions for a warm body to maintain thermal balance.

We will consider the hyperbaric thermal balance measurements of Raymond *et al.* (1968, 1975) and of Timbal *et al.* (1974). Each study used the approach of Epperson *et al.* (1966), in which the convective heat transfer term in the heat balance equation (14) was obtained by difference. In the first-mentioned studies, the subjects were pressurised in a hyperbaric chamber at an ambient temperature of 28–29°C. The $pO_2$ was 0·3 atm and the $pN_2$ was 1·1 atm with helium making up the rest of the gas pressure. The subjects wore light clothing and rested quietly in a stationary atmosphere. We will not go into the technical difficulties of making the required physiological measurements in a hyperbaric environment and can proceed straight to the results. Figure VII.17 summarises the data particularly clearly and needs little comment. Metabolism is fairly constant over the range of pressures used and, as expected, thermal balance is dominated by the increase in convective heat loss. Evaporative heat loss is slightly, and radiant heat loss is distinctly, reduced at high pressure, presumably because of the reduction in the temperature of the skin, which was concurrently measured (Fig. VII.18). Measurements of oxygen consumption gave no sign of an altered metabolic rate, although the subjects lost, on average, 5 per cent of their body weight.

The hyperbaric thermal balance study carried out by Timbal *et al.* (1974) in France adopted a similar approach to the previous one and the

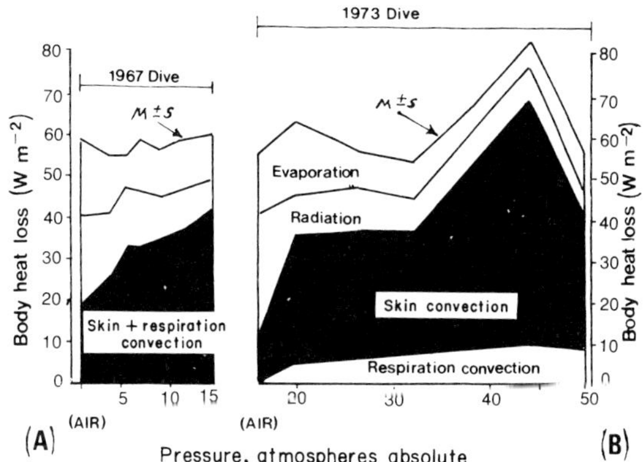


FIG. VII.17. Thermal balance in resting human subjects in hyperbaric helium. $M \pm S$, total heat loss which is equal to metabolic heat production, $I$, heat storage. Although $M$ appears to increase at high pressure it was associated with a temporary fall in core temperature during the measurements. ((A) Raymond *et al.*, 1968; (B) 1975)

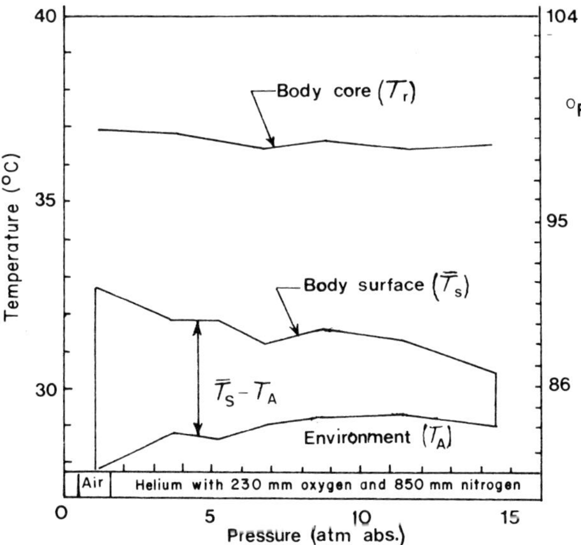


FIG. VII.18. The temperature gradient across the skin of human subjects is reduced in hyperbaric helium-oxygen although core (rectal) temperature is constant. Data from the same experiment are also plotted in Fig. VII.17(A). (Raymond *et al.*, 1968)

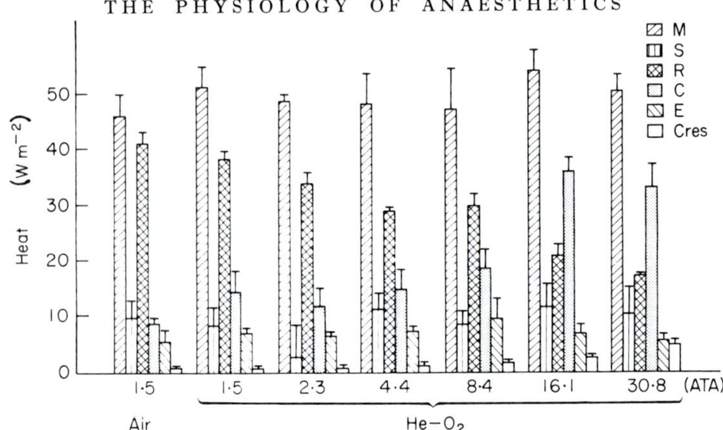


FIG. VII.19. Thermal balance in normal air and hyperbaric helium. $M$ = metabolism, $S$ = heat loss or gain, $R$ = radiant heat loss, $C$ = convective heat loss from the skin, $E$ = evaporative heat loss from the skin, $C_{res}$ = respiratory convective heat loss. (Timbal *et al.*, 1974)

agreement between the results is satisfactory (Fig. VII.19). Convective heat loss from the skin of subjects exposed to 30 atm helium was 30 W $m^{-2}$ in both cases, for example. The respiratory convective heat transfer and the total evaporative heat losses were also similar. Interesting discrepancies between the general increase in convective heat loss with increase in pressure are seen in both studies. Figure VII.17 shows a plateau over the pressure range 20–30 atm and Fig. VII.19 shows a levelling off at 16 atm. Respiratory heat loss increases steadily with increase in pressure in both, however. The decline in radiant heat loss in Fig. VII.19 is not due to a cooling of the skin (which was at a comfort temperature level at all pressures) but to the warming of the hyperbaric chamber walls.

The environment in which divers are liable to experience the most severe thermal stress in hyperbaric helium is the interior of a personnel transfer capsule (or diving bell). This is usually a steel sphere in which divers are transported down to the work site, and it is frequently used pressurised for transporting divers from the working depth to a hyperbaric chamber with no loss of pressure. Tauber *et al.* (1969) have calculated the quantity of heat which must be supplied to the individual diver in the transfer capsule to prevent harmful chilling.

For purposes of calculation the diver is taken to be equivalent to a 0·3 m diameter cylinder, 1·8 m tall, wearing a neoprene wet-suit supplemented with a hot-water suit on top. The gas is pure helium at 19·2 atm (equivalent to a depth of 190 m) and the gas temperature inside the transfer capsule is taken to be 4°C (40°F). Under these conditions we may write:

$$Q = H\,A\,\Delta T \qquad (15)$$

where $Q$ is the heat supplied to the diver's suit to maintain adequate body temperature, $H$ is the overall heat transfer coefficient, $A$ is the area of the body and $\Delta T$ the difference between the desired (normal) temperature of the skin and the environment. The biggest uncertainty in the calculation is in the estimation of $H$, but by taking a reasonable value for $H$ applicable to conditions in which gas within the capsule is circulating at a known velocity, $Q$ is equal to 1200 W. This simplified approach assumes the diver is resting, and it ignores respiratory and radiant heat losses, which are significant. The radiant heat loss, $H_R$, is given by

$$H_R = E\,S_0\,A[(460 + T)^4 - (500)^4]. \tag{16}$$

This is equation (11). In the above version, $E$ is the emissivity of the capsule and neoprene suit; $S_0$ is the Stefan-Boltzmann constant; and $A$ is the area of the radiant object, the diver.

$(460 + T)$ corresponds to the temperature of the outer surface of the neoprene suit and 500 is the temperature of the wall of the transfer capsule, both in degrees Rankine.

For a suited diver, Tauber *et al.* used equation (17) to describe the heat loss by convection and radiation:

$$\underset{\text{(radiation)}}{E\,S_0\,A[(460 + T)^4 - (500)^4]} + \underset{\text{(convection)}}{h_c A(T - 40)}. \tag{17}$$

T is the temperature of the external surface of the suit, 40 the gas temperature (°F) and *hc* the convective heat transfer coefficient The heat which must be supplied to the outer suit has to flow through the diver's neoprene suit and maintain a normal skin temperature of 34·4°C (94°F). This heat flow is

$$k\frac{A}{L}(94 - T) \tag{18}$$

where $k$ = thermal conductivity of the neoprene suit, $A$ = surface area of diver, $L$ = thickness of suit and, 94 the normal skin temperature.Tauber *et al.* used a trial and error method until the following equation was satisfied

$$k\frac{A}{L}(94 - T) = E\,S_0\,A[(460 + T)^4 - (500)^4] + H_c A(T - 40). \tag{19}$$

If $T$ is 23°C (73·5°F) the equation balances with 1320 W on either side. Thus a resting diver in a capsule filled with helium at 4°C (40°F) and 19 atm, insulated in the way described and exposed to a known gas velocity, requires at least 1320 W to offset convective and radiant heat transfer. Another 100 W or so can be added to offset the respiratory heat loss.

Conditions in a helium-filled diving bell may be very severe but once the diver enters the outside water his convective heat losses rise considerably, as water is an excellent heat transfer fluid. In the water the diver is liable to work fairly hard and ventilation will increase the respiratory heat loss, which can reach very high values indeed.

## 4.6 RESPIRATORY HEAT LOSS

### 4.6.1 *Normal conditions*

Heat is lost from the lungs by two of the four different mechanisms by which heat is transferred from the body as a whole. Gas is usually expired from the lungs at a higher temperature than it is inspired and thus body heat is carried away by convective means. Expired gas usually contains more moisture than when it is inspired, and therefore evaporative heat transfer occurs in the lungs and airways. Quantitatively these two processes can be expressed as follows:

$$RHL = \dot{V}_e \rho C_p (t_e - t_i) + \dot{V}_e L (W_e - W_i) \qquad (20)$$

where $RHL$ = respiratory heat loss, $\dot{V}_e$ = expired gas litres min$^{-1}$ STP dry, $\rho$ = density, $C_p$ = specific heat kcal g$^{-1}$ °C, $t_e$ and $t_i$ = expired and inspired gas temperature, $L$ = latent heat of vaporisation of water in prevailing conditions, and $W_e$ and $W_i$ = expired and inspired water content in g litre$^{-1}$. The first term in equation (20) is the convective heat transfer. Because of the tidal nature of pulmonary ventilation, convective heat loss may be simply calculated. It is the product of the temperature difference between inspired and expired air, the minute volume $\dot{V}$, and the term $\rho C_p$. It is therefore considerably influenced by the magnitude of ventilation and environmental temperature and pressure (Table VII.5). The second term in equation (20) is the evaporative heat loss and is normally much the larger part of the total respiratory heat loss. It, too, is affected by the rate of ventilation and, of course, by the humidity of the inspired air. Respiratory heat loss is normally a small part of the body's total heat loss although it is of importance in low temperatures experienced in polar conditions. Divers breathing helium mixture are liable to inspire cold gas, but the convective part of their respiratory heat loss is likely to be much greater than in cold air because of the properties of helium and its high density and specific heat at depth. In the following section we shall see how in extreme conditions the respiratory heat loss becomes a major component of the body's total heat loss.

### 4.6.2 *Hyperbaric helium–oxygen*

There are relatively few published reports of respiratory heat loss measurements in hyperbaric helium. However, a few accounts which are available

TABLE VII.5

Respiratory heat loss by convection and evaporation

| Air temperature (°C) | Conditions | Convective heat loss (W) | Evaporative heat loss (W) | Total respiratory heat loss (W) |
|---|---|---|---|---|
| 20 | Subjects resting $\dot{V}$ 6 litre STPD/dry min$^{-1}$ 45% humidity | 2 | 9 | 11·2[a] |
| 20 | $\dot{V}$ 10 litre STP, min$^{-1}$ 50% humidity | 3 2 | 14·3 | 17·5[c] |
| 20 | Subjects stand, resting 50% humidity | 2·8 | 10·9 | 13·7[b] |
| 0 | as above | 6·2 | 9·6 | 15·8[b] |
| −20 | as above | 9·7 | 8·07 | 17·77[b] |
| −40 | as above | 13·1 | 6·86 | 19·96[b] |
| −22 | $\dot{V}$ 11 litre STPD min$^{-1}$ } same subject | — | 12·5 | —[d] |
| −22 | $\dot{V}$ 42 litre STPD min$^{-1}$ } same subject | — | 64·7 | —[d] |
| 97 | $\dot{V}$ 10 litre STP, min$^{-1}$ 50% humidity | 0 | 9 | 9[c] |

[a]Hanson (1974); [b]Webb and [c]Day, both quoted in Goodman *et al.* (1971); [d]Brebbia *et al.* (1957).

are highly competent, demonstrating both the principles outlined above and the importance of accurate measurements on which to base calculations.

One such study was carried out by Webb and Annis (1966), who used subjects breathing various gas mixtures from an open-circuit Scuba demand valve while exercising gently at normal and high pressure. Measurements were made of inspired and expired gas temperature and humidity, and $\dot{V}_e$. Note that equation (20) contains no other terms which require experimental determination. Webb and Annis investigated respiratory heat loss in helium and in an oxygen–sulphur hexafluoride mixture, the latter providing a high-density breathing mixture without recourse to a hyperbaric chamber. The term $\rho C_p$ was calculated for the particular gas mixture used from the handbook data. Table VII.6 lists data for both pure gases and experimental mixtures. The heat capacity of a given gas mixture was calculated as follows: the molar heat capacity of the

TABLE VII.6

Properties of gases and gas mixtures at 27°C and 1 atm (From Webb and Annis, 1966)

| Gas | M.W. | Specific heat ($C_p$) (cal $g^{-1}$, °C) | Density ($\rho$) (g $litre^{-1}$) | $\rho C_p$ *cal* $litre^{-1}$°C |
|---|---|---|---|---|
| $O_2$ | 32 | 0·2198 | 1·3026 | 0·2863 |
| $N_2$ | 28 | 0·2486 | 1·1438 | 0·2843 |
| He | 4 | 1·242 | 0·1684 | 0·2092 |
| $SF_6$ | 146 | 0·159 | 6·0287 | 0·9586 |
| Mixtures | | | | |
| Air | — | 0·2402 | 1·1791 | 0·2832 |
| 20% $O_2$–80% He | — | 0·5608 | 0·3986 | 0·2235 |
| 20% $O_2$–80% $SF_6$ | — | 0·1621 | 5·0838 | 0·8241 |
| 4% $O_2$–96% He | — | 0·9320 | 0·2137 | 0·1992 |

pure gases were added up according to their mole fraction in the mixture. Then the average molecular weight of the mixture was computed, which, when divided into the sum of the molar heat capacities, yielded the heat capacity per unit mass of mixture. A useful range of $\rho C_p$ is apparent in Table VII.6, with the $O_2$–$SF_6$ mixture being more than four times the value of helium mixture. The latent heat term ($L$) in equation (20) was taken to be 0·58 kcal $g^{-1}$.

TABLE VII.7

Respiratory heat loss in different gases (Data of Webb and Annis, from Goodman *et al.* 1971)

| Gas mixture | Inspired temp. (°C) | $\Delta T$°C | Resp. heat loss (W) | Total heat loss (W) | Resp. heat loss (% total) |
|---|---|---|---|---|---|
| Air 1 atm | 20·9 | 13·4 | 39·0 | 383 | 9·5 |
| Air 4 atm | 17·0 | 16·3 | 62·7 | 334 | 17·7 |
| $SF_6$ 80%–$O_2$ 20% 1 atm | 20·2 | 13·7 | 49·4 | 348 | 13·5 |
| He 80%–$O_2$ 20% 1 atm | 22·7 | 11·1 | 33·4 | 376 | 9·0 |
| He 80%–$O_2$ 20% 4 atm | 15·3 | 15·4 | 33·4 | 216 | 16·3 |
| He 96%–$O_2$ 20% 8 atm | 16·5 | 16·0 | 62·7 | 264 | 25·1 |

At atmospheric pressure the subjects underwent a 5 min period of treadmill exercise followed by a similar rest period at temperatures between 5 and 15°C. In the hyperbaric chamber the subjects were immersed in water at a similar temperature, clothed in a full dry-suit.

The resultant values of respiratory heat loss are listed in Table VII.7. The dense gas $SF_6$ gives a high respiratory heat loss at 1 atm pressure, and doubling the pressure of helium from 4 atm to 8 atm doubles the respiratory heat loss. A similarity between air and helium at 1 atm is to be seen in both the respiratory heat loss and the fraction of the total heat loss which it comprises. The low density of helium offsets its high specific heat under these conditions (see Table VII.6).

The absolute amount of heat lost through the respiratory airways is also influenced by the inspired temperature (which was varied a little), humidity which was constant and low, and by the minute volume. Figure VII.20 shows the relationship between $\rho C_p$ and the respiratory heat loss per litre of gas ventilated.

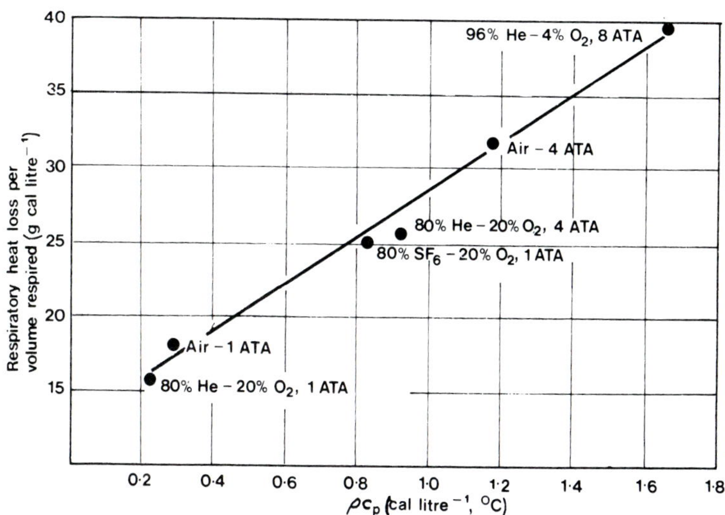


FIG. VII.20. Respiratory heat loss with subjects breathing gas mixtures of different values of $\rho C_p$. (Webb, 1970)

The respiratory heat loss which Webb and Annis measured in hyperbaric helium–oxygen was not particularly large, but extrapolation of the curve in Fig. VII.20 shows that respiration at much higher pressures will cause very large heat losses. These have been directly measured by Goodman *et al.* (1971) and Raymond *et al.* (1975) using a similar approach to that just described. In Goodman's study, the subjects wore heated suits and swam against a trapeze ergometer in a hyperbaric chamber filled with water. The respired gas was supplied through the apparatus shown in Fig. VII.21, which allowed $\dot{V}e$, $O_2$, $CO_2$ and expired moisture to be measured. Four pressures were used, approximately 14, 20, 26 and 30 atm, corresponding to diving depths of 137, 198, 259 and 304 m, which required the subjects to be confined to the hyperbaric chamber for several days.

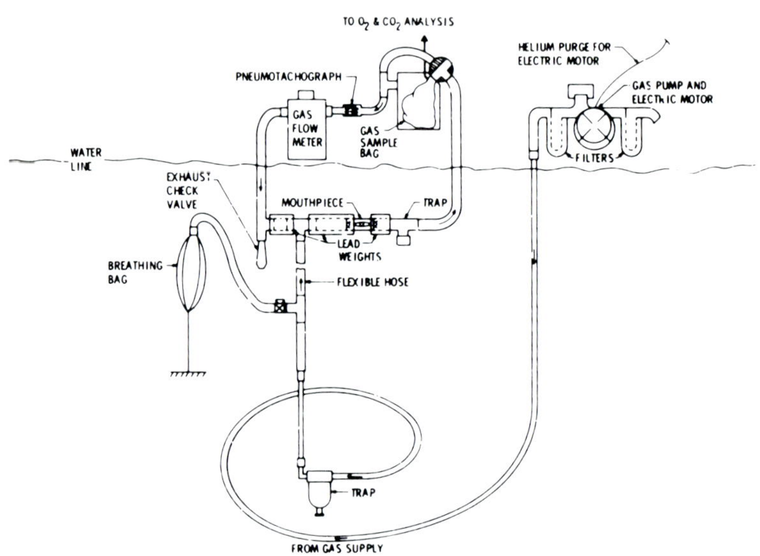


FIG. VII.21. Apparatus used to measure respiratory heat loss in subjects swimming in a hyperbaric chamber. (Goodman *et al.*, 1971)

The most arduous feature of the experiment was the low water temperature which determined the inspired gas temperature; 1·6, 7·2 and 12·7°C were selected. At the lowest temperatures, the subjects suffered a marked chilling of the chest and back and experienced severe shivering, despite wearing heated suits. Many experiments were cut short because of the harsh conditions. Also as a result of the cold, the subjects secreted copious amounts of fluid from the upper respiratory tract. The results obtained are a noteworthy achievement, providing accurate measurements of respiratory heat loss in helium mixtures at extreme pressure in conditions simulating those faced by commercial divers. Figure VII.22 shows the relationship between ventilation rate and respiratory heat loss which was calculated from the experimental data using equation (20) at different pressures and at two temperatures.

Respiratory heat loss was found to be proportional to pressure (Fig.

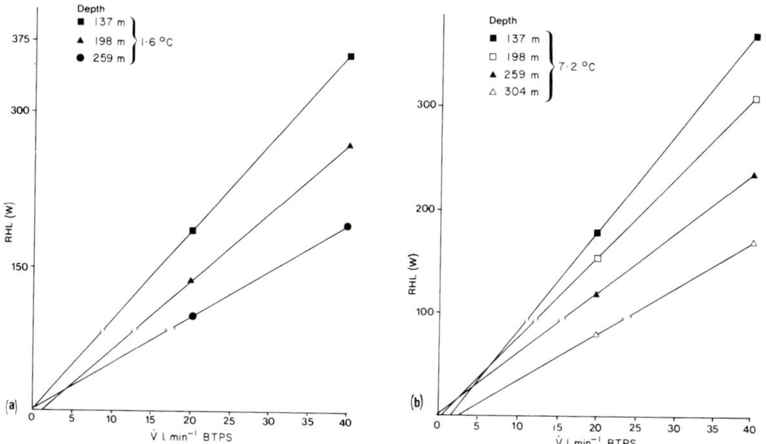


FIG. VII.22. Respiratory heat loss in subjects swimming at different pressures. The effect of variations in $\dot{V}$, plotted on the horizontal axis at 1·6°C (a) and 7·2°C (b) are shown. (Goodman *et al.*, 1971)

VII.23(a). Figure VII.23(b) from Tauber *et al.* (1969) shows rather different data calculated from first principles. The maximum respiratory heat loss measured by Goodman *et al.* (1971) was 400 W. Heat transfer in the airways is able to keep up with the increased heat capacity of the compressed gas because, unlike the skin, the airways are not able to reduce the temperature gradient at the tissue–gas interface. Evaporative heat loss is a decreasing fraction of the total respiratory heat loss as pressure increases.

The values for respiratory heat loss may be related to the body's thermal balance as a whole, using equation (14). As the subjects wore suits which were heated by circulating hot water, body heat loss was restricted to convective heat transfer ($H_c$), and thus a simplified equation was used

$$M = S + W + H_c + RHL \qquad (21)$$

where $M$ is metabolically produced energy derived from oxygen consumption, $W$, work energy and $S$ was obtained by calculation from body temperature changes as previously mentioned. $H_c$ could be calculated having obtained the respiratory heat loss $RHL$, or obtained from the relation

$$H_c = h_s A_b(T_r - T_{mws}) \qquad (22)$$

$h_s$ is the coefficient of heat transfer, $A_b$ the appropriate body area, $T_r$ and $T_{mws}$ the rectal and mean weighted skin temperatures, respectively. In the most extreme cases, respiratory heat losses amounted to over 50 per cent of the metabolically produced heat, and the subjects failed to maintain a normal body temperature. Goodman *et al.* (1971) concluded that divers with a 250 W respiratory heat loss breathing helium at 7°C, $\dot{V}$ 35 litre min$^{-1}$ at a depth of 260 m, for longer than 90 min run the serious risk of hypothermia Fig. VII.22(b). Under such conditions the inspired gas must be heated.

Let us examine the relationship between respiratory heat loss, ventilation and metabolic heat. In normal conditions exercise increases the metabolic rate and thereby increases heat production more than the increased ventilation increases heat loss, and so body temperature tends to rise. In helium at fairly high pressure the body is unable to increase its body temperature by exercising, because the increase in ventilation which is required causes an even greater loss of heat. This is an extraordinary thermal imbalance. At what combination of temperature and pressure does helium–oxygen achieve this? This is a question of practical significance in diving and of considerable theoretical interest.

Calculations carried out by Tauber *et al.* (1969) suggest that respiratory heat losses in helium at 4°C equal the heat produced metabolically in resting divers at a pressure of 14 atm, equal to 140 m depth. On increasing the pressure some advantage is gained by increasing the metabolic rate, but at a high work and ventilation rate, respiratory heat loss equals metabolic heat at 23 atm pressure, 230 atm in depth. According to Braithwaite's (1972) analysis, helium–oxygen at 0°C and 20 atm or 12°C and 30 atm, represents the maximum tolerable respiratory heat loss for working divers equipped with hot water suits.

### 4.6.3 *Convective character of helium*

The very useful and illuminating term convective character is used by Webb (1970) to score the cooling potency of helium and other fluids. Convective character is simply calculated from the physical properties of fluids which are important in heat transfer mechanisms: $\rho C_p K/\eta$. These terms are: $\rho$ (density), $C_p$ (specific heat), $K$ (thermal conductivity), $\eta$ (viscosity). The convective character of a particular fluid is the ratio of its convective constant to the convective constant of air at normal atmospheric pressure. Convective character thus allows hyperbaric heliox to be compared with any other gaseous environment or indeed with fluids such as water.

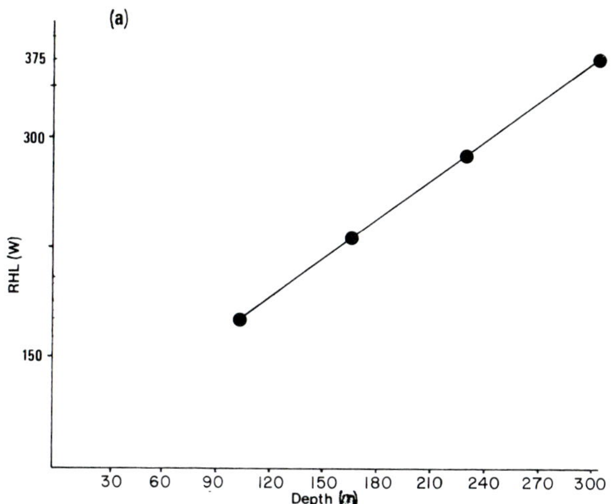


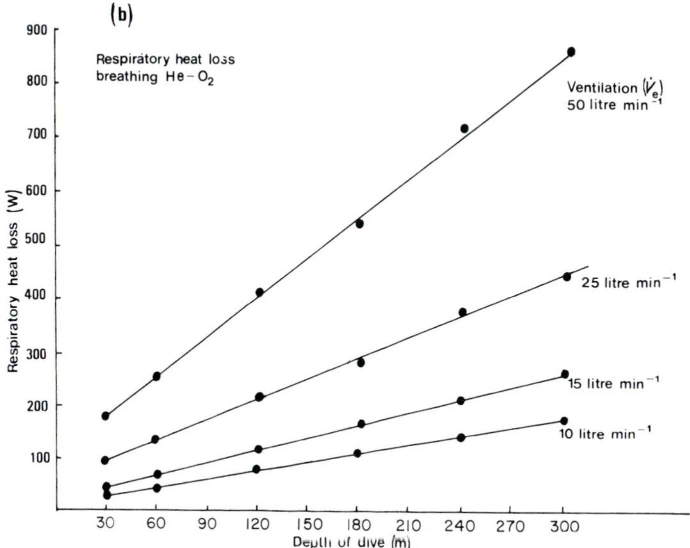


FIG. VII.23. (a) Respiratory heat loss in subjects swimming at 7·2°C at different pressures (Goodman *et al.*, 1971). (b) Calculated data, inspired temperature 4°C. (Tauber *et al* 1969)

We should not overlook the point that helium's very weak narcotic potency permits its use in diving and related experimental work at very high pressures. The inert gas whose narcotic potency resembles that of helium is neon. A neon–helium mixture promises to be a respirable gas which stresses the thermal balance of divers to a lesser extent than heliox (Schreiner *et al.*, 1972) but its high density, however, is disadvantageous in ventilation.

## 4.7 EXPERIMENTAL HYPOTHERMIA

When thermal regulation mechanisms are overwhelmed by heat loss to the environment, body temperature falls and the interesting condition of hypothermia sets in. In some mammals, e.g. the golden hamster (*Mesocricetus auratus*), hypothermia can be tolerated for a prolonged period. It is readily distinguishable from hibernation in which the animal lies curled up with the muscles tense and is able to rewarm its body given the appropriate stimulus. A hamster which is chilled experimentally lies limp and insensible and is unable to rewarm itself. One reason for its tolerance of hypothermia is probably its ability to maintain a ventilation rate sufficient to meet the needs of its depressed metabolism. Hypothermia is a common hazard to humans in cold climates and is also of clinical importance, being used in certain types of surgery.

Helium can be used to induce experimental hypothermia, as in the experiments with golden hamsters carried out by Fischer and Musacchia (1968). Groups of animals were kept at selected temperatures (23, 15, 12, 10, 5·5, 1°C) for a few hours in an atmosphere of either 80 per cent helium–20 per cent oxygen or 80 per cent $N_2$–20 per cent oxygen. A normal body temperature was maintained by the animals in $N_2$–$O_2$ irrespective of the temperature. In those animals subjected to helium–oxygen at 10°C or less, body temperature fell below the normal value of 38°C, and they became limp and insensitive to touch. Their respiratory frequency was low (19/min after 20 hr at 10°C in helium–oxygen, compared to 82/min in air) and they lost less weight than animals exposed to air.

Analysis of blood gases (Volkert and Musacchia, 1970) showed hypothermic animals to have a normal $pO_2$, $pCO_2$ and pH. Natural selection appears to have endowed hamsters with the ability to depress their metabolic rate in a balanced way, avoiding dangerous changes in blood pH in particular. Animals in helium at the intermediate temperature of 15°C were virtually normal in all respects while those at 12·5°C either showed symptoms of hypothermia or were normal. Clearly hamsters in helium at 12·5°C are at the limits of their thermal regulation capacity.

It is interesting to speculate on the way in which warm-blooded animals might have evolved if, in the evolution of the earth's atmosphere, helium had taken the place of nitrogen. In the present experiments the hamsters' inability to maintain a normal body temperature is brought about by increased convective heat losses aided by the reduced insulation caused by helium in the animals' fur. Presumably helium-adapted animals would have evolved like our present-day aquatic mammals, in being relatively large and insulated with fat rather than fur.

In certain types of surgery, particularly cardiac surgery, it is established procedure to cool the patient, stop the circulation temporarily and then to restore gradually normal body temperature. Improvements in the methods of inducing hypothermia in humans are of some importance. To this end Beran *et al.* (1975) employed a number of the heat transfers mechanisms to accelerate the heat loss in an anaesthetised rabbit. The animal respired a helium–oxygen mixture which was chilled to about 0°C and dehumidified Pulmonary ventilation was artificially sustained through an endotracheal tube. The combined effects of helium, low humidity, low temperature and by-passing the heat exchange section of the upper respiratory airways was considerable. The loss of body heat from animals cooled by the use of a conventional chilled wrap was half that obtained from animals additionally cooled by the cold helium method. The extension of these procedures to patients is not yet reported in the literature.

## 4.8 METABOLIC RESPONSES TO HELIUM–OXYGEN

It was first shown by Cook *et al.* (1951) that mice increased their metabolic rate when confined in helium at normal atmospheric pressure. It was subsequently shown that mice with a relatively low metabolic rate (as in large individuals or in thyroidectomised animals) responded to helium exposure to a greater extent than mice with a high metabolic rate (Young and Cook, 1953). Part, if not all, of the metabolic stimulus of helium is the accelerated heat loss. The data of Leon and Cook (1960) show this rather clearly in Table VII.8. This preliminary study was followed by a much larger one, in which many physiological measurements were made (Rhoades

TABLE VII.8

Stimulating effect of helium on the metabolic rate of rats (Leon and Cook, 1960)

| Ambient temperature: | 19·5°C | 25°C | 29·7°C |
|---|---|---|---|
| Increase in $\dot{Q}O_2$ over a period of 2 hr in helium, % control in air at the same temperature | 46% | 22% | 4·8% |

*et al.*, 1967). It was found that rats kept in helium eat more, respire oxygen faster, exhibit higher heart and respiratory rates than animals kept in air, provided the helium temperature is below about 33°C. In ambient temperatures greater than 33°C the body temperature tends to be normal and the rats' metabolism is indistinguishable from those with normal body temperature in air. In contrast, men in helium at atmospheric pressure seem to maintain an adequate body temperature and do not increase their metabolic rate (Bowers *et al.*, 1966; Raymond *et al.*, 1968, 1975). This is probably partly a consequence of man's large volume to surface area ratio which minimises the thermal stress introduced by helium. In general large mammals have a wider thermal–neutral zone than small animals and man is thermally less tightly coupled to his environmental temperature than a mouse or a rat.

Thermal measurement on rats exposed to helium at normal and high pressures have been made by numerous workers. Stetzner and De Boer (1972), for example, found that whereas a helium–oxygen mixture at normal pressure at 30°C was adequate to sustain a normal body temperature, 34·4°C was required at 41 atm. Table VII.9 gives a good measure of the metabolic stress which helium can indirectly bring about.

TABLE VII.9

$QO_2$ and weight loss in rats in heliox at 41 atm (Stetzner and De Boer, 1972)

| Temperature (°C) | $QO_2$, % control | Weight change, mean/rat |
|---|---|---|
| 30 | +35% | −15 gm |
| 34·4 | + 4% | − 6 gm |
| Controls | | |
| 30 in heliox at 1 atm | + 3% | + 2 gm |
| 24·4 in air at 1 atm | 0 | + 4 gm |

No observations of the effects of exercise in small mammals in helium seem to have been carried out, although Gillmore and Eicher (1974), who used telemetry to measure the body temperature of mice and guinea pigs in hyperbaric conditions, noted how unrestrained animals were better able to keep warm than animals tethered with a thermistor cable. No observations seem to have been made on the thermal effects of other inert gases.

Potent agents such as hyperbaric nitrogen or argon may directly affect thermoregulatory centres or metabolism by their anaesthetic action. There is no evidence that helium exerts an effect on thermal balance by such means.

### 4.9 CONCLUSIONS

When helium replaces the nitrogen in air it upsets the normal thermal balance of warm-blooded animals and demonstrates how their thermal regulation has evolved in harmony with the convective character of air. By accelerating heat transfer processes helium is able to indirectly modify the metabolism of test mammals.

Helium's exceedingly weak narcotic potency allows men and test animals to be subjected to high pressures without narcotic effects. Under these conditions helium's physiological potency derives from its bulk and transport properties.

## 5 General conclusions

In this chapter we have examined some non-narcotic effects of anaesthetics and inert gases. The latter in particular provide man and other higher animals with an artificial environment which can either be relatively normal and safe or it can be designed to stress thermal balance or gas transport mechanisms. What the inert gases may lack in anaesthetic potency they compensate for in their non-narcotic properties.

Life has evolved in equilibrium with inert gases, chiefly nitrogen, and it will be of interest to learn the full extent of their evolutionary impact. Undoubtedly nitrogen's properties have influenced the evolution of pulmonary mechanics (not just in mammals) and convective heat transfer in warm-blooded animals. The view has even been expressed that nitrogen in air at atmospheric pressure continually affects our mental performance, because its substitution by helium seems to stimulate mental processes (Winter *et al.*, 1975). The idea that we are permanently slightly anaesthetised by nitrogen has implications so enormous that it must be examined with great caution elsewhere. It does, however, illustrate the point previously made in Chapter I, that the physiology of anaesthetics and inert gases should be seen within the framework of environmental physiology.

## References

Bennett, L. J. and Miller, K. W. (1973). (Privately circulated data.)

Bennett, P. B and Elliott, D. M. (eds) (1975). "The Physiology and Medicine of Diving and Compressed Air Work". Baillière Tindall, London.

Beran, A. V., Proctor, K. G. and Sperling, D. R. (1975). Hypothermia and rewarming induced by surface and He–$O_2$ inhalate temperature control. *J. Appl. Physiol.* **39**, 337–340.

Birkebak, R. C. (1966). Heat transfer in biological systems. *Int. Rev. Gen. Zool.* **2**, 269–344.

BONURA, M. S., SNYDER, R. E. and WHITE, W. J. (1967). Thermal comfort zones for helium–oxygen atmospheres at reduced pressures. *Aerospace Medicine*, **38**, 912–916.

BOWERS, R. W. and FOX, E. L. (1967). Metabolic and thermal responses of man in various He–$O_2$ and air environments. *J. Appl. Physiol.* **23**, 561–565.

BOWERS, R. W., MATHEWS, D. K. and FOX, E. L. (1966). Metabolic and thermal responses of man during exposure to He–$O_2$ and air gaseous mixtures. *Physiologist*, **9**, 143. (Abstract.)

BRAITHWAITE, W. R. (1972). The calculation of minimum safe inspired gas temperature limits. Report RR–12–72, Navy Experimental Diving Unit, Washington, D.C.

BREBBIA, D. R., GOLDMAN, R. F. and BUSKIRK, E. R. (1957). Water vapour loss from the respiratory tract during outdoor exercise in the cold. *J. Appl. Physiol.* **11**, 219–222.

BÜHLMAN, A. A., FREI, P. and KELLER, H. (1967). Saturation and desaturation with $N_2$ and He at 4 atm. *J. Appl. Physiol.* **23**, 458–462.

COOK, S. F., SOUTH, F. E. and YOUNG, D. R. (1951). Effect of helium on gas exchange of mice. *Amer. J. Physiol.* **164**, 248–250.

ECKMAN, W. W., PHAIR, R. D., FENSTERMACHER, J. D., PATLAK, C. S., KENNEDY, C. and SOKOLOFF, L. (1975). Permeability limitation in estimation of local brain blood flow with $^{14}C$ antipyrine. *Amer. J. Physiol.* **229**, 215–221.

EGER, E. I. (1974). "Anesthetic Uptake and Action". Williams and Wilkins, Baltimore.

EPPERSON, W. L., QUIGLEY, D. G., ROBERTSON, W. G., BEHAR, V. S. and WELCH, B. E. (1966). Observations on man in an oxygen–helium environment at 380 mm Hg. Total pressure: III Heat exchange. *Aerospace Medicine*, **37**, 457–462.

EVERSOLE, U. H. (1938). The use of helium in anaesthesia. *J. Amer. Med. Ass.* **110**, 878–880.

FINK, B. R. (1955). Diffusion anoxia. *Anesthesiol.* **16**, 511–519.

FISCHER, B. A. and MUSACCHIA, X. J. (1968). Responses of hamsters to He–$O_2$ at low and high temperatures: induction of hypothermia. *Amer. J. Physiol.* **215**, 1130–1136.

FORKERT, L., WOOD, L. D. H. and CHERNACK, R. M. (1975). Effect of gas density on pulmonary compliance. *J. Appl. Physiol.* **39**, 906–910.

FOX, L. E., WEISS, H. S., BARTELS, R. L. and HIATT, E. P. (1966). Thermal responses of man during rest and exercise in a helium oxygen environment. *Arch. Environ. Health*, **13**, 23–28.

GAIT, D. J. and MILLER, K. W. (1973). Novel approach to improved submarine escape performance. *Aerospace Medicine*, **44**, 645–648.

GEORG, J., LASSEN, N. A., MELLEMGAARD, K. and VINTHER, A. (1965). Diffusion in the gas phase of the lungs in normal and emphysematous subjects. *Clin. Sci.* **29**, 525–532.

GILLMORE, J. D. and EICHER, M. (1974). Parabarosis and experimental infections. 2. Body temperatures of small animals; methods of observation and control. *Aerospace Medicine*, **45**, 249–253.

GLAUSER, S. C., GLAUSER, E. M. and RUSY, B. F. (1967). Gas density and the work of breathing. *Respir. Physiol.* **2**, 344–350.

GOODMAN, M. W., SMITH, N. E. and COLSTON, J. W. (1971). Hyperbaric respiratory heat loss study. Report N00014–71–C–0099, Westinghouse Electric Corp., Annapolis, Maryland.

GRAVES, D. J., IDICULA, J., LAMBERTSEN, C. J. and QUINN, J. A. (1973). Bubble formation in physical and biological systems: A manifestation of counter diffusion in composite media. *Science*, **179**, 582–584.

GRIFFITHS, H. B., MILLER, K. W., PATON, W. D. M. and SMITH, E. B. (1971). On the role of separated gas in decompression procedures. *Proc. Roy. Soc. (Lond.) B*, **178**, 389–406.

HALSEY, M. J. and EGER, E. I. (1973). Fluid shifts associated with gas-induced osmosis. *Science*, **179**, 1139–1140.

HANSON, R. DE G. (1974). Respiratory heat loss at increased core temperature. *J. Appl. Physiol.* **37**, 103–107.

HARDY, J. D. (1949). Heat transfer. In "Physiology of Heat Regulation" (L. H. Newburgh, ed.), p. 457. W. B. Saunders, Philadelphia.

HEMMINGSEN, E. A. (1975). Cavitation in gas-supersaturated solutions. *J. Appl. Phys.* **46**, 213–218.

HEMPLEMAN, H. V. (1960). The unequal rates of uptake and elimination of tissue nitrogen gas in diving procedures. Report U.P.S. 195, Med. Research Council R.N. Personnel Research Committee.

HEMPLEMAN, H. V. (1975). Decompression theory: British practice. *In* "The Physiology and Medicine of Diving and Compressed Air Work" (P. B. Bennett and D. H. Elliott, eds). Baillière Tindall, London.

HILLS, B. A. (1967). Diffusion versus blood perfusion in limiting the rate of uptake of inert non-polar gases by skeletal rabbit muscle. *Clin. Sci.* **33**, 67–87.

HILLS, B. A. (1970). Vital issues in computing decompression schedules from fundamentals. II. Diffusion versus blood perfusion in controlling blood: tissue exchange. *Int. J. Biometeor.* **14**, 323–342.

HILLS, B. A. (1971). Osmosis induced by nitrogen. *Aerospace Medicine*, **42**, 664–666.

HILLS, B. A. (1972). Gas-induced osmosis in the lung. *J. Appl. Physiol.* **33**, 126–129.

HILLS, B. A. (1977). The biophysical basis of prevention and treatment. *In* "Decompression Sickness" Vol. 1. Wiley Interscience.

JOHNSON, L. R. and VAN LIEW, H. D. (1974). Use of arterial $PO_2$ to study convective and diffusive gas mixing in the lungs. *J. Appl. Physiol.* **36**, 91–97.

KELLER, H. and BÜHLMAN, A. A. (1965). Deep diving and short decompression by breathing mixed gases. *J. Appl. Physiol.* **20**, 1267–1270.

KETY, S. S. (1951). The theory and applications of the exchange of inert gas at the lungs and tissues. *Pharmacol. Rev.* **3**, 1–41.

KETY, S. S. and SCHMIDT, C. F. (1945). The determination of cerebral blood flow in man by the use of nitrous oxide in low concentrations. *Amer. J. Physiol.* **143**, 53–66.

KETY, S. S., HARMEL, M. H., BROOMELL, H. T. and RHODE, C. B. (1948). The solubility of nitrous oxide in blood and brain. *J. Biol. Chem.* **173**, 487–496.

KYLSTRA, J. A., LONGMUIR, I. S. and GRACE, M. (1968). Dysbarism: Osmosis caused by dissolved gas? *Science*, **161**, 289.

LAMBERTSEN, C. J. (1971). Therapeutic gases; oxygen, carbon dioxide and helium. *In* "Drill's Pharmacology in Medicine" (J. R. Dipalma, ed.). McGraw-Hill, New York.

LAMBERTSEN, C. J. and IDICULA, J. (1975). A new gas lesion syndrome in man, induced by "isobaric gas counter diffusion". *J. Appl. Physiol.* **39**, 434–443.

LEON, H. A. and COOK, S. F. (1960). A mechanism by which helium increases metabolism in small mammals. *Amer. J. Physiol.* **199**, 243–245.

LONGMUIR, I. S. and GRACE, M. (1969). Physiological effect of the osmotic pressure of dissolved gas. *Fed. Proc.* **28**, 2580. (Abstract.)

MAIO, D. A. and FARHI, L. E. (1967). Effect of gas density on mechanics of breathing. *J. Appl. Physiol.* **23**, 687–693.

MEAD, J., TURNER, J. M., MACKLEM, P. T. and LITTLE, J. B. (1967). Significance of the relationship between lung recoil and maximum expiratory flow. *J. Appl. Physiol.* **22**, 95–108.

MORRISON, J. B. and FLORIO, J. T. (1971). Respiratory function during a simulated saturation dive to 1,500 feet. *J. Appl. Physiol.* **30**, 724–732.

PETERSEN, R. E. and WRIGHT, W. B. (1976). Pulmonary mechanical functions in man breathing dense gas mixtures at high ambient pressures—predictive studies III. *In* "Proceedings of the Fifth Symposium on Underwater Physiology" (C. J. Lambertsen, ed.) Fed. Am. Soc. Exp. Biol., Bethesda, Md.

RACKOW, H., SALANITRE, E. and FRUMIN, M. J. (1961). Dilution of alveolar gases during nitrous oxide excretion in man. *J. Appl. Physiol.* **16**, 723–728.

RADFORD, E. P. (1964). The physics of gases. *In* "Handbook of Physiology", vol. 1. American Physiol. Society, Washington, D.C.

RAYMOND, L. W., BELL, W. H., BONDI, K. R. and LINDBERG, C. R. (1968). Body temperature and metabolism in hyperbaric helium atmospheres. *J. Appl. Physiol.* **24**, 678–684.

RAYMOND, L. W., THALMANN, E., LINDGREN, G., LANGWORTHY, H. C., SPAUR, W. H., CROTHERS, J., BRAITHWAITE, W. and BERHAGE, T. (1975). Thermal homeostasis of resting man in helium–oxygen at 1–50 atm absolute. *Undersea Biomed. Res.* **2**, 51–67.

RHOADES, R. A., WRIGHT, R. A., HIATT, E. P. and WEISS, H. S. (1967). Metabolic and thermal responses of the rat to a helium oxygen environment. *Amer. J. Physiol.* **213**, 1009–1013.

ROTH, E. M. (1966). Gas physiology in space operations. *New England J. Med.* **275**, 144–153.

SALANITRE, E., RACKOW, H., GREENE, L. T., KLONYMUS, D. and EPSTEIN, R. M. (1962). Uptake and excretion of sub-anesthetic concentrations of nitrous oxide in man. *Anesthesiol.* **23**, 814–822.

SALANITRE, E., WOLF, C. L. and RACKOW, H. (1967). Pulmonary exchange of divinyl ether in man. *Anesthesiol.* **28**, 535–539.

SALZANO, J., OVERFIELD, E. M., RAUSCH, D. C., SALTZMAN, H. A., KYLSTRA, J. A., KELLEY, J. S. and SUMMITT, J. K. (1971). Arterial blood gases, heart rate and gas exchange during rest and exercise in men saturated at a simulated seawater depth of 1000 feet. *In* "Proceedings of the Fourth Symposium on Underwater Physiology" (C. J. Lambertsen, ed.). Academic Press, New York and London.

SALTZMAN, H. A., SALZANO, J. V., BLENKARN, D. and KYLSTRA, J. A. (1971). Effects of pressure on ventilation and gas exchange in man. *J. Appl. Physiol.* **30**, 443–449.

SCHEID, P. and PIIPER, J. (1975). Diffusion in alveolar gas exchange. *In* "Physiological Basis of Anaesthesiology" (W. W. Mushin *et al.*, eds). Piccin Medical Books, Padua, Italy.

SCHREINER, H. R., HAMILTON, R. W. and LANGLEY, T. D. (1972). Neon, an attractive new commercial diving gas. *Offshore Technology Conference*, Houston, Texas. 501–516. (Reprint No. 1561.)

STETZNER, L. C. and DE BOER, B. (1972). Thermal balance in the rat during exposure to helium-oxygen from 1 to 41 atmospheres. *Aerospace Medicine*, **43**, 306–309.

STRAUSS, R. H. and KUNKLE, T. D. (1974). Isobaric bubble growth: A consequence of altering atmospheric gas. *Science*, **186**, 443–444.

TAUBER, J. F., RAWLINS, J. S. P., and BONDI, K. R. (1969). Theoretical thermal requirements for the Mark II Diving system. Report Nav. Med. Res. Inst., Bethesda, U.S.A. M4306.

TIMBAL, J., VIEILLEFOND, H., GEUNARD, H. and VARERE, P. (1974). Metabolism and heat losses of resting man in a hyperbaric helium atmosphere. *J. Appl. Physiol.* **36**, 444–448.

VARENE, P., TIMBAL, J. and JAQUEMIN, C. (1974). Effect of different ambient pressures on airway resistance. *J. Appl. Physiol.* **22**, 699–706.

VOLKERT, W. A. and MUSACCHIA, X. J. (1970). Blood gases in hamsters during hypothermia by exposure to He–$O_2$ mixture and cold. *Amer. J. Physiol.* **219**, 919–921.

VOROSMARTI, J., BRADLEY, M. E. and ATHONISEN, N. R. (1975). The effects of increased gas density on pulmonary mechanics. *Undersea Biomed. Res.* **2**, 1–10.

WEBB, P. (1970). Body heat loss in undersea gaseous environments. *Aerospace Medicine*, **41**, 1282–1288.

WEBB, P. and ANNIS, J. F. (1966). Respiratory heat loss with high density gas mixtures. Report on contract 4965(00), Office of Naval Research, Washington, D.C.

WINTER, P. M., BRUCE, D. L., BACH, M. J., JAY, G. W. and EGER, E. I. (1975). The anaesthetic effect of air at atmospheric pressure. *Anesthesiol.* **42**, 658–661.

WOOD, L. D. H. and BRYAN, A. C. (1971). Mechanical limitations of exercise ventilation at increased ambient pressure. *In* "Proceedings of the Fourth Symposium on Underwater Physiology" (C. J. Lambertsen, ed.). Academic Press, New York and London.

YOUNG, D. R. and COOK, S. E. (1953). The effect of helium on the gas exchange of mice as modified by body size and thyroid activity. *J. Cell Comp. Physiol.* **42**, 319–325.

ZALTZMAN, G. L. and ZINOV'EVA, I. D. (1965). Comparative determination of permissible supersaturation value of the human body with different gases under difficult conditions. *In* "The Effect of the Gas Medium and Pressure on Body Functions" (M. P. Brestkin, ed.). NASA–TT–F–358, Washington, D.C.

# Subject Index